Lecture Notes in Artificial Intelligence 10357

Subseries of Lecture Notes in Computer Science

More information about this series at http://www.springer.com/series/1244

Petra Perner (Ed.)

Advances in Data Mining

Applications and Theoretical Aspects

17th Industrial Conference, ICDM 2017
New York, NY, USA, July 12–13, 2017
Proceedings

 Springer

Editor
Petra Perner
Institute of Computer Vision and Applied
  Computer Sciences
Leipzig, Sachsen
Germany

ISSN 0302-9743   ISSN 1611-3349   (electronic)
Lecture Notes in Artificial Intelligence
ISBN 978-3-319-62700-7   ISBN 978-3-319-62701-4   (eBook)
DOI 10.1007/978-3-319-62701-4

Library of Congress Control Number: 2017945723

LNCS Sublibrary: SL7 – Artificial Intelligence

Printed on acid-free paper

This Springer imprint is published by Springer Nature
The registered company is Springer International Publishing AG
The registered company address is: Gewerbestrasse 11, 6330 Cham, Switzerland

Preface

The 16th event of the Industrial Conference on Data Mining ICDM was held in New York (www.data-mining-forum.de) running under the umbrella of the World Congress on "The Frontiers in Intelligent Data and Signal Analysis, DSA 2017" (www.worldcongressdsa.com).

After the peer-review process, we accepted 26 high-quality papers for oral presentation. The topics range from theoretical aspects of data mining to applications of data mining, such as in multimedia data, in marketing, in medicine, and in process control, industry, and society. Extended versions of selected papers will appear in the international journal *Transactions on Machine Learning and Data Mining* (www.ibai-publishing.org/journal/mldm).

A tutorial on Data Mining, a tutorial on Case-Based Reasoning, a tutorial on Intelligent Image Interpretation and Computer Vision in Medicine, Biotechnology, Chemistry and Food Industry, and a tutorial on Standardization in Immunofluorescence were held before the conference.

We would like to thank all reviewers for their highly professional work and their effort in reviewing the papers.

We also thank the members of the Institute of Applied Computer Sciences, Leipzig, Germany (www.ibai-institut.de), who handled the conference as secretariat. We appreciate the help and understanding of the editorial staff at Springer, and in particular Alfred Hofmann, who supported the publication of these proceedings in the LNAI series.

Last, but not least, we wish to thank all the speakers and participants who contributed to the success of the conference. We hope to see you in 2018 in New York at the next World Congress on "The Frontiers in Intelligent Data and Signal Analysis, DSA 2018" (www.worldcongressdsa.com), which combines under its roof the following three events: International Conferences Machine Learning and Data Mining, MLDM, the Industrial Conference on Data Mining, ICDM, and the International Conference on Mass Data Analysis of Signals and Images in Medicine, Biotechnology, Chemistry and Food Industry, MDA.

July 2017   Petra Perner

Organization

Chair

Petra Perner   IBaI Leipzig, Germany

Program Committee

Ajith Abraham   Machine Intelligence Research Labs, USA
Brigitte Bartsch-Spörl   BSR Consulting GmbH, Germany
Orlando Belo   University of Minho, Portugal
Bernard Chen   University of Central Arkansas, USA
Jeroen de Bruin   Medical University of Vienna, Austria
Antonio Dourado   University of Coimbra, Portugal
Stefano Ferilli   University of Bari, Italy
Geert Gins   KU Leuven, Belgien
Warwick Graco   ATO, Australia
Aleksandra Gruca   Silesian University of Technology, Poland
Hartmut Ilgner   Council for Scientific and Industrial Research,
                                    South Africa
Pedro Isaias   Universidade Aberta, Portugal
Piotr Jedrzejowicz   Gdynia Maritime University, Poland
Martti Juhola   University of Tampere, Finland
Janusz Kacprzyk   Polish Academy of Sciences, Poland
Mehmed Kantardzic   University of Louisville, USA
Eduardo F. Morales   INAOE, Ciencias Computacionales, Mexico
Armand Prieditris   Newstar Labs, USA
Rainer Schmidt   University of Rostock, Germany
Victor Sheng   University of Central Arkansas, USA
Kaoru Shimada   Section of Medical Statistics, Fukuoka Dental College,
                                    Japan
Gero Szepannek   Santander Consumer Bank, Germany

Additional Reviewer

Juliane Perner   Cancer Research UK Cambridge Institute

Contents

Incorporating Positional Information into Deep Belief Networks for Sentiment Classification

Yong Jin[1], Harry Zhang[1], and Donglei Du[2]

[1] Faculty of Computer Science, University of New Brunswick, Fredericton, NB,
Canada, E3B 5A3
{yjin1,hzhang}@unb.ca,
[2] Faculty of Business Administration, University of New Brunswick, Fredericton, NB,
Canada, E3B 5A3
ddu@unb.ca

Abstract. Deep belief networks (DBNs) have proved powerful in many
domains including natural language processing (NLP). Sentiment classification has received much attention in both engineering and academic
fields. In addition to the traditional bag-of-word representation for each
sentence, the word positional information is considered in the input. We
propose a new word positional contribution form and a novel word-to-
segment matrix representation to incorporate the positional information
into DBNs for sentiment classification. Then, we evaluate the performance via the total accuracy. Consequently, our experimental results
show that incorporating positional information performs better on ten
short text data sets, and also the matrix representation is more effective than the linear positional contribution form, which further proves
the positional information should be taken into account for sentiment
analysis or other NLP tasks.

Keywords: Deep belief networks, Sentiment classification, Positional
information, Matrix representation

1 Introduction

Sentiment classification task is a popular research issue in the NLP community,
which is to determine the sentiment polarity of a text. The fast growing amount
of online opinion resources, such as product review sites and social media websites, has produced much interest on the task of sentiment classification [8,6,18].
Therefore, people in academic and engineering areas are paying attention to
develop an automatic system that can identify the polarities of the opinions.

Various techniques have been applied in sentiment analysis area, including
traditional machine learning approaches, language models [11,19], and the recently developed deep learning methods, such as deep neural networks [2], deep
belief networks [15], recursive matrix-vector model [16], and recursive deep network [17]. Pang *et al* use three traditional machine learning methods: naive
Bayes (NB), maximum entropy (ME) classification, and support vector machines

© Springer International Publishing AG 2017
P. Perner (Ed.): ICDM 2017, LNAI 10357, pp. 1–15, 2017.
DOI: 10.1007/978-3-319-62701-4_1

(SVMs) for movie sentiment classification [10]. Raychev and Nakov introduce a novel language independent approach to the task of determining sentiment polarities of the author's opinion regarding a specific topic in natural language texts [12]. It intakes the positional information into NB classifier. In particular, it introduces a method to measure the positional importance, that is, instead of value one, the occurrences of the words at different positions contribute different values to their frequency values in the sentence. The DBN [5] is an effective deep learning approach consisting of multiple layers of hidden variables, which are accommodated to abstract higher representations from the raw inputs. Sarikaya *et al* use additional unlabeled data for DBN pre-training and combine DBN-based learned features for text classification [15].

Some researchers consider the word order information into several different NLP tasks. Pahikkala *et al* introduce a framework based on a word-position matrix representation of text for natural language disambiguation tasks [9]. Specifically, in disambiguation tasks, each input example is consisting of a word to be disambiguated and its surrounding context words, then a kernel function is applied to map the features. Johnson and Zhang propose an effective way of bag-of-words conversion into convolutional layer to exploit the word order of text for text categorization using convolutional neural networks (CNNs) [7]. In detail, the region representation called *seq-CNN* is to embed text regions into low dimensional vector space. Meanwhile, considering the model becomes too complex and the training is too expensive, they provide an alternative way *bow-CNN* to perform bag-of-words conversion to make region vectors.

In this paper, we propose two ways to incorporate the positional information into DBNs for sentiment classification. Firstly, inspired by [12], we propose a new linear positional contribution form by scale normalization. Secondly, motivated by the work in [9,7], we propose a new word-to-segment matrix to denote a sentence, in which each sentence is divided into several segments (not convolutional). We do not consider the word order within each segment, but the segment order is represented using the word-to-segment matrix. Overall, this paper explores the effect of positional information on the sentiment classification task for short texts such as Twitter messages. Compared to the basic bag-of-word representation, the word positional information will capture some grammar information of a text.

The rest of the paper is organized as follows. In Section 2, we introduce three types to represent positional information; In Section 3, the experiments and results are described in detail; In Section 4, we come to the conclusion and discuss some future work.

2   DBN Incorporating Positional Information

2.1   Positional Contribution

Typically, a word is a grammar component in a sentence, and it has neighboring words resulting in different combinations of words, which means each word has a positional contribution value in the sentence. Basically, a word should be assigned a relatively higher weight as its position occurrence increases in the sentence. The reason is that when we read the first word, the sentence's sentiment polarity is not clear at all, but when we go to the last word after reading all previous words, the sentiment polarity becomes unambiguous.

The positional information has been introduced into the NB classifier [12] in the form of Eq. 1, in which a simple linear interpolation is used to measure the position-dependent information in a sentence. Suppose the vocabulary has N word attributes, W_j denotes the jth word (attribute) in the vocabulary, and it has value of x_j that is equal to zero when word W_j does not exist in the sentence, otherwise it is calculated via Eq. 1.

$$x_j = q_0 + q \cdot \frac{p}{n}, q_0 \geq 0, q > 0, j = 1, 2, ..., N, \tag{1}$$

where q_0 represents some constant value from the starting position of the sentence, and q is the position fractional weight, p is the position occurrence of word W_j in the sentence, n is the length of the sentence.

In order to normalize the values from Eq. 1 within the same scale, we force the value x_j to fall in the range of $[0, 1]$ (scale normalization). So we propose a new positional contribution form described in Eq. 2.

$$x_j = \theta + (1 - \theta) \cdot \frac{p}{n}, 0 \leq \theta \leq 1, \tag{2}$$

where p and n represent the same as those in Eq. 1, θ is the ratio between the word's presence and its positional contribution within the range of $[0, 1]$. When $\theta = 0$, the value only represents the positional contribution, while it means the traditional presence value if $\theta = 1$. Actually, the form of Eq. 2 is a special form of Eq. 1 just through the assumption of $q + q_0 = 1$, but it has two parameters. However, given that $0 < p \leq n$ is correct for each sentence because p represents the word's position in a sentence while n denotes the length of the sentence, this form offers the flexibility to adjust the ratio θ to assign the value into the interval of $[0, 1]$ that is fed into the DBNs. Each sentence incorporating the positional contribution is represented by an N-dimensional vector.

2.2   Matrix Representation

To better represent a sentence with the vocabulary words and incorporate the word order information into the model, also inspired by the linear transformation of the word-position matrix representation in the word disambiguation task [9], we try to use a matrix to represent a sentence. Intuitively, word-to-word matrix representation is introduced here. Suppose that a sentence is represented by an $N*N$ matrix M with the following two definitions:

- $M_{ii} = 1$, if the word W_i exists in the sentence, otherwise is 0, for $i = 1, 2, ..., N$;

- $M_{ij} = 1$,if the word W_i occurs before W_j with only one space (the two words are neighbors in the sentence), otherwise is 0, for i, $j = 1, 2, ..., N$ (i is not equal to j).

The word-to-word matrix representation can exactly describe the word order in each sentence. However, because there is normally a large vocabulary size for each training data set, the word-to-word matrix representation will consequently result in an extremely large size of input for training. Specifically, if the vocabulary size is N, the word-to-word matrix will be N^2, then the training time will be squarely increased, which will also cost too much memory of the computer.

In order to decrease the dimension size, we come up with an idea: each sentence is roughly divided into three segments in grammar order of the sentence. For example, the sentence *"the cat sat on that mat"* is divided into three segments as: segment 1 of {*"the cat"*}, segment 2 of {*"sat on"*}, and segment 3 of {*"that mat"*}. In this case, we do not consider the word order in a local region (within each segment), but we take into account the order information of the three segments in the sentence, which will not lose much information for short texts, and meanwhile the dimensional size is not too large.

Hence, we propose a simplified form of word-to-word matrix representation called word-to-segment matrix representation. But there is still some difference between word-to-word and word-to-segment: word-to-word is denoted by a 0-1 matrix, while word-to-segment would distinguish word's impact on different segments (detailed in the later matrix definitions). Based on the different impact relationship values, we propose two types of matrix representations. Considering the words in a sentence are normally organized in a logical order, and also to simplify the problem, the number of segments is not defined from the grammar view, but from the view that the number of words in each segment would be approximately similar. Specifically, let $nSeg$ denote the number of segments, n is the sentence length for a specific sentence. Then we define $nSeg$ as below:

- If $n < nSeg$, then each segment is at most one word in the sentence, for example, the sentence *"love it"* to be divided into three segments, then the first segment is {*"love"*}, the second segment is {*"it"*}, while the third segment is empty;
- If $n \geq nSeg$, the length of each segment $segLen$ (except for the last segment) is defined as the floor integer of n over $nSeg$. In detail, for the segment k through 1 to $nSeg - 1$, the position of segment k in the sentence is from $1 + (k-1) * segLen$ to $k * segLen$; while for the segment $nSeg$, the position is from $k * segLen + 1$ to n. For example, a sentence *"it is the best"* to be divided into three segments in order is {*"it"*}, {*"is"*}, and {*"the best"*}.

There are two types of impact definitions introduced here. Let the word-to-segment matrix denoted by M (each column represents each word in the vocabulary and each row denotes each segment), its element value is defined in the following two types accordingly. For both two types' definitions, $M_{ij} = 1$ when the word W_j occurs in the segment i for $i = 1,2,3$ and $j = 1,2, ..., N$, which means this word has full impact on its own segment. The word's impact on different segments is described below.

To clearly explain the word-to-segment matrix representations, we take the previously mentioned sentence *"the cat sat on that mat"* for example, and assume the vocabulary is {*"are"*, *"cat"*, *"mat"*, *"on"*, *"sat"*, *"take"*, *"that"*, *"the"*} with size $N = 8$. The first type is illustrated in Table 1, where the word-to-segment matrix M has size of 3*8 (same in Table 2, and also the empty cells are zeros). In Table 1, the words existing in the first segment(S1: *"the cat"*) have no impact on the latter two segments, the words in the second segment (S2: *"sat on"*) have impact value of a on the first segment, and the words in the third segment (S3: *"that mat"*) have impact on both the second and first segments with impact values a and b respectively. Given that the distance between the third segment and the first segment is larger than that between the third segment and the second segment, the impact values should satisfy the condition of $0 < b < a < 1$.

Table 1: Type one of word-to-segment matrix representation for one sentence (matrix1)

Segment	are	cat	mat	on	sat	take	that	the	words
S1		1	b	a	a		b	1	*"the cat"*
S2			a	1	1		a		*"sat on"*
S3			1				1		*"that mat"*

The second type is shown in Table 2. Another reasonable assumption is that a word in the segment not only has impact on its previous segment but also on its latter segment, but does not have impact on its remote segment (e.g., the first and the third segments). Furthermore, the three segments do not have the same impact on each other because they locate in different positions of the sentence. Specifically, the words in segment 1 (or segment 3) have impact on segment 2 with value c (or e), and the words in segment 2 have impact on both segment 1 and segment 3 with value d. The impact values have the restriction of $0 < c < d < e < 1$.

Table 2: Type two of word-to-segment matrix representation for one sentence (matrix2)

Segment	are	cat	mat	on	sat	take	that	the	words
S1		1		d	d			1	*"the cat"*
S2		c	e	1	1		e	c	*"sat on"*
S3			1	d	d		1		*"that mat"*

2.3 DBN settings

Through the definitions of the above two types of word-to-segment representations for each sentence, the input into the DBN model is transformed via the average operation over each column of the matrix. That is, each sentence is represented by an N-dimensional vector and each element value equals to the average of corresponding column of the three segments, which not only keeps the same size of input features as the original bag-of-words representation, but also contains the prior segment order information of the sentence.

Therefore, compared to traditional bag-of-words representation, there are overall four kinds of inputs: baseline bag-of-words representation, positional representation, matrix1 representation and matrix2 representation. The inputs are then fed into our DBNs respectively for further experimental comparison.

The DBNs introduced in our experiments are similar in [5], which includes RBMs to train the initial weights in an unsupervised way and then transmitted to ordinary neural networks for back propagation. In the DBNs for sentiment classification, from the visible input to the penultimate layer, we accommodate the widely used sigmoid function as the hidden neuron in Eq 3.

$$\varphi(v) = sigm(v) = \frac{1}{1 + e^{-v}}, \tag{3}$$

where v is the independent variable for sigmoid function φ.

Besides, there are some function options from the penultimate layer to output layer, such as softmax, and sigmoid functions. However, the sigmoid function proves to be more effective in both neuron models in this paper, which is written as Eq 4.

$$Label^S = sigm(w^{out} \cdot S + b^{out}), \tag{4}$$

where w^{out} is a weight matrix connecting the output layer and its previous hidden layer, b^{out} is the corresponding bias vector, and S represents the "Sentence" of the penultimate layer, and $Label^S$ is finally calculated as a C-vector (C is the number of classes) in which the largest value indicates its class.

The training process of DBNs is divided into two steps [5]: unsupervised RBM training and supervised neural network training. The first step follows the practical guide written by Hinton [4], and the second one is actually the traditional back propagation, a widely applied method to fine-tune the weights, introduced by Rumelhart *et al* [13]. Specially, for each data set, the same size of input is fed into the DBNs, so the running time difference lies in the computation of different input transformations that will not cost much time compared to DBN training. Therefore, here we focus on their classification performance on total accuracy.

3 Experiments and Results

In this section, we design a variety of experiments to verify the power of positional information of sentences in the DBNs. Then the experimental results are compared from different angles to analyze the effect and sensitivity of positional information for sentiment classification.

3.1 Data Collection and Pre-processing

We mainly focus on short text sentiment classification since the positional representation and word-to-segment matrix representation described above will lose some important effect for long text sentiment analysis. Therefore, several short text data sets, e.g. Twitter messages, are selected here for implementations.

(1) STS-T: Stanford Twitter Sentiment (STS) Test Set [14], a manually annotated data set of STS with positive and negative labels.
(2) STS-G: a gold data set extracted from STS with positive and negative labels [14].
(3) SST: Sentiment Strength Twitter data set [14], including three sentiment labels (positive, negative and neutral). We will use the data set as two, one is tri-class data set and the other is binary class removing neutral tweets.
(4) HCR: Health Care Reform (HCR) Twitter data set [14], including three classes (positive, negative and neutral). Similar as (3), the data set is used a tri-class set and a binary class set.
(5) FT: Full Twitter data [1],including three classes (positive, negative and neutral). Similar as (3), the data set is used of tri-class data set and binary class data set.
(6) GT: Game Tweets regarding the video games, are real-time collected and labeled by us with three labels (positive, negative and neutral). Also, this data set is used for tri-class and binary data sets.
(7) HCR2, SST2, FT2, GT2: These four data sets are respectively derived from HCR, SST, FT and GT with neutral tweets removed for binary classification.

Each raw data set needs to be pre-processed for training the model. Firstly, all characters are converted to lowercase since upper case and lower case have no differences for sentiment polarities; Secondly, the URLs in the data set are removed, because they do not make much sense for the sentiment; Thirdly, we transform some acronyms and abbreviations to their completely expanded forms. For example, "*i've*" is replaced by "*i have*", "*can't*" or "*cannt*" is "*can not*", "*won't*" or "*wont*" is "*will not*", "*shouldn't*" or "*shouldnt*" is "*should not*", with details in Table 3. In this way some meaningful words especially the negation word "*not*" are kept as they are essential for sentiment. Finally, some punctuations, such as @, /, |, $, are deleted as well since they contribute little to the text sentiment.

After the above pre-processing, we need to obtain initial trainable data that can be directly used in DBNs. Since Twitter texts are all limited to 140 characters long, each word in one Twitter text occurs only once for most of the time, so the vocabulary of each data set is extracted as attributes, each word token is denoted by its presence or not, and then each sentence (training example) can be represented by a vector where the element is one if the word exists in it,

Table 3: Corresponding expanded forms for abbreviations

abbrev.	expanded forms	abbrev.	expanded forms
i've	i have	won't/wont	will not
i'm	i am	wouldn't/wouldnt	would not
i've	i have	shouldn't/shouldnt	should not
i'll	i will	can't/cannt	can not
it's	it is	couldn't/couldnt	could not
let's	let us	isn't/isnt	is not
she's	she is	wasn't/wasnt	was not
he's	he is	aren't/arent	are not
she'll	she will	weren't/werent	were not
he'll	he will	don't/dont	do not
you've	you have	doesn't/doesnt	does not
there're	there are	didn't/didnt	did not
there's	there is	haven't/havent	have not

otherwise zero (filtered by the *StringToWord* in Weka 3.6.12 [3]). Besides, the sentence's label is denoted by a vector with value one at corresponding position and zeros elsewhere. For example, here is an example with positive label in a binary classification task, then its label vector (output) can be defined as $(1, 0)$ in which the first element denotes positive while the second element is negative.

Table 4: Summary statistics of data sets used in the experiments

Data set	size	N	avgL	maxL	C	class sizes	mini-batch
STS-T	495	2067	14.5	32	3	179/139/177	11
STS-G	2000	1172	16.4	33	2	632/1368	50
HCR	2300	1018	18.8	32	3	541/400/1359	46
HCR2	1900	1084	19.1	32	2	541/1359	38
SST	4000	985	16.6	37	3	1251/1800/949	50
SST2	2200	1030	17.6	37	2	1251/949	44
FT	5000	1066	14.0	34	3	1664/1664/1672	50
FT2	3250	1162	15.2	34	2	1625/1625	50
GT	12000	929	13.8	33	3	3983/4013/4004	50
GT2	8000	984	14.3	33	2	3984/4016	50

[*]size: number of examples; N: number of words extracted as the vocabulary; avgL: average text length; maxL: maximal text length; C: number of classes; class sizes: number of examples for positive/neutral/negative if $C = 3$, otherwise for positive/negative if $C = 2$; mini-batch: the size of mini-batch for each data set in the experiments.

3.2 Results and Analysis

In this paper we implement the experiments based on the following four model variations:

(1) DBN-presence: DBN model with the basic bag-of-words representation (presence / non-presence) as input with $\theta = 1$ in Eq. 2.

(2) DBN-position: DBN model with the vocabulary incorporating the word's presence and positional contribution value as input with $0 \leq \theta < 1$ in Eq. 2.

(3) DBN-matrix1: DBN model with the average of first type word-to-segment matrix representation as input. Each word attribute has value one if existing in its own segment of the sentence (full impact on its own segment), while it only has impact on its all preceding segments. That is, the words existing in the first segment have no impact on the other two segments, the words in the second segment have some impact (value a) on the first segment, and finally the words in the third segment have impact on both the second and first segments with decreasing impact values of a and b respectively in Table 1, with the restriction of $0 < b < a < 1$.

(4) DBN-matrix2: DBN model with the average of second type word-to-segment matrix representation as input. Each word attribute has value one if it exists in its own segment of the sentence, and also has impact on its nearest segment(s) including its preceding and posterior segments with the impact values of c, d, e with $0 < c < d < e < 1$ in Table 2.

We perform the above four models on ten data sets listed in Table 4. The DBN structure is manually set to consist of two hidden layers. Specifically, it is: input (visible units) $\rightarrow$ 400 hidden units $\rightarrow$ 100 hidden units $\rightarrow$ output layer (class labels). Meanwhile, the NB classifier is also performed for reference comparison. On the other hand, to speed up the experiments in this paper, each data set is performed using five-fold cross validation (four for training and one for testing) to obtain an average accuracy. For some hyper-parameters in our experiments, firstly in RBM unsupervised training, the momentum is 0.1 and learning rate is 0.1, the number of epochs is set as 50. Secondly, for supervised BP training, the mini-batch sizes are listed in Table 4, the sparsity penalty parameter is 0.1, and the maximum number of epochs is 500 with 10^{-6} as the convergence control based on early stopping rule.

We firstly manually set the parameters for the classification results and give a comparison among different DBN model variations. Each accuracy value in the following tables is average total accuracy of cross validation. Then, we perform a range of experiments to analyze the effect of parameters with respect to θ, (a, b), and (c, d, e), investigating whether there exist some hidden patterns for these parameters in different data sets or whether they are robust to the classification performance. The DBNs are performed on the platform of MATLAB 2014a and the NB model is performed in Weka 3.6.12 in the PC of 64-bit OS, Intel Core i5-5200U, CPU 2.20GHz, and RAM 8.0GB.

Classification Results. To obtain the classification results, we set the parameters for each data set listed in Table 5 for experimental results through a number of tests.

Table 5: Positional parameters of each data set used in the experiments

Model	Position (θ)	Matrix1 (a, b)	Matrix2 (c, d, e)
STS-T	0.3	(0.6, 0.1)	(0.3, 0.4, 0.6)
STS-G	0.6	(0.8, 0.3)	(0.3, 0.4, 0.8)
HCR	0.9	(0.6, 0.5)	(0.2, 0.5, 0.7)
HCR2	0.6	(0.8, 0.3)	(0.2, 0.5, 0.8)
SST	0.8	(0.4, 0.3)	(0.2, 0.4, 0.7)
SST2	0.3	(0.8, 0.1)	(0.2, 0.5, 0.7)
FT	0.4	(0.6, 0.1)	(0.3, 0.5, 0.8)
FT2	0.9	(0.8, 0.3)	(0.2, 0.5, 0.8)
GT	0.7	(0.6, 0.5)	(0.2, 0.5, 0.8)
GT2	0.7	(0.8, 0.3)	(0.2, 0.4, 0.7)

The classification results of different model variations on the ten data sets are shown in Table 6. Note that the NB classifier is performed based on the presence/non-presence word features of each data set. The accuracy values of the three models (DBN-position, DBN-matrix1, DBN-matrix2) are respectively compared with DBN-presence using a two-tailed t-test. The numbers followed by the sign * indicate the accuracy passes the t-test at the significance level of 95% for each data set. The results indicate that all the four DBN models perform better than the NB model (average 63.74%). Meanwhile, DBN-matrix2 has a relatively highest average accuracy (68.26%) for all the data sets, and DBN-presence performs worse than the other three DBN models.

In Table 7, an explicit comparison among the four model variations is summarized, where $w/l/t$ indicates that the model wins in w data sets, loses in l data sets, and ties in t data sets. In Table 7, the three rows (DBN-position, DBN-matrix1, and DBN-matrix2) show their comparisons with DBN-presence respectively. Especially, because the DBN-presence only has ties or losses with the other three models, it is omitted in Table 7. To summarize more clearly, the models are compared in three aspects: all data sets (10), tri-class data sets (5) (Note: five data sets with three labels: STS-T, HCR, SST, FT, and GT), and binary data sets (5) (Note: five data sets with only two labels: STS-G, HCR2, SST2, FT2, and GT2.).

Table 7 shows that DBN-matrix2 performs best on all the ten data sets for one-by-one comparisons (seven wins and three ties). Besides, the DBN-matrix1 and DBN-position also have four wins and five wins respectively, while the DBN-presence only has loses or ties. Consequently, it is obvious that the positional information of sentences exactly provides positive effect on the sentiment classification issues, and also the second type matrix representation is more effective than the positional contribution form. On the other hand, the positional infor-

Table 6: Experimental results on classification accuracy (%)

Model	NB	DBN-presence	DBN-position	DBN-matrix1	DBN-matrix2
STS-T	66.9	61.01	64.65 *	65.66 *	66.06 *
STS-G	79.7	83.50	84.85	82.60	83.25
HCR	63.1	64.91	66.40 *	65.61	66.33 *
HCR2	74.4	78.11	78.66	78.26	78.89
SST	51.7	53.40	54.45	54.80	55.30 *
SST2	66.5	70.00	71.59 *	71.64 *	72.77 *
FT	47.0	51.16	52.70 *	52.56 *	53.44 *
FT2	63.3	68.18	68.71	68.95	68.92
GT	54.4	57.69	59.93 *	60.73 *	60.71 *
GT2	70.4	75.39	75.83	75.46	76.94 *
average	63.74	66.34	67.78	67.61	**68.26**

Table 7: Summary of classification accuracy comparisons

Model	all data sets (10)	tri-class data sets (5)	binary data sets (5)
DBN-position	5/0/5	4/0/1	1/0/4
DBN-matrix1	4/0/6	3/0/2	1/0/4
DBN-matrix2	7/0/3	5/0/0	2/0/3

mation seems more effective for tri-class data sets, as DBN-matrix2 has five wins for all the five tri-class data sets, and DBN-position and DBN-matrix1 have four wins and three wins out of five respectively. While the results for binary data sets are not so good as tri-class data sets. It is probably because the word position and the word order will account more exact information for the sentiment. So the more positional information is integrated, the better for multi-class sentiment classification.

In essence, the positional contribution form and word-to-segment matrix representation are different ways to describe the word positional information of a sentence. For positional contribution form, the contribution values are linearly augmented as the position increases; for matrix1 representation, the words have relatively larger weights in the latter segments, but they are the same in the same segments; while for matrix2 representation, the words in the middle segments have largest weights since they play impact on both its previous and latter segments. To summarize, the word positional representations improve the word presence features for short text sentiment classification.

Effect of Parameters. In our experiments, there are some important parameters which need to be set manually. Whether the results are sensitive to the parameters needs further investigation. Hence, we perform a variety of experiments in DBNs to examine the effect of parameters.

Firstly, we investigate the effect of position parameter θ in Eq. 2. Here the values are set from 0.0 through 1.0 with the interval of 0.1. Each data set is performed with five-fold cross validation for each parameter. The results are shown in Table 8.

Table 8: Classification accuracy of four data sets vs. position parameter

θ	0.0	0.1	0.2	0.3	0.4	0.5	0.6	0.7	0.8	0.9	1.0
STS-T	63.03	58.18	63.03	**64.65**	61.21	61.01	60.81	61.62	57.37	57.37	61.01
STS-G	81.65	82.30	82.75	83.00	83.50	83.50	**84.85**	82.85	84.25	83.75	83.50
HCR	64.30	64.96	64.74	66.00	64.83	65.91	65.78	65.00	64.61	**66.43**	64.91
HCR2	77.89	77.32	76.95	77.95	78.42	78.05	**78.66**	78.11	78.16	78.00	78.11
SST	52.05	51.80	53.85	53.33	53.78	53.73	53.80	54.33	**54.45**	52.50	53.40
SST2	69.95	70.82	71.00	**71.59**	71.41	70.82	71.05	71.09	70.45	69.91	70.00
FT	49.88	49.96	51.28	52.18	**52.70**	51.10	52.28	51.42	51.44	51.90	51.16
FT2	66.83	68.68	67.69	68.43	68.52	67.91	68.09	68.55	67.57	**68.71**	68.18
GT	56.18	58.85	57.74	58.93	59.66	59.81	59.10	**59.93**	59.58	59.34	57.69
GT2	74.00	74.59	75.84	74.15	74.76	75.05	75.43	**75.83**	75.58	74.53	75.39

Table 8 shows the effect of position parameter θ on the average classification accuracy of the ten data sets. In this table, the last data column for each data set with position ratio at 1.0 represents the traditional bag-of-words representation (presence), and the bold number is the highest accuracy value in each row. It means that different ratio values between the word's presence and its positional contribution have different contributions to the text's sentiment. Normally, if θ is relatively bigger, less positional information is integrated, while the word contributes more information if θ is smaller. It indicates that, the parameter value of θ needs a careful investigation for each data set. In Table 8, the θ value corresponding with the bold number for each row is selected for the comparison on the previous classification results (similar in Table 9 and Table 10).

Secondly, the matrix1 representation is investigated for the values of $a = 0.2$, 0.4, 0.6, 0.8 and $b = 0.1, 0.3, 0.5$, with the restriction of $0 < b < a < 1$. Table 9 shows the results of accuracy on each set of (a, b). For instance, the data set of STS-T has the highest average accuracy of 65.66% at $(0.6, 0.1)$ among all the nine parameter settings for matrix1 representation. It gives a picture of the sensitivity for the values of (a, b) on the performance and provides evidence to choose good parameters.

Finally, the matrix2 representation is performed with the parameter values of $c = 0.1, 0.2, 0.3, d = 0.4, 0.5$ and $e = 0.6, 0.7, 0.8$. The accuracy results of each set of (c, d, e) are given in Table 10. For example, the parameter setting $(c = 0.3, d = 0.5, e = 0.8)$ reaches the highest accuracy 53.44% for FT data set.

Table 9: Classification accuracy of four data sets vs. matrix1 parameters

(a, b)	(2,1)	(4,1)	(6,1)	(8,1)	(4,3)	(6,3)	(8,3)	(6,5)	(8,5)
STS-T	64.44	60.00	**65.66**	64.85	61.21	65.45	64.65	65.25	63.84
STS-G	68.40	68.40	72.15	71.40	73.35	74.20	**82.60**	77.85	81.90
HCR	62.13	65.48	65.26	65.13	64.74	64.17	65.09	**65.61**	64.57
HCR2	72.63	73.53	77.53	77.26	77.42	76.89	**78.26**	78.21	77.63
SST	54.23	53.78	52.83	51.10	**54.60**	52.70	53.40	53.25	53.78
SST2	62.27	70.64	70.09	**71.64**	70.82	71.05	70.77	69.41	69.91
FT	45.82	49.84	**52.56**	51.48	50.42	51.72	52.12	49.78	52.36
FT2	48.92	52.55	58.49	65.02	52.00	65.26	**68.95**	67.45	68.28
GT	46.17	48.72	55.51	59.79	51.37	50.78	60.05	**60.73**	59.97
GT2	64.08	64.99	70.78	75.16	74.75	74.91	**75.46**	71.26	75.44

*E.g.(2,1) indicates that (a, b) is $(0.2, 0.1)$, others are similar.

The results of different parameter settings not only show how to choose the parameters, but also demonstrate these parameters play a significant role in DBN models, because the performance results are not very robust to those parameters. Furthermore, it is possible that some other better parameters are not included in our experiments since it is difficult to perform all experiments with exhaustive parameter searching. However, it is really necessary to point out that the positional information (positional contribution and word-to-segment) affects the sentiment classification positively if we set the right parameters.

4　Conclusions and Future Work

In this paper, we propose several ways to incorporate the positional information of texts into DBNs with four model variations for sentiment classification and perform a variety of experiments to verify the effect of positional information towards the sentence sentiments. By choosing the good parameters, the experiments reveal that the word position and word order do provide positive effect on the classification performance. The results indicate that the traditional bag-of-words representation can be improved by incorporating some positional information represented by the word positional contribution form and the word-to-segment matrix representation. Also, it can be seen that positional information works more effective for tri-class classification compared to binary classification. In other words, each word attribute in a sentence has different effect for sentence sentiment.

In the future, this work can be improved from the following views: (1) For the positional representation, the linear positional transformations implemented in this paper seem too simple, and some more sophisticated curve functions (e.g. logit function or symmetrical function) probably perform relatively better; (2)

Table 10: Classification accuracy of four data sets vs. matrix2 parameters

(c, d, e)	(1,4,6)	(1,4,7)	(1,4,8)	(1,5,6)	(1,5,7)	(1,5,8)	(2,4,6)	(2,4,7)	(2,4,8)
STS-T	61.62	65.66	65.86	64.04	60.81	62.02	65.25	65.05	59.80
STS-G	69.20	68.40	82.75	71.35	82.25	83.10	78.60	73.70	79.95
HCR	65.78	65.43	64.43	65.09	64.70	65.35	64.65	65.22	65.17
HCR2	77.00	77.95	77.95	77.00	77.11	76.68	77.68	77.16	76.68
SST	54.15	53.98	52.98	52.90	53.45	53.15	53.15	**55.30**	53.23
SST2	70.41	71.45	71.41	71.36	71.45	71.14	72.14	72.32	71.64
FT	49.80	51.92	52.18	50.14	50.96	53.00	52.82	50.58	52.44
FT2	67.66	68.43	68.40	63.66	68.09	68.28	66.95	65.02	64.28
GT	58.40	55.41	54.12	57.57	55.90	58.96	56.08	56.79	58.34
GT2	71.94	76.06	76.35	70.89	75.59	75.15	76.15	**76.94**	71.40
(c,d,e)	(2,5,6)	(2,5,7)	(2,5,8)	(3,4,6)	(3,4,7)	(3,4,8)	(3,5,6)	(3,5,7)	(3,5,8)
STS-T	63.23	62.42	64.04	**66.06**	64.04	63.64	61.41	64.85	61.21
STS-G	81.20	83.10	83.15	82.20	79.30	**83.25**	83.15	83.20	77.70
HCR	63.91	**66.33**	65.22	65.57	64.22	65.30	65.30	64.87	65.00
HCR2	77.79	77.47	**78.89**	77.00	75.21	76.63	77.42	76.74	77.68
SST	54.13	53.65	53.60	54.39	53.08	53.40	54.05	53.28	53.13
SST2	72.36	**72.77**	71.36	71.14	68.32	70.14	72.05	72.14	71.68
FT	52.24	52.86	52.42	53.14	51.76	52.96	52.72	52.40	**53.44**
FT2	66.03	67.23	**68.92**	64.89	67.91	67.26	66.31	68.43	68.03
GT	58.40	56.28	**60.71**	43.08	60.21	60.31	59.19	56.75	56.06
GT2	75.54	75.99	76.29	76.15	75.80	75.74	76.20	76.30	75.26

[*]E.g. (1,4,6) means (c, d, e) is (0.1, 0.4, 0.6). Similar to other numbers.

Try to extend the matrix representation into a row vector for each sentence (even though it will be an extremely large sparse matrix and cost much memory during the training, this is an approach), letting each element value in the matrix be a single feature into the model; (3) Only three segments are introduced here, more segments (e.g. four or more segments) for a sentence (or long text) may be more reasonable, which will probably account more word order information into the model. These will be some of our future research directions.

References

1. Agarwal, A., Xie, B., Vovsha, I., Rambow, O., Passonneau, R.: Sentiment analysis of twitter data. In: Proceedings of the workshop on languages in social media. pp. 30–38. Association for Computational Linguistics (2011)
2. Collobert, R., Weston, J.: A unified architecture for natural language processing: Deep neural networks with multitask learning. In: Proceedings of the 25th international conference on Machine learning. pp. 160–167. ACM (2008)

3. Hall, M., Frank, E., Holmes, G., Pfahringer, B., Reutemann, P., Witten, I.H.: The weka data mining software: an update. ACM SIGKDD explorations newsletter 11(1), 10–18 (2009)
4. Hinton, G.: A practical guide to training restricted boltzmann machines. Momentum 9(1), 926 (2010)
5. Hinton, G., Osindero, S., Teh, Y.W.: A fast learning algorithm for deep belief nets. Neural computation 18(7), 1527–1554 (2006)
6. Hu, X., Tang, J., Gao, H., Liu, H.: Unsupervised sentiment analysis with emotional signals. In: WWW 2013 - Proceedings of the 22nd International Conference on World Wide Web. pp. 607–617 (2013)
7. Johnson, R., Zhang, T.: Effective use of word order for text categorization with convolutional neural networks. In: NAACL HLT 2015 - 2015 Conference of the North American Chapter of the Association for Computational Linguistics: Human Language Technologies, Proceedings of the Conference. pp. 103–112 (2015)
8. Liu, B.: Sentiment analysis and opinion mining. Synthesis lectures on human language technologies 5(1), 1–167 (2012)
9. Pahikkala, T., Pyysalo, S., Boberg, J., Jrvinen, J., Salakoski, T.: Matrix representations, linear transformations, and kernels for disambiguation in natural language. Machine Learning 74(2), 133–158 (2009)
10. Pang, B., Lee, L., Vaithyanathan, S.: Thumbs up?: sentiment classification using machine learning techniques. In: Proceedings of the ACL-02 conference on Empirical methods in natural language processing-Volume 10. pp. 79–86. Association for Computational Linguistics (2002)
11. Ponte, J.M., Croft, W.B.: A language modeling approach to information retrieval. In: Proceedings of the 21st annual international ACM SIGIR conference on Research and development in information retrieval. pp. 275–281. ACM (1998)
12. Raychev, V., Nakov, P.: Language-independent sentiment analysis using subjectivity and positional information. In: RANLP. pp. 360–364 (2009)
13. Rumelhart, D.E., Hinton, G.E., Williams, R.J.: Learning representations by back-propagating errors. Cognitive modeling 5 (1988)
14. Saif, H., Fernandez, M., He, Y., Alani, H.: Evaluation datasets for twitter sentiment analysis: a survey and a new dataset, the sts-gold (2013)
15. Sarikaya, R., Hinton, G.E., Deoras, A.: Application of deep belief networks for natural language understanding. IEEE Transactions on Audio, Speech and Language Processing 22(4), 778–784 (2014)
16. Socher, R., Huval, B., Manning, C.D., Ng, A.Y.: Semantic compositionality through recursive matrix-vector spaces. In: Proceedings of the 2012 Joint Conference on Empirical Methods in Natural Language Processing and Computational Natural Language Learning. pp. 1201–1211. Association for Computational Linguistics (2012)
17. Socher, R., Perelygin, A., Wu, J.Y., Chuang, J., Manning, C.D., Ng, A.Y., Potts, C.: Recursive deep models for semantic compositionality over a sentiment treebank. In: Proceedings of the conference on empirical methods in natural language processing (EMNLP). vol. 1631, p. 1642. Citeseer (2013)
18. Tang, D., Wei, F., Yang, N., Zhou, M., Liu, T., Qin, B.: Learning sentiment-specific word embedding for twitter sentiment classification. In: 52nd Annual Meeting of the Association for Computational Linguistics, ACL 2014 - Proceedings of the Conference. vol. 1, pp. 1555–1565 (2014)
19. Zhai, C., Lafferty, J.: A study of smoothing methods for language models applied to information retrieval. ACM Transactions on Information Systems (TOIS) 22(2), 179–214 (2004)

Tracking Multiple Social Media for Stock Market Event Prediction

Fang Jin[1], Wei Wang[2], Prithwish Chakraborty[2], Nathan Self[2], Feng Chen[3], and Naren Ramakrishnan[2]

1 Department of Computer Science, Texas Tech University
2 Discovery Analytics Center, Department of Computer Science, Virginia Tech
3 Department of Computer Science, University at Albany, SUNY

Abstract. The problem of modeling the continuously changing trends in finance markets and generating real-time, meaningful predictions about significant changes in those markets has drawn considerable interest from economists and data scientists alike. In addition to traditional market indicators, growth of varied social media has enabled economists to leverage micro- and real-time indicators about factors possibly influencing the market, such as public emotion, anticipations and behaviors. We propose several specific market related features that can be mined from varied sources such as news, Google search volumes and Twitter. We further investigate the correlation between these features and financial market fluctuations. In this paper, we present a Delta Naive Bayes (DNB) approach to generate prediction about financial markets. We present a detailed prospective analysis of prediction accuracy generated from multiple, combined sources with those generated from a single source. We find that multi-source predictions consistently outperform single-source predictions, even though with some limitations.

Keywords: Market prediction; Multiple social media; Features combination; Google trends; Twitter burst; News sentiment.

1 Introduction

Predictions concerning financial markets are complicated by their inherent volatility. Capturing signals of this volatility and providing proper estimates about 'market flips' is of prime interest to economists. This problem has attracted great interest from researchers in diverse disciplines such as economics, statistics and data science. Consequently this has led to a wide variety of methods aimed at modeling stock markets [11, 16, 17, 20, 23].

In most of the traditional approaches, researchers characterize the stock market by the historical records of prices and try to find signatures that indicate rising or falling prices based on this historical time series. However, such financial time series methods are generally incognizant of human indicators, such as public reaction, and have frequently been found wanting in their accuracy at predicting sudden, large changes in market value [4]. Recently, with the pervasive growth of social media [6, 14] which allow individuals to readily express their

© Springer International Publishing AG 2017
P. Perner (Ed.): ICDM 2017, LNAI 10357, pp. 16–30, 2017.
DOI: 10.1007/978-3-319-62701-4_2

sentiments [21], views and concerns, real-time mining of such factors has become possible. Furthermore, different aspects of public sentiment can now be extracted by analyzing multiple social networks. In this paper, we collect and analyze global search trend data from Google, archived news articles from Bloomberg News and relevant tweets from Twitter. Using unsupervised methods, we extract features from these publicly available data sources. Using these features, we design a set of experiments to investigate the correlations between human behavior and market fluctuations in South American markets. With this analysis, we propose models that predict large changes (events) in market value using the most informative extracted factors. To be specific, given these three data sources at day d and historical stock prices for a market, our proposed models attempt to predict the stock market value of at least day $d + 1$.

The key contributions of this paper are:

- We propose a systematic analysis of Google Search Trends, Bloomberg News and Twitter to gather information about market trends and quantify these social media trends.
- We identify burst features from Twitter and further group burst features into burst events. We also investigate and find the correlations of these burst events with market trends.
- We present Delta Naive Bayes Model to predict finance market fluctuations by fusing multiple social media sources. Though there have been earlier attempts at investigating combinations of sources for finance related applications [12, 19], most of the work has focused on surveying datasets. In that respect, to the best our knowledge, our work is the first not only to compare sources but also to propose predictive models that leverage multiple sources.
- Finally, we present our findings about the underlying cross-correlation among market indicators extracted from multiple social media sources and extensively analyze the information from each data source.

2   Related Work

In this section we review a few works related to the fields of financial time series analysis, modeling of financial markets and extraction of features from online data streams.

Financial Time Series Analysis Financial time series analysis has been one of the most popular approaches to market modeling. Generalized Autoregressive Conditional Heteroskedasticity (GARCH) models [5] have been widely applied in the financial domain since the 1980s. Clustering algorithms have been applied to redescribe time series [11] and identify temporally correlated stocks [1], methods which we have employed in our data processing. More recently, [17] proposed information dissipation length as a leading indicator to measure global instability.

Correlation with Social Media With the recent development and prevalence of big data platforms [25, 24], adopting data mining tasks on online social networks has been shown to produce state-of-the-art results. There have been a number of exploratory analyses correlating social media with stock markets. [16] found a correlation between transaction volumes of top companies and Google search volumes of those companies' names. [12] investigated the correlation of search query volume and the Dow Jones Industrial Average (DJIA) and found that a higher search volume of certain finance terms indicates lower DJIA prices. Further, [15] found that trading strategies based on the volumes of 98 keywords from Google Search Trends outperformed random investment with respect to overall turnover. Recently, [13] study the "co-movements" between stock prices and news articles for stock market prediction. [4] calculated public sentiment from Twitter and found the "calm" sentiment curve has an especially strong correlation with DJIA values. [20] found that the number of connected components in a constrained subgraph within time-constrained graphs has high correlation with traded volume. In this paper, we build on the research that suggests that aggregate search volume and sudden changes in sentiment across social networks are correlated with financial market performance by combining these factors in a unified prediction framework.

Feature Fusion With respect to fusion methods, [10] proposed the popular Kalman Filter approach for linear filtering and prediction problems which measures multiple sequential sensors to estimate a system's dynamic states. The naive Bayes classifier has been recognized as an effective model which can estimate class labels for multi-dimensional features based on maximum posterior probabilities. Hidden Markov models have been applied successfully for temporal pattern recognition in areas such as sequential images [27]. In our experiment, we employ and revise the classic naive Bayes model, for feature fusion and finance prediction.

3   Feature Extraction

The first step in our method is to extract features from social medias. As shown in Figure 1, each data source requires processing to become a time series which we use for prediction. For articles from Bloomberg News, we use topic modeling, natural language processing and sentiment analysis to estimate values for economic indicators. For Google data, we employ Lasso regression to determine which terms from our custom set of finance related terms are the most informative. For Twitter, we chain "burst features" into "burst events" and consider the ones with the greatest hotness degree.

3.1   Google Search Trends

We consider Google search volumes a source for surrogate information about public feelings towards certain stock indices and concerns about financial futures. These dynamic trends are potentially important factors which can influence a

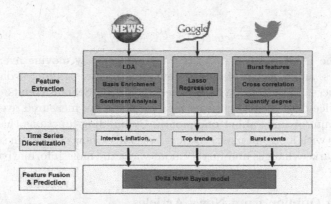

Fig. 1. Overall System Framework.

finance market analysis framework. Several authors have used Google search volumes to predict stock value changes [12, 15, 16]. However, since their approaches compute search volumes for keywords and phrases using cumulative returns over a fixed period of time, they are prone to miss the dynamic contribution of keywords over a search cycle as the importance of those words changes. To find the most informative terms for each cycle, we use L1-regularization, or Lasso [22] to filter our dictionary of finance terms. This approach makes our method robust enough to easily incorporate expert feedback by adding keywords as needed.

Formally, for each of our K keywords we can compute its search volume, $x^k(t)$, at time t. We propose that this volume influences future stock prices, $P(t+1)$, at time $t+1$. We compare the change in market price, $\Delta P(t+1)$, at time $t+1$ with the change in search volume of the keywords at time t, $\Delta x^k(t)$. We express this correlation as a linear equation as shown in Equation 1. a_k is each term's weight.

$$\Delta P(t+1) = \sum_{k=1}^{K} a_k \Delta x^k(t) \qquad (1)$$

Weights are computed for each term using Equation 2 where the best keywords are the ones with highest, non-zero weights. We fix λ from empirical evaluations based on the accuracy of the final predictions.[1]

$$\begin{aligned} a &= (a_1, a_2, \ldots, a_K) \\ &= argmin_a(\sum_t (\Delta P(t) - \sum_{k=1}^{K} a_k \Delta x^k(t))^2 + \lambda \sum_k |a_k|) \end{aligned} \qquad (2)$$

Since Google search volumes are reported only weekly, they can mask the general tendency of public queries. We address this problem by considering the $z - score(4)$ of the Google search volumes for each week as the representative value for every day of that week. For market prices, which change every day, we use $z - score(30)$ of each day's closing price as the value for that day. A $z - score$

---

[1] For the results reported in this paper we use $\lambda = 0.8$.

is defined by:

$$z - score(n) = (X - \mathcal{M})/\Sigma \tag{3}$$

where X is the 1-day difference, $\mathcal{M}$ is the trailing n-day moving average of 1-day differences, and Σ is the standard deviation of those trailing n-day moving 1-day differences. This step aligns the weekly Google Search Trends data with daily stock price data. It also ensures that variances are measured over roughly the same amount of time: about one month. By building functions that connect Google trends with financial market values, we can see how changes in search volume reflect market changes and furthermore select the most informative features for prediction.

3.2 Mining Opinion from News Articles

To make these predictions, we are interested in economic features such as interest rates, inflation rates, GDP, consumer confidence, and foreign investment. Unfortunately, official values for these features are not immediately available which makes them unusable for real-time economic forecasting. Instead of relying on values from state-run or research institutions, we employ natural language and statistical processing on economic news stories published online by Bloomberg News and consider the processed content a surrogate for the actual, unavailable data.

We collected 401,923 Bloomberg articles (freely available using their API), from April 2010 to June 2013. To filter this collection to only articles about financial news, we employ the popular Latent Dirichlet Allocation [2] topic modeling algorithm to identify interesting topics. We randomly chose 25,041 articles as a training set, fixed the number of topics at 30 and computed a set of topics. We then manually selected the topics most relevant to finance and economics based on an analysis of the top 30 keywords of each topic. Any document in the corpus composed of any of the selected topics is considered relevant for financial prediction.[2]

Next, we use a suite of natural language processing tools developed by Basis Technology to perform lemmatization, sentence boundary detection, part of speech tagging and noun phrase detection on each article. From there, we use a custom country level dictionary to bin articles by country. This natural language processing allows us to detect passages that refer to a particular country since articles may discuss multiple countries. Finally, we use a customized economic feature dictionary along with noun phrase detection outputs to calculate sentiment scores for each feature for each country. These sentiment scores serve as inputs for our economic predictions.

3.3 Twitter Burst Detection

With the explosion in social media networking, Twitter has quickly evolved as an invaluable source for reflecting social movements and exhibiting the complex emotions of individuals. Research [4, 20] shows that financial markets are

---

[2] A single document can comprise of at most five topics.

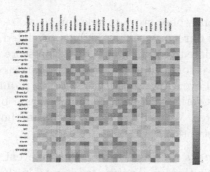

Fig. 2. Feature cross correlation from Twitter burst events.

closely tied with social movements and, furthermore, driven by human emotions. Using Twitter, we can detect the most recent dramatic societal movements in real-time which we represent as Twitter bursts [7]. Analyzing the correlation between these bursts and market turbulence, we investigate how they correlate with financial markets. Our entire framework for the Twitter burst detection involves the following steps:

1. Build the network of burst features: For a given window of time, we employ TF-IDF to identify burst features and assign each feature with a burst score.
2. Cluster burst features into burst events: Each pair of nodes that are correlated are connected by an edge with weight equal to their correlation score.
3. Calculate surrogate data: We use Twitter volume and sentiment scoring to measure each burst event's influence.

Identifying Burst Features Since we are interested in predictions for country-wide stock markets, we need to determine from which country a tweet was generated. Though tweets can contain user supplied latitudes and longitudes, most tweets do not have this data and must be geolocated by other means. To circumvent this, we applied another geo-enriching algorithm to our dataset by using the geocoding tool described in [18]. Once we have a set of tweets for a country of interest we calculate the 'burst score' for each of its terms over time. Each term's burst score is calculated using 'term frequency-inverse document frequency'(TF-IDF). At given time t, when a term's $|z - score(30)|$ is greater than a threshold, it was labeled as a burst feature.

Grouping into Burst Events After calculating burst scores for each term, we have a set of burst feature vectors denoted by $B = \{b_0, b_1, \ldots, b_l\}$. We intend to build a graph, $G = (B, E, C)$, such that nodes are connected by an edge in E if there exists a strong enough correlation C between the two nodes [26]. We use Equation 4 to calculate correlation scores where f^* denotes the complex conjugate of f and m is the time lag. The results can be seen in Figure 2.

$$(f * g)[n] = \sum_{m=-\infty}^{\infty} f^*[m]g[n + m] \tag{4}$$

We only consider vectors whose cross-correlation score is greater than a predefined threshold, 0.1 for our purposes, as highly correlated. We create an edge, e, between highly correlated vectors with correlation score c. Note that because only highly correlated vectors (burst features) are connected by edges, grouping burst features into burst events is transformed into the problem of detecting communities in Twitter networks with each community as a burst event composed of a set of burst features. We employ the Louvain method [3] for community detection in Twitter networks.

Measure Event *Hotness* Degree We consider an event that takes place over a short period of time with high Twitter volume to be *hot*. For each event E, we take the list of burst features f_i and their associated Twitter volumes and calculate the event's hotness degree. Combining public reaction to this event, we label an event with optimistic and pessimistic quantifiers. Mathematically, for a topic with m burst events, Twitter volume v_i and negative sentiment accounts s_{ni}, its hotness degree, $H_e(t)$, at time slot t can be computed according to Equation 5.

$$H_e(t) = \sum_{i=1}^{m} \frac{v_i * s_{ni}}{t} \tag{5}$$

4 Ensemble and Prediction

Google search volume is generally considered to be a weather vane of consumer confidence and public expectations about economic situations, which can serve as a background detector for market predictions [15, 16]. Internet articles from companies like Bloomberg News can be a good source for insider opinions, expert views, important event reports and situation analysis, factors containing more information and relatively more professional and accurate signals compared to other social media sources [8]. Finally, Twitter is one of the more popular social media networks for capturing emerging social movements. Especially for breaking news, Twitter moves much faster than traditional media [9]. To make full use of these sources, we construct a feature fusion framework to combine the features extracted from each: news factors (F_n), Google search volume factors (F_g), and Twitter event factors (F_t). We expect to obtain an improved performance by combining these three sources in an intelligent way.

4.1 Feature Fusion

Since our observation features come from three different sources, we consider them to be independent of each other. As shown in Figure 3, we denote $Y = \{y_n\}$ as the label of the historical stock price class, where y_n represents the stock price class on day n. We also denote the observation features from Google Search Trends, Bloomberg News and Twitter as F_g, F_n and F_t, respectively. We apply a strategy for combining these features, which can be expressed as $x_n = (F_g, F_n, F_t)_n$, where x_n is the combination of the three feature observed on day $n - 1$. In addition, normalization is required before feature fusion since F_g, F_n and F_t might be of different scales.

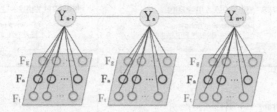

Fig. 3. Feature fusion: Y_n is the state of the market. F_n is news factor. F_t is Twitter factor. F_g is Google trends factor.

4.2   Stock Discretization

To ensure that periods of high volatility and changes that occur over many weeks are captured accurately, we calculate $z - scores$ over a time window of 30 days. Based on $z - score$ thresholds, we define stock status as following. When $z - score(30) \leq -1$, we call it as **'plunge'** status, when $z - score(30) \geq 1$, we call it as **'surge'** status, otherwise, when $|z - score(30)| < 1$, we call it **'flat'** status.

4.3   Delta Naive Bayes Model

Naive Bayes is a well known classifier which, assuming conditional independence of features, computes the posterior probability of labels for an input feature set. Unfortunately, the basic naive Bayes model is suboptimal for financial market predictions because of the highly skewed class distribution. In our training process, we found that 78.6% of our features were assigned to the same class. The result is reasonable considering the stock market values do not often have large, abrupt value changes.

Algorithm 1 Delta Naive Bayes Model

INPUTS: historical dataset $\{(x_{n-L}, y_{n-L}), ..., (x_n, y_n)\}$, training window size L, and probabilty threshold t, and most recent observation $\{x_{n+1}\}$
OUTPUTS: $\hat{y}_{n+1}$
  1. Using Naive Bayes Model, compute conditional probability matrix $P(Y|X)$ from $s_0 = \{(x_{n-L}, y_{n-L}), ..., (x_{n-1}, y_{n-1})\}$;
  2. Using Naive Bayes Model, compute conditional probability matrix $Q(Y|X)$ from $s_1 = \{(x_{n-L+1}, y_{n-L+1}), ..., (x_n, y_n)\}$;
  3. Compute $\Delta P(Y|X)=\text{abs}(Q(Y|X) - P(Y|X))$;
  4. $\bar{p}_{max} = \max_{y_j} \Delta P(Y = y_j|X = x_{n+1})$;
  5. Return $\hat{y}_{n+1} = \arg\max_{y_j} \Delta P(Y = y_j|X = x_{n+1})$ when $\bar{p}_{max} > t$; otherwise return $\hat{y}_{n+1}$ as **'flat'**.

To overcome this disadvantage, we propose the Delta Naive Bayes Model in Algorithm 1. Unlike the basic naive Bayes model, this classifier is based on the

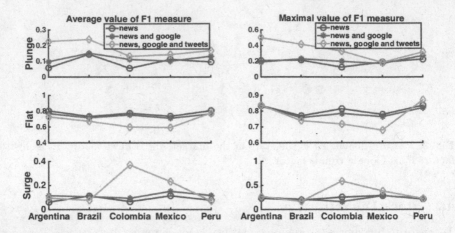

Fig. 4. Performance with single news, two sources and three sources respectively on multiple countries. The upper row shows the plunge status prediction performance, with the left one present average F_1 measure, and the right one present maximal F_1 measure. The middle row shows the flat status performance. The bottom row shows the surge status performance.

incremental change of the posterior probability instead of the posterior probability itself. For the nth time step in the discretized series, we compute $P(Y|X)$ and $Q(Y|X)$ on the historical datasets $s_0 = \{(x_{n-L}, y_{n-L}), \ldots, (x_{n-1}, y_{n-1})\}$ and $s_1 = \{(x_{n-L+1}, y_{n-L+1}), \ldots, (x_n, y_n)\}$, respectively. The difference between s_0 and s_1 is represented by $\Delta P(Y|X) = Q(Y|X) - P(Y|X)$. For a new observation x_{n+1}, we don't use $P(Y|X)$ or $Q(Y|X)$ directly as the classifier. Instead, we first check the maximal value of the conditional probability, $\bar{p}_{max} = \max_{y_j} \Delta P(Y = y_j | X = x_{n+1})$. If $\bar{p}_{max}$ is larger than some predefined threshold t, y_{n+1} will be predicted as:

$$\hat{y}_{n+1} = \arg\max_{y_j} \Delta P(Y = y_j | X = x_{n+1}).$$

Otherwise, y_{n+1} will be predicted as **'flat'** state.

As we can see, the only difference between dataset s_0 and s_1 is that the point (x_{n-L}, y_{n-L}) is replaced with the most recent point (x_n, y_n). Hence the difference of the two conditional probability matrices $P(Y|X)$ and $Q(Y|X)$ should be quite small. This helps explain why the basic naive Bayes model nearly always assigns the same class. In addition, the influence of the most recent point (x_n, y_n) on our prediction is attenuated by the other $L-1$ points. It is important to note that using L points helps us avoid the effects of a few noisy data points. The choice of L is also important. Using a low value reduces computation complexity of computing the conditional probability matrix $P(Y|X)$ but risks greater interference from noisy points. On the other hand, using a high value for L leads to better masking of interference but greater computational complexity.

Table 1. Delta Naive Bayes model performance on Colombia stock index COLCAP. N denotes source of news, G represents Google trends source and T means Tweets source. t shows transition threshold, p represents precision and r denotes recall.

Source	t	Plunge			Flat			Surge		
		p	r	F_1	p	r	F_1	p	r	F_1
N	0.02	0.147	0.074	0.097	0.69	0.78	0.731	0.146	0.132	0.138
	0.04	0.165	0.048	0.069	0.683	0.873	0.764	0.072	0.045	0.052
	0.06	0.133	0.028	0.044	0.686	0.909	0.78	0.088	0.035	0.047
	0.08	0.145	0.023	0.038	0.688	0.93	0.79	0.105	0.035	0.049
	0.1	0.195	0.018	0.033	0.691	0.949	0.799	0.112	0.03	0.043
NG	0.02	0.156	0.097	0.118	0.697	0.742	0.718	0.142	0.153	0.147
	0.04	0.178	0.097	0.124	0.702	0.817	0.754	0.138	0.094	0.11
	0.06	0.17	0.083	0.109	0.692	0.839	0.757	0.104	0.063	0.075
	0.08	0.186	0.083	0.112	0.688	0.856	0.762	0.117	0.054	0.07
	0.1	0.19	0.071	0.1	0.688	0.881	0.772	0.149	0.049	0.068
NGT	0.02	0.135	0.169	0.149	0.582	0.499	0.535	0.325	0.418	0.366
	0.04	0.134	0.146	0.139	0.593	0.594	0.588	0.379	0.418	0.394
	0.06	0.128	0.124	0.123	0.583	0.63	0.6	0.405	0.418	0.407
	0.08	0.15	0.124	0.134	0.587	0.681	0.627	0.374	0.358	0.36
	0.1	0.175	0.091	0.118	0.584	0.74	0.65	0.355	0.302	0.322

5   Empirical Evaluation

Our goal is to predict significant fluctuations in the value of a stock exchange. For this challenge, we focus on South American countries including Argentina, Brazil, Colombia, Mexico, and Peru. We first describe our evaluation criteria then present a detailed performance analysis.

5.1   Evaluation Criteria

In the absence of a standard performance matrix, we define an evaluation criteria based on 'hit rate' as follows. For a single day there are only three status: plunge or flat or surge. To account for the accuracy of each status, only when the prediction status is exactly the same, we consider the prediction to be correct.

Recall, precision and F_1 measure are widely used metric employed in classification problem. A formal definition of recall is $r = \frac{TP}{TP+FP}$, precision is $p = \frac{TP}{TP+FN}$ and F_1 measure is $F_1 = \frac{2rp}{r+p}$, where TP denotes true positive, FP shows false positive, and FN means false negative. While it is fairly straightforward to compute precision and recall for a binary classification problem, it can be quite confusing as to how to compute these values for a multi-class classification problem. In our case, we are 3-class multi classification problem. For an individual class, the false positives are those instances which were classified as that class, but in fact aren't, and the true negatives are those instances which are not that class, and were indeed classified as not belonging to that class (regardless of whether they were correctly classified).

5.2 Performance Analysis

Our experiment dataset is from January 1, 2012 to July 31, 2013. We set training window size L vary from 40 to 200 days, to test the efforts in shaping quality distribution. We present an exhaustive evaluation of Delta Naive Bayes model against multiple aspects as follows:

How do our model perform for the 5 countries of interest and how well does the prediction for plunge, flat and surge status? We plot the overall prediction performance for each country regard the three stock status in Figure 4. We can see while predict plunge status, Brazil and Argentina achieve prominent performance. When predicting surge status, Colombia is superior than the rest countries, with F_1 measure as high as 0.6. While predicting the flat status, Mexico and Peru exhibit outstanding and stable performance.

Single source or multiple sources, which performance is better at capturing surge/plunge status? As we know, most of the status of stock market is flat status, which makes it even challenging to predict the surge or plunge status. In our experiment, we pay special attention to surge and plunge status prediction, which is shown as upper row and bottom row, respectively in Figure 4. We can see, three sources prediction persistently outperforms two sources prediction, while two sources prediction is greater or equal to single feature prediction moderately. Another clear evidence can be found in Table 1, where shows detailed prediction performance of Colombia, in recall, precision and F_1 measure. We can see as for the surge status, the combined three-sources prediction F_1 measure can be as high as 0.407, while the two-sources prediction maximal F_1 measure is 0.147, and single news prediction maximal score as low as 0.138.

How does the transition threshold t help shape our quality distribution, in recall and precision? For Flat state, a larger t often generates a better F_1. While Plunge and Surge state, a larger t is often accompanied by a lower F_1. So the selection of t is a tradeoff among plunge, surge and flat $F_1 measure$.

How does training size window L vary with F_1 measure scores? We find that given transition threshold, different training size window lead to different F_1 measure depending on fused sources, as can be seen in Figure 5(a). In the case of Colombia, as the three sources prediction, the best performance happens at training window size of 70, both in the plunge status and surge status. However, the relationship are not always consistent across different fusing sources. Thus a general statement about the efficacy of the training window sizes would be hasty since even though the combined sources may show some trends as discussed above, the relationships may be different for a particular example.

5.3 Case Study

In this section we further explore the importance of sources by looking at a few specific examples as case studies.

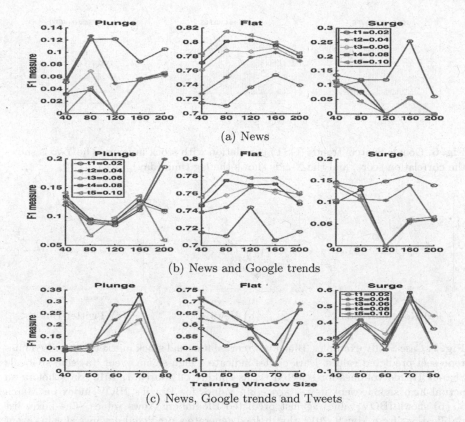

(a) News

(b) News and Google trends

(c) News, Google trends and Tweets

Fig. 5. Prediction performance of Colombia, with various threshold and training window size. The X axis denotes training window size and Y axis shows F_1 measure. (a)Prediction using single source of news. (b)Prediction using news and Google trends. (3)Prediction using news, Google trends and Tweets.

Google Search Trends Google Search Trends of a specific country trends to have a high correlation with country level stock index. We plot four examples of Google Search Trends(GST) terms with country level stock index in Figure 6. We can see the correlation score between GST and stock index can be as high as close to 0.8. Google Search Trends also proved to be a good leading indicator. Figure 7(a) shows the correlation of the tracked Google search trends feature with the $z - score(30)$ of IBOV stock index. As we can see, GST showed an appreciable peak just days before the actual peak on March 27, 2013.

News Opinion News opinions were found to be great indicators and for some cases are quite good at detecting sudden changes in the stock market. As can be seen in Figure 7(b), with the help of a series of news articles from May 9, 2012, including "*Brazil Bulls Capitulate as State Intervention Spurs Outflows,*" the news mining approach successfully predicted that the IBOV index would be

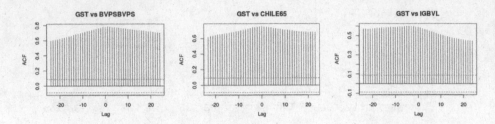

Fig. 6. Google Search Trends (GST) correlation with stock indexes. The Y-axis shows the correlation score, and the X-axis shows the lag time (day).

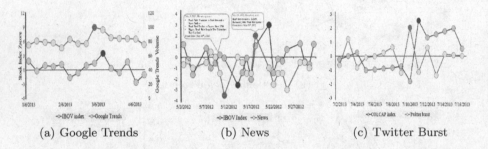

(a) Google Trends   (b) News   (c) Twitter Burst

Fig. 7. Case Study examples. Blue lines represent actual stock market values. Red lines represent predicted values. Dark circles indicate a high-sigma event. (a) shows Google Search Trends volumes for selected terms spiking on March 1, 2013, days before an actual high sigma event in $z - score(30)$ values for Brazil's IBOV index on March 5. (b) shows IBOV values against predicted Bloomberg News values. The boxes list publications from May 9, 2012, that helped generate a predicted downward value event on May 14 and from May 20 that were used in the prediction of the uptick on May 21. (c) shows values for Twitter bursts surging on July 10, 2013, before a high sigma event in $z - score(30)$ values for Colombia's COLCAP index on July 12.

adversely affected. Similarly, the approach was also able to predict an uptick on May 21 based on the news entitled *"Brazil Says Economic Growth Recovering After Weak First Quarter,"* published on May 20. The key takeaway from a few such other examples was that our news mining method is very effective for market prediction, especially when high quality news is available.

Twitter Burst Events At the same time our case studies indicate that Twitter is also an invaluable source of information. News moves through Twitter much faster than traditional news media. Figure 7(c) shows an example highlighting how Twitter bursts are leading indicators of stock movements and, although noisy, if properly processed can be an invaluable real-time indicator for financial markets. In our experiments, the burst event detection method has been able to capture such extraordinary movements in Twitter chatter and effectively detect bursts of market-influencing events.

6 Discussion

In this paper, we propose a novel approach to anticipate financial market fluctuations by combining multiple social media sources. We employ language modeling, topic clustering and sentiment analysis to extract opinions from news articles, Lasso regression on Google search volume data to select the most informative features, and burst detection and event grouping techniques on Twitter data to identify market indicators from Twitter data. Finally, we present the results using Delta Naive Bayes model. In the process, we also demonstrate that the combination of multiple data sources achieves better prediction performance than any of these media sources alone.

7 Acknowledgment

Supported by the Intelligence Advanced Research Projects Activity (IARPA) via Department of Interior National Business Center (DoI/NBC) contract number D12PC000337, the US Government is authorized to reproduce and distribute reprints of this work for Governmental purposes notwithstanding any copyright annotation thereon. Disclaimer: The views and conclusions contained herein are those of the authors and should not be interpreted as necessarily representing the official policies or endorsements, either expressed or implied, of IARPA, DoI/NBC, or the US Government.

References

1. Basalto, N., Bellotti, R., De Carlo, F., Facchi, P., Pascazio, S.: Clustering stock market companies via chaotic map synchronization. Physica A: Statistical Mechanics and its Applications 345(1), 196–206 (2005)
2. Blei, D.M., Ng, A.Y., Jordan, M.I.: Latent dirichlet allocation. JMLR 3, 993–1022 (2003)
3. Blondel, V.D., Guillaume, J.L., Lambiotte, R., Lefebvre, E.: Fast unfolding of communities in large networks. Journal of Statistical Mechanics: Theory and Experiment 2008(10), P10008 (2008)
4. Bollen, J., Mao, H., Zeng, X.: Twitter mood predicts the stock market. Computational Science 2(1), 1–8 (2011)
5. Bollerslev, T.: Generalized autoregressive conditional heteroskedasticity. Journal of econometrics 31(3), 307–327 (1986)
6. He, W., Guo, L., Shen, J., Akula, V.: Social media-based forecasting: A case study of tweets and stock prices in the financial services industry. Journal of Organizational and End User Computing (JOEUC) 28(2), 74–91 (2016)
7. Jin, F., Khandpur, R.P., Self, N., Dougherty, E., Guo, S., Chen, F., Prakash, B.A., Ramakrishnan, N.: Modeling mass protest adoption in social network communities using geometric brownian motion. In: Proc. KDD'14. pp. 1660–1669. ACM (2014)
8. Jin, F., Self, N., Saraf, P., Butler, P., Wang, W., Ramakrishnan, N.: Forex-foreteller: Currency trend modeling using news articles. In: Proc. KDD'13 Demo Track. pp. 1470–1473. ACM (2013)

9. Jin, F., Wang, W., Zhao, L., Dougherty, E., Cao, Y., Lu, C.T., Ramakrishnan, N.: Misinformation propagation in the age of twitter. Computer 47(12), 90–94 (2014)
10. Kalman, R.: A new approach to linear filtering and prediction problems. Journal of basic Engineering 82(1), 35–45 (1960)
11. Lavrenko, V., Schmill, M., Lawrie, D., Ogilvie, P., Jensen, D., Allan, J.: Mining of concurrent text and time series. In: KDD'00 Workshop. pp. 37–44 (2000)
12. Mao, H., Counts, S., Bollen, J.: Predicting financial markets: Comparing survey, news, twitter and search engine data. Quantitative Finance Papers 1112(1051) (2011)
13. Ming, F., Wong, F., Liu, Z., Chiang, M.: Stock market prediction from wsj: Text mining via sparse matrix factorization. In: Data Mining (ICDM), 2014 IEEE International Conference on. pp. 430–439. IEEE (2014)
14. Piñeiro-Chousa, J., Vizcaíno-González, M., Pérez-Pico, A.M.: Influence of social media over the stock market. Psychology & Marketing 34(1), 101–108 (2017)
15. Preis, T., Moat, H.S., Stanley, H.E.: Quantifying trading behavior in financial markets using Google Trends. Scientific reports 3 (2013)
16. Preis, T., Reith, D., Stanley, H.E.: Complex dynamics of our economic life on different scales: insights from search engine query data. Phil Trans Math Phys Eng Sci 368(1933), 5707–5719 (2010)
17. Quax, R., Kandhai, D., Sloot, P.M.: Information dissipation as an early-warning signal for the Lehman Brothers collapse in financial time series. Scientific reports 3 (2013)
18. Ramakrishnan, N., Butler, P., Muthiah, S., Self, N., Khandpur, R., Saraf, P., Wang, W., Cadena, J., Vullikanti, A., Korkmaz, G., et al.: 'beating the news' with embers: forecasting civil unrest using open source indicators. In: Proc. KDD'14. pp. 1799–1808. ACM (2014)
19. Rao, T., Srivastava, S.: Modeling movements in oil, gold, forex and market indices using search volume index and twitter sentiments. In: Proc. WebSci'13. pp. 336–345 (2013)
20. Ruiz, E.J., Hristidis, V., Castillo, C., Gionis, A., Jaimes, A.: Correlating financial time series with micro-blogging activity. In: Proc. WSDM'12. pp. 513–522 (2012)
21. Sul, H.K., Dennis, A.R., Yuan, L.I.: Trading on twitter: Using social media sentiment to predict stock returns. Decision Sciences (2016)
22. Tibshirani, R.: Regression shrinkage and selection via the lasso. Journal of the Royal Statistical Society. Series B (Methodological) pp. 267–288 (1996)
23. Veiga, A., Jiao, P., Walther, A.: Social media, news media and the stock market. News Media and the Stock Market (March 29, 2016) (2016)
24. Wang, J., Yao, Y., Mao, Y., Sheng, B., Mi, N.: Omo: Optimize mapreduce overlap with a good start (reduce) and a good finish (map). In: IPCCC'15
25. Wang, J., Yao, Y., Mao, Y., Sheng, B., Mi, N.: Fresh: Fair and efficient slot configuration and scheduling for hadoop clusters. In: CLOUD'14. pp. 761–768. IEEE (2014)
26. Weng, J., Lee, B.S.: Event Detection in Twitter. In: Proc. ICWSM'11 (2011)
27. Yamato, J., Ohya, J., Ishii, K.: Recognizing human action in time-sequential images using hidden Markov model. In: Proc. CVPR '92. pp. 379–385 (1992)

Ensemble Sales Forecasting Study in Semiconductor Industry

Qiuping Xu[1] and Vikas Sharma[2]

Intel Corporation. {qiuping.xu, vikas.sharma}@intel.com

Abstract. Sales forecasting plays a prominent role in business planning and business strategy. The value and importance of advance information is a cornerstone of planning activity, and a well-set forecast goal can guide sale-force more efficiently. A forecasting usually depends on many factors such as the product feature, supply chain constrain, market demand, market share, promotion strategy, competition, macroeconomics condition and others. However, most of those data is hard or even impossible to collect. In this paper CPU sales forecasting of Intel Corporation, a multinational semiconductor industry, was considered. We consolidated the available data resource and forecasting requirement, matched them against the optimal methodology. Past sale, future booking, exchange rates, Gross domestic product (GDP) forecasting, seasonality and other indicators were innovatively incorporated into the quantitative modeling. Benefit from the recent advances in computation power and software development, millions of models built upon multiple regressions, time series analysis, random forest and boosting tree were executed in parallel. The models with smaller validation errors were selected to form the ensemble model. To better capture the distinct characteristics, forecasting models were implemented at lead time and lines of business level. The moving windows validation process automatically selected the models which closely represent current market condition. The weekly cadence forecasting schema allowed the model to response effectively to market fluctuation. Generic variable importance analysis was also developed to increase the model interpretability. Rather than assuming fixed distribution, this non-parametric permutation variable importance analysis provided a general framework across methods to evaluate the variable importance. This variable importance framework can further extend to classification problem by modifying the mean absolute percentage error(MAPE) into misclassify error. This forecast output now helps formulate part of the input provided to public and investors as guideline for the following quarter during Intel's quarterly earning release.

1 Background and Motivation

Sales forecasting is the foundation for planning various phases of the company's business operations. A sales forecast predicts the value of sales over a period of time. Marketing and other managerial functions need different types of forecasting horizons because each directly affects a different business function. The work

© Springer International Publishing AG 2017
P. Perner (Ed.): ICDM 2017, LNAI 10357, pp. 31–44, 2017.
DOI: 10.1007/978-3-319-62701-4_3

presented in this paper was focusing on short term forecast. The forecast was made for tactical reasons that included production planning, sale target setup, short-term cash requirements and adjustments that needed to be made for sales fluctuations.

For the sale forecasting to be accurate, all (not limited to) of the following factors need to be considered: historical perspective, economic conditions, expected market share, total available market, manufacturing constrains, efficiency of distribution channel. The sensitivity of those parameters also needs to be considered and incorporated. However, due to the complexity of the markets and low visibility of data resources, alternative data needs to be considered and methodology needs to be developed to better describe market's dynamic. Thus, the sale forecasting effort is still an active field that attracts extensive attention from both industry and academia.

Research in sales forecasting can be traced back to 1950s [2]. Since then a large number of sales forecasting papers have been published, in which various forecasting techniques have been proposed. The most commonly used techniques for sales forecasting include statistically based techniques like time series [20, 3, 11] regression techniques [13], artificial neural networks [9, 12, 18] and other hybrid models [6, 7].

No surprise that each type of the models has its particular advantages and disadvantages compared to other approaches. In this paper, we investigated and compared the performance of different techniques on Intel's weekly CPU sale forecast. In particular, Extreme Gradient Boosting algorithm, Random forest, ensembled linear regression and ensembled autotegrreive integrated moving average models were considered. Given the fact that neural network performs generally well for sales forecasting if the demand is not seasonal and quite non fluctuating [21], and taking interpretability and data size into account, we left the direct implementation of neural network out the discussion of this paper. The comprehensive comparisons among those methods were illustrated across product segments and lead times. Undoubtedly, this research will not only provide an accurate basis of demand forecast, but also greatly rich the study on selecting and benchmark forecasting techniques for practical applications .

The rest of this paper is organized as follows: Sec.2 includes the data description of this work; Sec.3 describes forecasting models, moving windows validation process and variable importance extraction methodology; Empirical results and insights are included in Sec.4; Finally, Sec.5 concludes the paper with a summary of our results and potential areas for future work.

2   Data and Feature Engineering

As the leading company in Semiconductor industry, Intel combines advanced chip design capability with a leading-edge manufacturing capability and be able to satisfy customers' needs along different product lines. By considering maturity levels of business segments, the CPU prediction problem was further divided into three sub problems - Desktop CPU, Notebook CPU and Server CPU sale forecast. One usage of this model was to be incorporated into the forecast pipeline to help generate the outlook for the next quarter as included in Intel's earning report Fig.1 to guide company's execution.

Business Outlook

Intel's Business Outlook does not include the potential impact of any business combinations, asset acquisitions, divestitures, strategic investments and other significant transactions that may be completed after July 20.

The acquisition of Altera was completed in early fiscal year 2016. As a result of the Altera acquisition, we have acquisition-related charges that are primarily non-cash. Our guidance for the third quarter and full-year 2016 include both GAAP and non-GAAP estimates. Reconciliations between these GAAP and non-GAAP financial measures are included below.

Q3 2016	GAAP	Non-GAAP	Range
Revenue	$14.9 billion	$14.9 billion ^	+/- $500 million
Gross margin percentage	60%	62%	+/- a couple pct. pts.
R&D plus MG&A spending	$5.1 billion	$5.1 billion ^	approximately
Amortization of acquisition-related intangibles included in operating expenses	$90 million	$0	approximately
Impact of equity investments and interest and other, net	$75 million net loss	$75 million ^ net loss	approximately
Depreciation	$1.5 billion	$1.5 billion ^	approximately

Fig. 1: Part of outlook for Q3'16 from Intel's Q2'16 earning report

In this project, we were provided weekly sales data from 2012. This data represented a time-stamped weekly total sale of Intel's CPU by line of business. In addition, we were provided the historical back-then booking data, which showed a screen shot of booking information at a time in the past. Historical average selling price was also provided.

Besides internal data, we also derived variables from the following data.

- Foreign exchange outlook (back-then and forward look quarterly forecast): the forecast exchange rate between RMB(Currency of China) and US dollar, EURO and US dollar.
- GDP outlook (back-then and forward look quarterly forecast): GDP quarterly YoY[1] forecast at world wide level and some key regions (Mainland China and Europe).
- Seasonality: varies from 1 to 4 to indicate the four seasons of a year.
- WeekofQuarter: varies from 1 to 13. This variable was used to indicate in quarter week.
- Special events: Chinese New Year (CNY), indicates the effect of CNY (0-1). We spread the CNY effect evenly to the +/-5 days around Lunar new year, and the summarized effect was computed weekly.
- Time-stamp: the quarters in this dataset were assigned to a list of consecutive numbers start from 1. This variable was used to capture the general growth/shrink of the business.

Measuring GDP requires adding up the value of what is produced, net of inputs, across a wide variety of business lines, weighting each according to its

---

[1] The rate of change between current quarter and the same quarter of previous year

importance in the economy. Thus, GDP statistics are so prone to constant and substantial measurement errors and subjective bias. Similar problem exists in foreign exchange outlooks. To overcome this difficulty, instead of taking the absolute numbers from the outlook, we construed variables as the differences among consecutive outlooks to reduce the subjective bias and measurement errors. The negative change in consecutive GDP outlooks indicated a loss of confidence in the quarter of interest.

Figure 2 showed the world wide GDP YoY with the Intel revenue YoY [2] for same period of time (Q1'12 to Q2'16). In order to visualize those two series together, those two series were normalized to have mean of zero and standard deviation of one. This picture revealed the fact that GDP plays an important role in revenue forecasting, especially for the points of inflection.

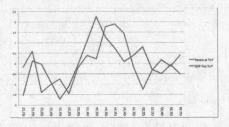

Fig. 2: World wide GDP YoY (Red) with Intel revenue YoY (Blue) from Q1'12 to Q2'16.

3   Methods and Technical Solutions

In this section, we started with our approach for dealing with multicollinearity. The forecast methodologies were briefly discussed afterwards. To incorporate the nature of time dependence in the data, a dynamic moving windows validation method was proposed. A general framework of variable importance was also implemented to better interpret the model output.

3.1   Multicollinearity

Multicollinearity is a phenomenon in which two or more predictor variables in a multiple regression model are highly correlated. Multicollinearity generally does not reduce the predictive power, it only affects calculations regarding individual predictors. In order to deal with this problem without giving up the interpretability, we took the approach of grouping similar numerical variables together and selecting the most representative variable from each group into the predictive model.

---

[2] 14 weeks Q1'16 has been normalize to 13 weeks by a factor of $\frac{13}{14}$

The similarity metric was defined as the absolute correlation coefficients between pair of variables. To cluster the variables, multiple dimensional scaling (MDS) [16] and K-means [17] were used. A pair-wise distance matrix among the variables was construed as 1-the pair-wise absolute correlation coefficients matrix. MDS algorithm was then applied to find the best representation in N-dimensional space such that the pair-variable distances are preserved as well as possible. The N in our experiment was selected to be the maximal to preserve as much information as possible. Then the mapping of MDS became the input of the K-means algorithm to form clusters. The number of clusters was determined by the ratio between 'between cluster variance' and 'total variance' (in the experiments, we set this number to be 80%). Within each cluster, the variable had the maximal absolute correlation coefficient (with response variable) was selected into the modeling steps. One example of before and after the multicolinearity treatment was shown in Fig. 3. In this example, the variance inflation factor was reduced from 166 to 4.3.

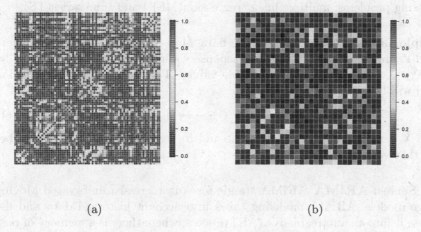

(a)   (b)

Fig. 3: The heat map of absolute correlation coefficients between pairs of variables before (a) and after(b) multicolinearity treatment.

3.2 Extreme Gradient Boosting

XGBoost [5] is short for Extreme Gradient Boosting. This is an efficient and scalable implementation of gradient boosting framework proposed by Friedman [8]. XGBoost constructs an additive model while optimizing a loss function. The loss functions not only accounts for the inaccuracy of the prediction (size of the residuals) but also considers a regularization term that controls the complexity of the model. The regularization helps to avoid over-fitting and has shown good performance in various Kaggle challenges. Empirical evidence shows that the boosting approach with small learning rate yields more accurate results[5]. In the current project, we used 0.01 as learning rate and 1000 as the number of trees for one set of boosting experiment.

3.3   Random Forest

Random forest (RF), which was first proposed by Breiman [4], is an ensemble of B trees $f_1(X), \cdots, f_B(X)$, where $X = X_1; \cdots; X_n$ is variable matrix of n observations, and X_i is the p-dimension feature vector of each observation. B is a predefined hyperparameter in RF algorithm. Unlike boosting tree, in random forest each tree is built on a bootstrap data set, independent of the other trees. RF also benefits from random feature selection to decorrelating the trees, thereby making the average of the resulting trees less variable and hence more reliable [14]. This 'small' tweak provides RF a large improvement over pure bagged trees.

3.4   Parametric Methods

In this section, we would like to briefly sketch two long standing methods in forecasting problems: multiple linear regression(MLR) and time series(TS).

Multiple Linear Regression Given data $D = (X_1, Y_1); \cdots ; (X_n, Y_n)$ where X_i and Y_i are the feature vector and response variable, respectively, where $X_i = [X_{i1}, X_{i2}, \cdots, X_{ip}]$ represents p variable values. The multiple linear regression (MLR) will take the following form:

$$f(X) = \beta_0 + \beta_1 X\_1 + \cdots + \beta_p X\_p + \epsilon \qquad (1)$$

Where X_j represents the j th variable and β_j quantifies the association between that variable and the response.

Time Series: ARIMA ARIMA stands for Autoregressive Integrated Moving Average models. ARIMA modeling takes into account historical data and decompose it into an autoregressive (AR) process, where there is a memory of past events; and Intergrated (I) process, which accounts for stabilizing or making the data stationary and ergodic, making it easier to forecast; and a Moving Average (MA) of the forecast errors, such that the longer the historical data, the more accurate the forecast will be, as it learns over time. ARMIA models therefore have three model parameters, one for the AR(p) process, one for the I(d) process, and one for the MA(q) process, all combined and interacting among each other and recomposed into ARIMA(p, d, q) model. In the experiments, the optimal ARIMA model for a given dataset was selected according to Akaike Information Criterion (AIC) [1]. That is, AIC was used to determine if a particular model with a specific set of p, d and q parameters was a good statistical fit.

Another potential advantage of time series forecasting is the ability to predict future values based solely on previously observed values. This method will reveal more usage when we move our forecast effort to newly emerging business sector with limited data visibility.

3.5   Model Selection and Ensemble Process

In forecast models, the response variables were the weekly CPU billings in each line of business, e.g. Desktop CPU, Notebook CPU and Server CPU. Since the existence of competitor and complimentary relationship among those sectors, for instance, to a certain extent notebook can be considered as substitute product of Desktop, the independent variables were used across line of business. We left around 35 (numerical and categorical) variables for each model which were derived from the categories of variables described in Sec.2 and past the multicolinearity treatment described in Sec.3.1.

Under the consideration of the ratio between numbers of observations and variables, we faced a problem of performing regression in a high-dimensional space. Sugguested by empirical results, using all the variables would result in over fitting. Traditional approach to tackle this problem was to use step-wise subset selection (forward, backward or hybrid approach) as long as that optimizes the measurement (AIC, BIC, or adjusted R^2). One well-known problem from those approaches was that the solutions were not always optimal [22, 10]. Thanks to the increasing of the computation power and development of parallel execution in data science tools [19], we were able to perform best subset selection using brute force technique. The models trained and validated on different subset of variables in parallel. After training, the models were ranked by their performance on validation set. The measurement we used was mean absolute percent errors (MAPE) because of its sensitivity to magnitude changes of the response variable. In validation set MAPE was defined as the average MAPE over all validation samples.

$$\text{MAPE} = \frac{100\%}{m} \sum_{i=1}^{m} \frac{|\hat{Y}_i - Y_i|}{Y_i} \tag{2}$$

To take full advantage of the millions of models, we took top M models, instead of a single best model, to form the ensemble model. This number M was determined by a combination of error threshold and change point algorithm proposed in [15]. In Fig.4, the curve shows sorted MAPE on validation set, the vertical line was determined by change point algorithm to show the number of models which performance similar on the validation set and better to be incorporated into the final ensemble stage. In this specific case, the number of models was 57.

Each model among the best M models was denoted by $\hat{f}_i$, the final prediction was given by averaging all the predictions

$$\hat{f}(X) = \frac{1}{M} \sum_{i=1}^{M} \hat{f}_i(X) \tag{3}$$

This ensemble approach was applied to all methods. Although XGBoost and random forest are already ensemble methods, we still added in another layer of ensemble. From this point on, we will denote the ensemble XGBoost, RF, MLR and ARIMA as EXGBoost, ERF, ELR and ETS respectively.

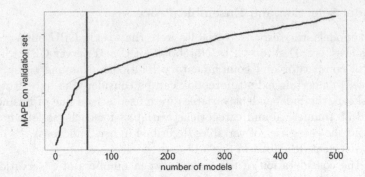

Fig. 4: Number of models detected by change point algorithm on MAPE on validation set

3.6   Moving Window Validation

Our goal is to forecast weekly CPU sale volume in short term (within 16 weeks). The approach was to build separate model for different lead times. e.g. standing at today time t, a model $\hat{f}_{t+1}$ to forecast the next week sale was built independently with the model $\hat{f}_{t+2}$ for forecasting the sale the week after next week. This approach fully considered the unique characteristic of forecasting with various lead times, also led to as large of a reduction in variance as averaging many uncorrelated quantities when we were more interested in monthly or quarterly total. In particular, this approach led to a substantial reduction (in percentage) variance over a multi-week window.

For each lead time model $\hat{f}_{t+j}$, we split the total data set into training data (2 years), validate data (1 year) and test data (1 year). For a fix leading time, to test the performance on next observation in the test set, the entire training, validation and test set were moved forward one week and the process was repeated, as depicted in Fig.5. This moving window framework considered the time-dependence of the data and would automatically select the models which represented back-then up to date market condition.

3.7   Variable Importance

Ensemble models typically result in improving accuracy and reducing variance versus a single model. However, Ensemble models are more difficult to interpret. In practical application, the lack of interpretability made the model adoption extremely hard. To overcome this, we borrowed of idea of permutation test and proposed a generic method to obtain an importance summary of variables.

The method was summarized as follow: For the M models $\hat{f}_i, i = 1, \cdots, M$ selected into the ensemble process, we recorded the $MAPE_0$ defined as in 2 of the validation set. For each variable $X_{\_j}$, repeat the following steps for number of times.

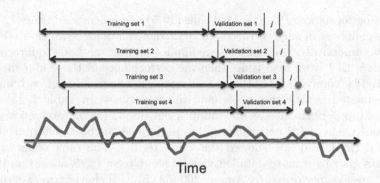

Time

Fig. 5: A illustration of training, validation and test(red dots) sets for model $\hat{f}_{t+j}$ with j weeks lead time.

1. Permute or "shuffle" the value of $X_{\_j}$ over all the data (e.g., by assigning different values to each observation from the set of actually observed values) without replacement.
2. Repeat the model training on the training set
3. Generate forecast result for observation in validation set
4. Calculate the MAPE_j defined as 2 on the validation set and record MAPE_j.

Over a sufficient amount of iterations (we used 100, although empirical result suggested results stabilized after 60 - 70 iterations.), the average MAPE_j was calculated and denoted as $\overline{\text{MAPE}}_j$. The difference between $\overline{\text{MAPE}}_j - \text{MAPE}_0$ was used as the importance measurement of variable j, while a large value of $\overline{\text{MAPE}}_j - \text{MAPE}_0$ indicated a more important variable.

Since the variable importance analysis only needed to run on M models, the computation was efficient. Without assuming certain distribution, this non-parametric permutation variable importance analysis provided a general framework across methods to evaluate the variable importance and to increase model interpretability. This approach could also be extended to classification problem by modifying the MAPE into measurement of misclassify error.

4 Empirical Evaluation

We would dive deep into the performance comparison of ELR, ERF, ETS and EXBoost for lead time (1-16 weeks) in this section. For various lead times, we fixed the test set as the weekly CPU sale of 2015 (Intel Calendar). The training data started from 2012. The training, validating and testing followed the moving windows described in Fig.5.

Models were tested on 52 target weeks in 2015 with 16 different lead times for different lines of business. The performance of those models was compared at the target week and lead time level. The detailed ranking counts were shown in Table 1. Besides the performance of 1 - 16 weeks of lead time, we also reported the model performance of shorter lead time (1 - 5) and longer lead time (12 - 16). The three numbers in each of the table show the counts for lead time 1-16, 1-5 and 12-16, respectively. For instance, we applied all four methods to forecast 1st week's sale with 1 week lead time, then the performance of those four methods were ranked by comparing MAPE on test samples. The ranking of each method was accumulated for the test weeks and the lead times. In Table 1, take first row of desktop performance as an example, out of 832 (= 52 forecast weeks × 16 lead time weeks) ELR experiments, 185 performed the best, 199 ranked 2nd, 221 ranked third, and 227 showed the worst results. The rank counts in the parentheses gave us a more detailed view of short-term ($\leq$ 5 weeks) and long-term ($\geq$ 12 weeks) performance. Among 260 (52×5) ELR short-term experiments for desktop, 58 achieved the best results, while 58, 71 and 73 ranked 2nd - 4th place respectively. The following conclusion could be drawn from this table: for different lines of business, the forecasting performance generated by different forecasting models were mixed, for instance, ERF performed better for desktop CPU forecasting, ELR generated better results for server, and EXBoost showed bipolar performance in notebook and server; another observation was that even for the same model, the lead time could have large effect on forecasting results, for instance, ETS performed better for shorter lead time forecasting on desktop.

Performance Ranking for DT	1	2	3	4
ELR	185 (58, 65)	199 (58, 86)	221 (71, 55)	227 (73, 54)
ERF	231 (73, 64)	206 (72, 54)	211 (59, 77)	184 (56, 65)
ETS	216 (75, 56)	206 (60, 61)	218 (62, 71)	192 (63, 72)
EXBoost	200 (54, 75)	221 (70, 59)	182 (68, 57)	229 (68, 69)
Performance Ranking for MB	1	2	3	4
ELR	211 (73, 63)	201 (66, 69)	212 (69, 49)	208 (52, 79)
ERF	174 (49, 61)	259 (75, 78)	227 (72, 68)	172 (64, 53)
ETS	201 (59, 63)	218 (61, 69)	204 (63, 67)	209 (77, 61)
EXBoost	246 (79, 73)	154 (58, 44)	189 (56, 76)	243 (67, 67)
Performance Ranking for SVR	1	2	3	4
ELR	227 (63, 72)	232 (71, 70)	203 (72, 60)	170 (54, 58)
ERF	177 (57, 50)	221 (75, 65)	233 (68, 77)	201 (60, 68)
ETS	196 (58, 73)	229 (63, 76)	221 (75, 65)	186 (64, 46)
EXBoost	232 (82, 65)	150 (51, 49)	175 (45, 58)	275 (82, 88)

Table 1: Performance ranking for models on weekly sale prediction for Desktop, Notebook and Server. This table showed the counts of each method for target weeks and lead times. The three numbers in each cell represented counts for all lead times, counts for lead time $\leq$ 5 and counts for lead time $\geq$ 12 .

To better understand the model performance on various lead times, we calculated the average MAPE across all the test weeks by lead time. Figure 6 showed the change of MAPE along lead time. Surprisingly, the MAPE did not show monotonically increase as lead time increased, although lead time of 1 week showed the lowest MAPE. This indicated the high dynamic nature of the market. Insights could be drawn from those figures, for instance, in desktop forecasting shown in Fig.6a, ELR showed better results than other methods in longer lead time forecast; in server forecast as in Fig.6c, combination of ERF for shorter forecasting and EXBoost for longer forecasting would outperform other methods.

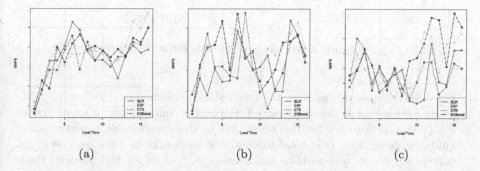

Fig. 6: Average MAPE on test set of models along lead time(in weeks) on Desktop(a), Notebook(b) and Server(c).

As we proposed in Sec.3.7, the variable importance was examined for each ensemble model. To show a clearer comparison among those methods, we fix the target week (1st week of 2015), lead time (1 week) and line of business (DT). The top 5 most important variables for each method were shown in Fig.7. In those figures, the x-axis showed the MAPE increase on the validation set by permuting the value of certain variable. All methods identified DT_backlog_1 [3] and ww_season_index [4] as the top two important variables, with the DT_backlog_1 as the dominant one. This showed that in the shorter term, the forward booking had strong predictive power, while the constant appearance of ww_season_index indicated the weekly sale had strong in-quarter-selling-pattern. For this specific example, the effect of top 2 variables was more dominant in ELR, ETS and EXBoost, while the effect of top variables in ERF more spread out. We also noticed the appearance of cross sectors variables, which implied the success of using cross sectors variables. For example, the SVR_w_3 [5] and MB_asp_sofar [6] showed up as important variables in this desktop forecast example.

Comparing all the empirical results from various methods, the following conclusion can be drawn:

---

[3] The booking amount for back then next week
[4] Week index in a quarter
[5] The Server CPU sale 3 weeks ago
[6] The average selling price of notebook

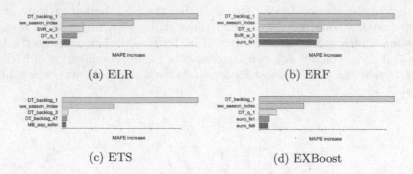

(a) ELR   (b) ERF

(c) ETS   (d) EXBoost

Fig. 7: Top 5 important variables of models for a Desktop example

1. For different lines of business, the forecasting performance generated by different forecasting models were mixed. For current quarter performance tracking (the total forecast for the weeks left in this quarter) and next quarter guidance generating (the total forecast for the weeks in next quarter), our current approach was to take the average output of all the models. Back track test results and live testing indicated that this approach was superior with less effort than the traditional bottom-up approach.
2. For different lead times, the forecasting error was not monotonically increasing as lead time increased.
3. For different lines of business and lead times, the combination of various models might yield better granular results. For example, averaging model outputs with inverse MAPE weighting is one possible approach. We will further investigate the performance of this approach.
4. Important variable analysis suggested strong correlation between certain variables and response. However, this correlation effect did not necessarily lead to concrete conclusion regarding the existence or the direction of causality relationship.

5 Significance and Impact

In this paper, we shared our work on the development, implementation and comparison of a sale forecasting tool to predict Intel's weekly CPU sale by lines of business. The novel feature engineering helped to reduce the subjective bias and measurement errors, especially for economic indicators. The time sensitive training, validating and testing framework allowed the model to effectively reflect changes in the environment and to continuously be evaluated and revised to maintain its credibility. The comparison among those models showed mixed results. This suggests that no general conclusion drawn as to which is the best forecasting technique to employ, but it is certainly true that a forecast should consider multiple models by taking into account the available data resources, characteristic of business and the requirement of the forecast. Our variable importance algorithm boosted the interpretability of the ensemble models and enabled stakeholders examining and giving timely feedback. Those positive aspects of our model led to the recent model adoption. Further work will be 1) To break the forecasting problem into finer granularity and to forecast at geography level and customer level; 2) To develop hybrid techniques. The current approach considers each model individually. In order to break the model boundaries, we plan to take the multi-stage approach. One model can serve as the pretreatment of the data or the feature engineering of variables, thus, to prepare better features for other models. This is similar to the idea of deep belief networks; 3) To combine the intelligence of human and the computing power of machine to capture other (can not observe in data) factors.

6 Acknowledgment

We want to thank Aziz Safa, Vice President of Information Technology, Mary Loomas, the Sr. Controller of World Wide Revenue and Aaron Smith, the manager of World Wide Revenue for their continuing support and sharing valuable business expertise through discussions. We also thank numerous other executives and employees for their assistance and support throughout this project.

References

1. H. Akaike. Information theory and an extension of the maximum likelihood principle. *In Proc. 2nd Int. Symp. Information Theory (eds B. N.Petrov and F.Csáki)*, pages 267–281, 1973.
2. J. Boulden. Fitting the sales forecast to your firm. *Business Horizons*, 1:65–72, 1958.
3. G. Box and G. Jenkins. *Time series analysis forecasting and control*. Prentice Hall, Englewood Cliffs, 1969.
4. L. Breiman. Random forests. *Machine Learning*, 45(1):5–32, 2001.
5. T. Chen and C. Guestrin. Xgboost: A scalable tree boosting system. In *Proceedings of the KDD*, San Francisco, California, 2016.
6. M. P. Clements and D. Hendry, editors. *A Companion to Economic Forecasting*. Backwell Publishing Ltd, Malden, MA, 2002.
7. K. Ferreira, B. Lee, and D. Simchi-Levi. Analytics for an online retailer: demand forecasting and price optimization. *Manufacturing & Service Operations Management*, 18:69–88, 2015.
8. J. H. Friedman. Greedy function approximation: A gradient boosting machine. *The Annals of Statistics*, 29(5):1189–1232, 2001.
9. C. L. Giles, S. Lawrence, and A. C. Tsoi. Noisy time series prediction using recurrent neural networks and grammatical inference. *Machine Learning*, 44(1):161–183, 2001.
10. A. Goodenough, A. Hart, and R. Stafford. Regression with empirical variable selection: Description of a new method and application to ecological datasets. *PLoS ONE*, 7(3):e34338. doi:10.1371/journal.pone.0034338, 2012.
11. G. K. Groff. Empirical comparison of models for short range forecasting. *Manag Sci*, 20:22–31, 1973.

12. G. Huang, Q. Zhu, and C. Siew. Extreme learning machine: a new learning scheme of feedforward neural networks. In *Proceedings of the international joint conference on neural network*, pages 25–29, Budapest, 2004.

13. C. Jain. Benchmarking forecasting software and systems. *J Bus ForecastMethods Syst*, 26:30–33, 2007.

14. G. James, D. Witten, T. Hastie, and R. Tibshirani. *An Introduction to Statistical Learning, with Applications in R.* Springer, New York, NY, 2013.

15. R. Killick and I. Eckley. Changepoint: an r package for changepoint analysis. *J Stat Software*, 58:1–19, 2014.

16. J. B. Kruskal and W. Myron. *Multidimensional Scaling.* Sage, everly Hills, CA, 1978.

17. J. B. MacQueen. Some methods for classification and analysis of multivariate observations. *Proceedings of 5-th Berkeley Symposium on Mathematical Statistics and Probability*, (1):281–297, 1967.

18. R. Sharma and A. K. Sinha. Sales forecast of an automobile industry. *International Journal of Computer Applications*, 53(12):25–28, 2012.

19. S. Weston and R. Calaway. Getting started with doparallel and foreach. `https://cran.r-project.org/web/packages/doParallel/vignettes/gettingstartedParallel.pdf`, 2005.

20. P. Winters. Forecasting sales by exponential weighed moving averages. *Manag Sci*, 6:324–342, 1960.

21. W. Wong and Z. Guo. A hybrid intelligent model for medium-term sales forecasting in fashion retail supply chains using extreme learning machine and harmony search algorithm. *Int J Prod Econ*, 128:614–624, 2010.

22. M. Yuan and Y. Lin. Model selection and estimation in regression with grouped variables. *J R Stat Soc Ser B Stat Methodol.*, 68:49–67, 2006.

Towards an efficient method of modeling "Next Best Action" for Digital Buyer's journey in B2B

Anit Bhandari[1], Kiran Rama[1], Nandini Seth[2], Nishant Niranjan[1], Parag Chitalia[1], Stig Berg[1]

[1] VMware Inc.
{anitb, rki, nniranjan, pchitalia, stigb}@vmware.com
[2] Indian Institute of Management – Bangalore
{nandini.seth15}@iimb.ernet.in

Abstract. The rise of Digital B2B Marketing has presented us with new opportunities and challenges as compared to traditional e-commerce. B2B setup is different from B2C setup in many ways. Along with the contrasting buying entity (company vs. individual), there are dissimilarities in order size (few dollars in e-commerce vs. up to several thousands of dollars in B2B), buying cycle (few days in B2C vs. 6-18 months in B2B) and most importantly a presence of multiple decision makers (individual or family vs. an entire company). Due to easy availability of the data and bargained complexities, most of the existing literature has been set in the B2C framework and there are not many examples in the B2B context. We present a unique approach to model next likely action of B2B customers by observing a sequence of digital actions. In this paper, we propose a unique two-step approach to model next likely action using a novel ensemble method that aims to predict the best digital asset to target customers as a next action. The paper provides a unique approach to translate the propensity model at an email address level into a segment that can target a group of email addresses. In the first step, we identify the high propensity customers for a given asset using traditional and advanced multinomial classification techniques and use non-negative least squares to stack rank different assets based on the output for ensemble model. In the second step, we perform a penalized regression to reduce the number of coefficients and obtain the satisfactory segment variables. Using real world digital marketing campaign data, we further show that the proposed method outperforms the traditional classification methods.

Keywords. Multi-class classification, literature survey, ensemble, regression, digital, robustness, non-negativity constraint, B2B, next-best action.

1 Introduction

VMware (VMW) is a virtualization, end user computing and cloud company with annual revenues of USD 6,571 million (as of 2015) and a market capital of USD 34 BB as of 2017. VMware sells products in the Software Defined Data Center (vSphere, VSAN, NSX for computing, storage & network virtualization respectively), end user computing (Air-watch, Horizon, and Fusion/Workstation) and cloud. The company exclusively caters to business customers – i.e. B2B.

© Springer International Publishing AG 2017
P. Perner (Ed.): ICDM 2017, LNAI 10357, pp. 45–56, 2017.
DOI: 10.1007/978-3-319-62701-4_4

VMW is characterized by 100% digital supply chain which means that all products are downloadable from the website (www.vmware.com). The company also promotes them online. Different individuals across companies worldwide visit the site to familiarize themselves with the products before making a purchase decision. Along with the overview of the product, there are various customer- interaction digital assets that are show to the VMW audience including and not limited to:

- Hands-on Labs (HoL): Here the visitor can evaluate a VMW product before making a decision to buy. HoLs provide a virtual environment where a visitor/email id can acquaint himself with the product by using it first-hand.
- Eval: Here the visitor can download a version of the software for his/her personal use
- Whitepapers: These downloadable papers cover a wide range of topics related to the product - such as usage, features, vs. competition summaries, Gartner research reports etc.
- Seminar & Webinar: Here the visitor can register for a seminar or a webinar

Fig. 1. Digital Assets example – Case Study Download, Webinar, Hands-on-Lab, Eval (in clockwise)

The digital buyer journey goes through 4 stages (Awareness → Interest → Trial → Action) which VMW would like to personalize. Of the digital assets available on the website, HoL is more of an Action, Eval is more of a Trial, and Whitepaper is more of Awareness whereas Seminar/Webinar are expressions of interest. To personalize the buyer's journey, it is imperative to identify the appropriate digital assets that need to be pressed to the consumers and an optimal ordering for the same. An example for digital assets is shown in Fig 1.

Traditionally, the models built for B2B interactions focus all their marketing effort assuming the company to be a unified entity. While this is a prudent assumption to design a "Propensity to buy" model, it is not appropriate for the "propensity to respond" model we are trying to propose. A company as a united entity may have a propensity to buy a product but the diverse individuals within the company will have varying propensities to respond. There are numerous individuals involved in the decision making process who go through various phases of nurture program before they complete a purchase. The traditional response models attempt to predict the likelihood of response for a marketing effort around a particular digital asset. However, these models do not take into account the past consumption event of the individual with the particular asset. The method presented here eliminates this drawback and work towards targeting customers more effectively.

1.1   Objective

In consultation with the Digital Marketing team, the Advanced Analytics & Data Sciences team came up with the following objectives:

- Determine the right order of digital assets to display to an individual email id
- Since the website would like to target groups of email ids, come up with a set of segment rules to identify top individuals for a digital asset and to target them with personalization on the website

2   Literature Review

2.1   Digital Marketing Literature

Digital Marketing is the process of fostering customer relationships through online activities to expedite the exchange of ideas, products, and services that satisfy the twofold objectives of the customer as well as the seller [1]. Lately, the use of internet to expand marketing efforts has intensified. With the internet serving as a computer facilitated marketplace in which customers and sellers can access each other, they can easily execute trade functions like sales, distribution and marketing [2]. Customers go through a compound decision making process before they make their final consumption. As McKinsey [3] rightly pointed out, "consumers don't want to feel subjected to the hard sell— they expect marketers to engage them, not dictate to them." As a result, traditional marketing strategies have reformed drastically leading to additional customer value, improved targeted marketing & escalated company profits [4]. Due to the comparative novelty and an ever-growing exigency, digital marketing is an exciting area for research – not just academically but in the industry. Academicians and practitioners have emphasized on the significance of digital marketing to deal with marketing mixes, which include global accessibility, convenience in updating, real-time information services, interactive communications features, and unique customization and personalized capabilities [5]. Plenty of attention has been focused on the tremendous opportunities digital marketing presents, with very little attention on the real challenges companies are facing with analyzing and interpreting digital. These challenges are discussed at length in the recent paper by Leeflang et al [8]. A number of books have been published in the recent years concentrating on B2B digital marketing (see [6, 7]). Even though these books are wide-ranging, there is still a great deal of understanding that is required to formulate models on digital data, especially in a B2B framework.

2.2   Machine Learning Models Literature

In the past decade, striving attempts have been made to design frameworks and models to derive value out of digital data using multi-class classification methods like logistic regression, random forests, gradient boosting etc. Random forest and gradient boosting models have been extensively used on digitally generated data in analytical Customer Relationship Management (CRM) to develop churn-prediction models [9, 10, 11, 22]. Random forest has been lucratively used in answering other marketing questions like predicting subject line open rates for targeted emails [12], privacy preserving data mining [13], defect prediction [14], building different attribution models [15, 16] etc. One of the principal application of gradient boosting in digital marketing has been in obtaining better recommender systems by improved unified search system [23] and adaptive advertising [24]. Various application of Regularized Logistic Regression can be found in the marketing literature such as for better targeting of display ads [19], multi-touch attribution model [20], analyzing the effectiveness of an ad [21] etc. Irrespective of the sizeable literature research done on digital marketing, very little has been said about identification and targeting of customers in a B2B framework. In an antecedent to the work present here, ensemble decision tree method has been used in a B2C setup to improve personalized advertising and behavioral targeting [18]. Ensemble methods use numerous learning algorithms to achieve improved predictive capabilities than what could be obtained from any of the featuring algorithms. Ensemble methods are popular for superior performance than other algorithms in most cases [17]. A current study [25] presents an ensemble method approach to developing a more accurate and implementable recommender system for B2B clients using collaborative filtering and gradient boosting.

3   Solution Framework

3.1   Integrating Online & Offline Features

The dataset for the model was created at an email address level using Pivotal Greenplum and Hadoop. In Greenplum which is an enterprise data warehouse, we created dataset with all historic bookings information. Similarly in the Hadoop environment from the clickstream data, we created dataset containing individual's online behavior. We then combined both these datasets to obtain the single view of an individual's behavior across offline and online features. We ended up with 800+ explanatory features and one target variable comprising of 5 classes. Fig 2 presents a block diagram declaring all the features that were used to build the model. We used time based cross validation to examine the performance of the created models. Time Based Cross Validation is the right technique in digital marketing scenarios where we try to predict the future based on what we observe in the past. Such models lead to generalization and limit overfitting. To form the training set, we took the information for explanatory variables from FY13Q1 (01/01/13) to FY15Q2 (06/30/15) and the target variable from FY15Q3 (07/01/15) to FY15Q4 (12/31/15). Similarly for validation of the model, we took explanatory variables from FY13Q1 (01/01/13) to FY15Q4 (12/31/15) and the target variable was from FY16Q1 (01/01/16) to FY16Q2 (06/30/16).

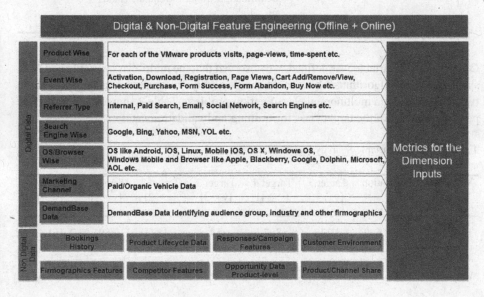

Fig. 2. Feature Bucket Diagram

3.2   Multiple Classification methods

Since we had to find the best possible asset for better targeting of an email address, the target variable was assigned based on the order of action taken among the assets: HoL, Eval, Seminar / Webinar, Download or other events. Table 1 shows how the target variable was defined based on order of action.

Table 1. Target Variable Definition

Criterion	Details	Target
Hands-on-Lab < Any Action	Hands – on – Lab was registered before doing one or more actions	5
Any Action < Hands – on – Lab	Hands – on – Lab was registered after doing one or more actions	4
Seminar / Webinar < Any Action	Seminar / Webinar was signed up before doing one or more actions	3
Download < Any Action	Any download event occurred before doing one or more actions	2
All remaining actions	Any other action done other than download, seminar / webinar, Hands – on – Lab	1

Once the target variable was defined, we observed that the multi-class variable had high sparsity. We ran algorithms which could handle highly imbalanced data and at the same time can solve for a multinomial classification problem. Table 2 shows the numbers of positives in each of the target class for training and validation datasets.

Table 2. Target Variable Sparsity in datasets

Models	Population	Target 5 [y = 5]	Target 4 [y = 4]	Target 3 [y = 3]	Target 2 [y = 2]	Target 1 [y = 1]	Target 0 [y=0]
Training	1,468,690	3,275	897	54	112	2,891	1,461,461
Validation	1,876,064	4,150	1,074	52	125	3,014	1,867,649

Once the target variable was defined, we executed multiple algorithms designed to solve a multinomial classification problem.

a) Since the data was highly imbalanced, the first approach we undertook was random forest [26] [28] with an under-sample method. The random forest method relies on an autonomous, pseudo-random procedure to select a small number of dimensions from a larger feature space [27] - in our case is 800+ features. We performed this for multiple iterations with different number of trees and under-samples of target class.

b) The next attempted model was the L2 regularized logistic regression using LIBLINEAR: A library for Large Linear Classification [29]. It has been shown that LIBLINEAR is very efficient on large sparse data [30]. We tried L2 logistic implementation for different costs and arrived at the best cost to obtain the maximum AUC. We also ran an iteration of LIBLINEAR with assigned weights for each class in the model.

c) The final model we tried was the extreme gradient boosting (xgboost) [31] [32] with two different approaches - optimizing eta, depth and number of trees. In the first trial we considered a higher depth of 9, with eta = 0.2 and no of rounds of trees = 200 which we call big_xg_boost and in the second trial, we decrease the depth to 7, eta = 0.5 and no of rounds = 15, called small_xg_boost.

We observed the AUC for all these five models for the multi-class target to compare their performances. Table 3 lists the AUC obtained for these models.

Table 3. Model Performance measure based on AUC for traditional methods

Models	Target 5	Target 4	Target 3	Target 2	Target 1
Random Forest Under-sample	0.78	0.84	0.83	0.85	0.89
L2 Regularized Logistic Regression	0.62	0.68	0.72	0.72	0.75
Weighted L2 Regularized Logistic Regression	0.73	0.80	0.66	0.78	0.74
XGBoost (Big)	0.72	0.75	0.74	0.72	0.85
XGBoost (Small)	0.78	0.85	0.88	0.87	0.89

3.3 Proposed Ensemble method using outputs of traditional methods

Now that we got the results from each of the multi-class classification models - To improve the AUC of the output, we propose an ensemble using the output of the models that were ran in the previous iteration along with enforcing a non-negativity constraint [33] in predicting digital action. In this ensemble method , for each target class we took the probability output from random forest under-sample, L2 regularized logistic regression, weighted L2 regularized logistic regression, XGBoost implementation as inputs into Lawson-Hanson NNLS implementation of non-negative least squares method in R [34] which results in the ensemble of models and is better performing than existing traditional methods. Table 4 shows how the ensemble method performs better then each of the individual models in each of the target class.

Table 4. Model Performance measure based on AUC including Model Ensemble results

Models	Target 5	Target 4	Target 3	Target 2	Target 1
Random Forest Under-sample	0.78	0.84	0.83	0.85	0.89
L2 Regularized Logistic Regression	0.62	0.68	0.72	0.72	0.75
Weighted L2 Regularized Logistic Regression	0.73	0.80	0.66	0.78	0.74
XGBoost (Big)	0.72	0.75	0.74	0.72	0.85
XGBoost (Small)	0.78	0.85	0.88	0.87	0.89
Ensemble Method	0.82	0.87	0.90	0.89	0.91

4   Conclusion

4.1   Results

As presented in Table 4, we observe that the ensemble technique presented in this paper performs better than the other ensemble methods – i.e. the ones obtained by combining identical classification methods. Intuitively, this improvement can be credited to the combination of the regularized logic regression method and the gradient boosting method. While the regularization takes care of the sparseness, boosting tends to increase the predictability of the model.

4.2   Uniqueness of the approach

In varied ways, the current research is innovative. The novelty of the research lies in the following particulars.

(i)   The model attempts to study the digital behavior of B2B customers which has been inadequately explored in both academia and industry as of yet. In a B2B scenario, the decision maker is not an individual (as in the B2C case). It is possible that a targeted company is highly diverse in its structure and needs. In addition to that, each individual who is a part of the decision-making process might respond differently to the target asset leading to increased complexity in the model.

(ii)   Disparate to most existing models, the current research presents the use of an ensemble of various dissimilar multi-class classification models (random forest, L2 regularized logistic regression- weighted and non-weighted and XGBoost- small and big). We demonstrate the superiority of this ensemble of models approach in mining highly imbalanced digital datasets in B2B which is forward-looking and highly sophisticated as compared to the existing literature.

(iii)   The model goes beyond conventional propensity to buy models and contributes by providing a Propensity to Respond model. This alteration leads to tremendous managerial implication which have been provided below in 4.3

(iv)   The model also extends in identifying high propensity to respond individuals by rules which can be implemented to do website personalization and give an engaging experience to the individual. This is particularly useful in the web analytics world where many organizations might not have the tools to enable targeting at an individual email id or cookie level. Translating the propensity model output into segments allows organizations the ability to do 1:M personalization where M is the number of segments in cases where the technology stack of the organization is not equipped for 1:1 marketing.

4.3   Managerial Implication

The ability to predict a customer's response to a digital action has remarkable value. Along with identifying probable customers who will choose a particular action; we can determine an order in which digital actions should be pressed to targeted customers. A predictive model based on the digital behavior and historic data of the existing customers can help identify potential sales leads and enhance customer experience by improved personalized marketing. The output of the model can be directly used to create a list of probable leads for varied marketing channels, re-targeting and social targeting. Further, the derived rules can be used to target customers who are most likely to consume specific digital content. This leads to enhanced segmentation and better understanding of the customized behavior for each segment. This in turn indicates to better identification of potential buyers or raw leads. These refined ways of identification and targeting advance to an increased efficiency due to a considerable reduction in cost and unnecessary marketing efforts.

4.4   Future Extensions

The method presented here can be used as a foundational framework to design the Next Best Digital Action (NBDA) using which B2B company should target prospects. Given the sparse literature around B2B, this work can be a basis for a system for the B2B digital buyer journey. Future research areas could include design of new metrics to analyze the digital buyer journey and development of models to optimize this metric directly. The multinomial problem could be framed as a recommendation problem as well and compared vs. existing models. A very interesting research area will be development of an algorithm to generate discriminating segments of users from the propensity model output that are business-user discernible and can be used by digital business sites that do not have the ability for individual email id/user level targeting.

References

1. Imber, J., & Toffler, B. A. (2008). *Dictionary of marketing terms*. Barron's snippet.
2. Farhoomand, A. F., & Lovelock, P. (2001). Global e-Commerce: Text and Cases Plus Instructor's Manual.
3. Edelman, D. C. (2010). Four ways to get more value from digital marketing. *McKinsey Quarterly, 6*.
4. Strauss, J. (2016). *E-marketing*. Routledge.
5. Kian Chong, W., Shafaghi, M., Woollaston, C., & Lui, V. (2010). B2B e-marketplace: an e-marketing framework for B2B commerce. *Marketing Intelligence & Planning, 28*(3), 310-329.
6. Miller, M. (2012). *B2B digital marketing: Using the web to market directly to businesses*. Que Publishing.
7. Järvinen, J., Tollinen, A., Karjaluoto, H., & Jayawardhena, C. (2012). DIGITAL AND SOCIAL MEDIA MARKETING USAGE IN B2B INDUSTRIAL SECTION. *Marketing Management Journal, 22*(2).
8. Leeflang, P. S., Verhoef, P. C., Dahlström, P., & Freundt, T. (2014). Challenges and solutions for marketing in a digital era. *European management journal, 32*(1), 1-12.
9. Burez, J., & Van den Poel, D. (2007). CRM at a pay-tv company: Using analytical models to reduce customer attrition by targeted marketing for subscription services. *Expert Systems with Applications, 32*(2), 277-288.
10. Nafis, S., Makhtar, M., Awang, M. K., RAHMAN, M. N. A., & DERIS, M. M. (2015). FEATURE SELECTIONS AND CLASSIFICATION MODEL FOR CUSTOMER CHURN. *Journal of Theoretical & Applied Information Technology, 75*(3).
11. Xie, Y., Li, X., Ngai, E. W. T., & Ying, W. (2009). Customer churn prediction using improved balanced random forests. *Expert Systems with Applications, 36*(3), 5445-5449.

12. Balakrishnan, R., & Parekh, R. (2014, October). Learning to predict subject-line opens for large-scale email marketing. In *Big Data (Big Data), 2014 IEEE International Conference on* (pp. 579-584). IEEE.

13. Szűcs, G. (2013). Decision trees and random forest for privacy-preserving data mining. *Research and Development in E-Business through Service-oriented Solutions. IGI Global, Hershey, PA, USA*, 71-90.

14. Pushpavathi, T. P., Suma, V., & Ramaswamy, V. (2014). Defect Prediction in Software Projects-Using Genetic Algorithm based Fuzzy C-Means Clustering and Random Forest Classifier. *International Journal of Scientific & Engineering Research*, 5(9).

15. Sinha, R., Saini, S., & Anadhavelu, N. (2014, October). Estimating the incremental effects of interactions for marketing attribution. In *Behavior, Economic and Social Computing (BESC), 2014 International Conference on* (pp. 1-6). IEEE.

16. Yadagiri, M. M., Saini, S. K., & Sinha, R. (2015, November). A non-parametric approach to the multi-channel attribution problem. In *International Conference on Web Information Systems Engineering* (pp. 338-352). Springer International Publishing.

17. Opitz, D., & Maclin, R. (1999). Popular ensemble methods: An empirical study. *Journal of Artificial Intelligence Research*, *11*, 169-198.

18. Koh, E., & Gupta, N. (2014). An empirical evaluation of ensemble decision trees to improve personalization on advertisement. In *Proceedings of KDD 14 Second Workshop on User Engagement Optimization*.

19. Perlich, C., Dalessandro, B., Raeder, T., Stitelman, O., & Provost, F. (2014). Machine learning for targeted display advertising: Transfer learning in action. *Machine learning*, *95*(1), 103-127.

20. Shao, X., & Li, L. (2011, August). Data-driven multi-touch attribution models. In *Proceedings of the 17th ACM SIGKDD international conference on Knowledge discovery and data mining* (pp. 258-264). ACM.

21. Farahat, A., & Shanahan, J. (2013, February). Econometric analysis and digital marketing: how to measure the effectiveness of an ad. In *Proceedings of the sixth ACM international conference on Web search and data mining* (pp. 785-785). ACM.

22. Lu, N., Lin, H., Lu, J., & Zhang, G. (2014). A customer churn prediction model in telecom industry using boosting. *IEEE Transactions on Industrial Informatics*, *10*(2), 1659-1665.

23. Wang, J., Zhang, Y., & Chen, T. (2012, December). Unified recommendation and search in e-commerce. In *Asia Information Retrieval Symposium* (pp. 296-305). Springer Berlin Heidelberg.

24. Addicam, S., Balkan, S., & Baydogan, M. (2015). Adaptive Advertisement Recommender Systems for Digital Signage.

25. Zhang, W., Enders, T., & Li, D. (2017, January). GreedyBoost: An Accurate, Efficient and Flexible Ensemble Method for B2B Recommendations. In *Proceedings of the 50th Hawaii International Conference on System Sciences*.

26. Ho, Tin Kam (1995). Random Decision Forests (PDF). Proceedings of the 3rd International Conference on Document Analysis and Recognition, Montreal, QC, 14–16 August 1995. pp. 278–282.

27. Ho, Tin Kam (1998). "The Random Subspace Method for Constructing Decision Forests" (PDF). IEEE Transactions on Pattern Analysis and Machine Intelligence. 20 (8): 832–844. doi:10.1109/34.709601

28. Breiman, L. (2001). Random forests. Machine learning, 45(1), 5-32

29. C.-C. Chang and C.-J. Lin. LIBSVM: a library for support vector machines, 2001. Software available at http://www.csie.ntu.edu.tw/~cjlin/libsvm.

30. R.-E. Fan, K.-W. Chang, C.-J. Hsieh, X.-R. Wang, and C.-J. Lin. LIBLINEAR: A library for large linear classification Journal of Machine Learning Research 9(2008), 1871-1874.

31. 29, No. 5, 1189-1232. 1999 REITZ LECTURE. Friedman, Jerome. GREEDY FUNCTION APPROXIMATION: A GRADIENT BOOSTING MACHINE'.

32. Friedman, Jerome; Hastie, Trevor; Tibshirani, Robert. Additive logistic regression: a statistical view of boosting (With discussion and a rejoinder by the authors). Ann. Statist. 28(2000), no. 2,337--407.doi:10.1214/aos/1016218223.
33. Chen, Donghui; Plemmons, Robert J. (2009). Nonnegativity constraints in numerical analysis. Symposium on the Birth of Numerical Analysis. CiteSeerX: 10.1.1.157.9203.
34. https://cran.r-project.org/web/packages/nnls/nnls.pdf

Association Rule-based Classifier
Using Artificial Missing Values

Kaoru Shimada[1,2], Takaaki Arahira[2] and Takashi Hanioka[2]

[1] Fukuoka Nursing College
[2] Fukuoka Dental College
2-15-1, Tamura, Sawara-ku, Fukuoka, 814-0193, Japan
{shimada, arahira, haniokat}@college.fdcnet.ac.jp

Abstract. In this paper, we propose a rule-based classification method that uses artificial missing values to improve the effectiveness and precision of medical data analysis. We apply artificial missing values to avoid the sharp boundary problem encountered when discretizing continuous variables. In discretization, we treat attribute values near the boundary as missing values. We evaluated the performance of the proposed artificial missing value-based classification method and our experimental results using medical data show this method to be effective for classification. The proposed method can reduce the number of rules required to build a classifier. It may also be able to control the relation between a false positive and true positive in rule-based classifiers.

Keywords: association rule, classification, evolutionary computation, incomplete data, missing value

1 Introduction

Association rule mining is the discovery of association relationships or correlations among a set of attributes (itcms) in a database [1, 2]. Association rule-based applications have been used for analysis in biomedical and health informatics fields and have demonstrated effective performance [3–5]. A class association rule (CAR) in the form "If X then *Class label* $(X \rightarrow Class\ label)$" is interpreted as "instances having the set of attributes X are likely to be classified to the *Class label*." Whereas effective CAR mining methods and application techniques have been proposed [6, 7], previous approaches cannot handle incomplete databases, i.e., databases in which there are missing values.

In biomedical and health informatics fields, datasets are likely to have many missing values due to incomplete personal information or failed experiments. Data can also be incomplete when multiple databases are joined and the attributes in the joined databases differ. Conventional CAR mining methods consider databases to be complete or else disregard instances that have missing values. Generally, instances with missing data are deleted for the purposes of rule mining or are completed using mean values or frequent categories [8, 9]. However, the discovery rule for biomedical data is such that these techniques

© Springer International Publishing AG 2017
P. Perner (Ed.): ICDM 2017, LNAI 10357, pp. 57–67, 2017.
DOI: 10.1007/978-3-319-62701-4_5

Table 1. Example of incomplete database and available instance.

(a)

ID	A_1	A_2	A_3	A_4	C
1	1	1	1	m	1
2	m	0	1	1	0
3	0	m	m	1	0
4	1	m	m	0	0
5	1	1	0	0	1

(b)

$(A_1=1) \wedge (A_2=1) \wedge (A_3=1)$
satisfied (available)
not satisfied (available)
not satisfied (available)
cannot judge
not satisfied (available)

(c)

$(A_3=1) \wedge (A_4=1)$
cannot judge
satisfied (available)
cannot judge
not satisfied (available)
not satisfied (available)

are difficult to apply. In addition, the classification of incomplete data should also be considered. In the conventional associative classifier, the first matching rule usually makes the prediction. Therefore, the order of the classifier rules affects the classification accuracy. When missing values exist in both the training and testing data, a multiple matched rules-based method may be appropriate.

We have previously proposed association rule mining tools for incomplete databases using an evolutionary computation method [10, 11]. In this paper, we leverage the advantages of rule extraction methods for incomplete databases and propose an artificial missing values-based method to improve the effectiveness and precision of rule-based classification for medical datasets. We apply artificial missing values (AMV) to avoid the sharp boundary problem encountered when discretizing continuous values. In discretization, we treat attribute values near the boundary as missing values, which means that certain areas near the boundary are not used for rule mining and classification. The proposed method can obtain more reliable results. Unlike fuzzy methods [12], the proposed method does not use probabilities to transform continuous variables which can also be beneficial in biomedical and health informatics fields. We previously presented the basic concept of the proposed method as a rule-based continuous value prediction method [13].

This paper is organized as follows. In section 2, we present related concepts and explain the CARs and classification for incomplete databases. We describe the proposed method in which AMV are used in section 3, present our experimental results in section 4, and draw our conclusions in section 5.

2　Rules and Classification Method

2.1　CARs for an Incomplete Database

Let A_i be an attribute (item) in the database. To clearly describe the algorithm, we indicate instances of attribute values as either 1 or 0, as shown in Table 1(a) [10], and missing values as "m." This means that the absence of item A_j is described as $A_j=0$ and a lack of information for A_j is indicated by '$A_j=m$'. Let C be the class label. Note that the database has no missing class labels. When the data has two classes, the class labels are denoted $C=i$ ($i=0,1$). X and Y denote the combination of attributes, e.g., $X=(A_j=1) \wedge \cdots \wedge (A_k=1)$. X is

Table 2. Contingencies of X and C for an incomplete database.

	$C = 1$	$C = 0$	$\sum_{row}$
X	$Y(1)$	$Y(0)$	$Y(1) + Y(0)$
$\neg X$			
$\sum_{col}$	$N(1)$	$N(0)$	$N(1) + N(0)$

represented briefly as $A_j \wedge \cdots \wedge A_k$. Some examples of the above are described in [11].

For CAR mining from an incomplete database, the number of instances required to calculate measurement differs with different rules [10]. Tables 1(a) and 1(b) illustrate the available instances feature. For example, let $X = (A_1 = 1) \wedge (A_2 = 1) \wedge (A_3 = 1)$. Instance $ID = 1$ in Table 1(a) satisfies X even though the value for A_4 is missing. When at least one attributes value of A_1, A_2, or A_3 is 0, the instance does not satisfy X. $ID = 2$ and 3 are available to determine X even though they have missing values. These instances are available for calculating rule measurements. $ID = 4$ is unavailable because, due to the missing values, we cannot determine whether the instances satisfy X. For $X = (A_3 = 1) \wedge (A_4 = 1)$, the combination of available instances differs from that of the previous case, as shown in Tables 1(b) and 1(c).

We exclude instances whose attribute values for a candidate rule equal 1 or m, except where all attribute values equal 1 [10]. We use the M and Y values introduced in [10] for the rule-by-rule measurement calculation as follows. The M value represents the number of instances whose attribute values for the rule equal 1 or m, and the Y value represents the number of instances where all the attribute values for the rule equal 1. The N value, which represents the number of available instances, is also defined in the rule measurement. In this paper, we use a contingency table (Table 2), which is related to $X \to (C = 1)$. $M(i)$, $Y(i)$, and $N(i)$ are used as the M, Y, and N values for $i = 1, 0$, respectively. These values satisfy the following formula: $N(i) = N_T(i) - (M(i) - Y(i))$, where, $N_T(i)$ is the total number of instances for $C = i$ in the database.

We define the measurements for CAR as follows [11]:

$$support(X \to (C = i)) = \frac{Y(i)}{N(1) + N(0)},$$

$$confidence(X \to (C = i)) = \frac{Y(i)}{Y(1) + Y(0)},$$

$$support(C = i) = \frac{N(i)}{N(1) + N(0)},$$

$$\chi^2(X \to (C = i)) = \frac{N \cdot (\frac{Y(i)}{N} - \frac{Y}{N} \cdot \frac{N(i)}{N})^2}{\frac{Y}{N} \cdot \frac{N(i)}{N}(1 - \frac{Y}{N})(1 - \frac{N(i)}{N})},$$

where, $N = N(1) + N(0)$ and $Y = Y(1) + Y(0)$.

2.2   CAR Mining Method

Details of the CAR mining method for incomplete databases are provided in [10] and [11]. This method extracts rules directly without constructing the frequent item sets used in previous rule mining approaches. Instead, we use available instances of attribute values including missing values to calculate the rule measurements. We developed this method using an evolutionary computation technique that adopts a new strategy to accumulate rules via its evolutionary process. In the experiments described in section 4, we used the same parameter settings for the CAR mining as those used in [10].

In this paper, we define important CARs as satisfying the following:

$$support(X \rightarrow (C = i)) \geq sup_{min}, \tag{1}$$

$$\chi^2(X \rightarrow (C = i)) > \chi^2_{min}, \tag{2}$$

$$confidence(X \rightarrow (C = i)) \geq support(C = i), \tag{3}$$

where, sup_{min} and χ^2_{min} are the minimum support and χ^2 value provided by users in advance. Equation (3) is required for positive associations of Eq. (2). For example, instances in the medical field sometimes include different characteristics for individuals. Therefore, important rules do not always have high confidence values. In the classification method based on matched multiple rules, both strict rules with a high confidence value and rules having a high χ^2 value even if they have a low confidence value are recommended.

2.3   Building a Multi-rules-based Classifier

Typically, rule-based classification involves two stages: training and testing. In the training stage, important CARs are generated for classification. In the testing stage, the obtained rules are applied to estimate the test data. We use an unordered rule-based model to build a classifier, which compares the accuracy or score of all the classes obtained from the multiple matched rules. The class having the highest accuracy or score is used for classification.

In this work, we use the method for building a classifier described in [10] and an incomplete data set as follows. We defined the *available rule* as the rule that can judge whether the new instance satisfies the antecedent of the rule.

[**Input**] A set of CARs and an instance to be classified

[**Output**] Class predicted by the classifier

[**Method**]   1. $available(i)$: Compute the total number of available rules satisfying $C = i$ in the classifier ($i = 0, 1$)

2. $match(i)$: Compute the number of rules in the classifier, whose antecedent match the new data and satisfy $C = i$

3. $score(i) = \frac{match(i)}{available(i)}$

If $available(i) = 0$ then $score(i) = 0$

4. The instance is predicted as $C = \arg \max score(i)$

5. If $score(1) = score(0) > 0$ then $C = 1$

6. If $score(1) = score(0) = 0$ then no classification of C is returned.

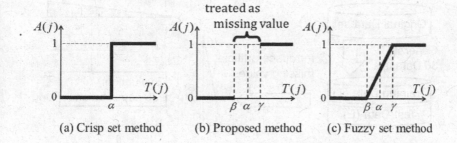

Fig. 1. Example of continuous variable transformation.

3 AMV

We propose the use of AMVs to improve the effectiveness and precision of rule-based classification systems, as originally proposed in [13]. Figure 1(a) shows an example of the discretization of a continuous attribute. Here $T(j)$ is transformed into $A(j)$, and α is a breakpoint of $T(j)$ for the crisp set. Generally, rule-based classification algorithms using discretization of continuous attributes encounter the sharp decision boundary problem. In this study, we apply AMVs to avoid the sharp boundary problem in the discretization of continuous variables. In discretization, we treat attribute values near the boundary as missing values, as shown in Fig. 1(b). This means that uncertain areas are not used for rule mining and classification. Unlike fuzzy methods, as shown in Fig. 1(c), the proposed method does not use probabilities in the discretization process, but rather constructs a new decision boundary treatment rather than a crisp or fuzzy decision boundary. The width of the area containing AMVs can be set attribute by attribute. Optimization of the AMV settings is a challenge, however.

In the CAR mining stage, the proposed method extracts useful rules that are overlooked by conventional approaches. We modify CAR measurements using AMVs and use a set of distinct instances for each candidate rule, but in the classification stage, the proposed method automatically selects useful rule sets for classifying each test data entry. This classifier can avoid the over-fitting problem encountered in rule matching. The proposed method realizes human-like classification because it looks for reliable classification rules. In the biomedical and health informatics fields, there are many essential continuous attributes in their databases. The proposed AMV-based method differs from probability-based classification schemes, such as fuzzy methods, and is therefore suitable for application in these fields. One of the limitations of this method may be an ethical decision not to use available values. However, from the point of view of big data analysis, we need to develop new techniques for the selection of effective values in databases and thereby improve the attribute selection methods.

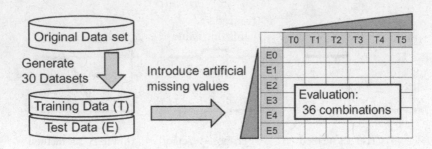

Fig. 2. Experimental setting.

4 Experimental Results

4.1 Experimental Settings

We used the *Pima Indians Diabetes Database* [14] from the UCI ML Repository
[15] for our evaluation. A summary of this database is as follows:
• Classification of signs of diabetes ($C = 1$ (tested positive for diabetes) and
$C = 0$)
• Eight continuous attributes ($T(j)$, $(j = 0, \ldots, 7)$) (Number of times pregnant,
Plasma glucose concentration at 2 hours in an oral glucose tolerance test, Dias-
tolic blood pressure (mm Hg), Triceps skin fold thickness (mm), 2-hour serum
insulin (mu U/ml), Body mass index, Diabetes pedigree function, and Age)
• 768 instances. (268 in $C = 1$, 500 in $C = 0$)
• Including missing attribute values
In [14], when forecasting the onset of diabetes mellitus, the sensitivity and specificity of the ADAP learning algorithm used by the authors was 76%. The difficulty of this classification is moderate, therefore, this data set was considered to
be a good one for evaluating the improved effectiveness of the classification and
precision of the analysis.

Figure 2 illustrates how we prepared the datasets for the experiments. We
independently generated 30 combinations of training and test data. We then
randomly selected ten percent of the instances (77 instances) as test data and
used the remaining 90% for training. We obtained our experimental results using
the average value of 30 data sets. We discretized the continuous attribute values
($T(j)$) to the sets of attributes $A(2j)$ and $A(2j + 1)$, for which the values were
1, 0, or missing. We performed the discretization of the values as follows:
1) Calculate the averaged value m_j and standard deviation s_j of $T(j)$ ($j =
0, \ldots, 7$).
2) Discretize the values of $T(j)$ using m_j and s_j. We defined the values of $A(2j)$
and $A(2j + 1)$ as shown in Fig. 3(a). This transformation has no scientific relevance, and is used only to generate a dataset for the purposes of estimation.
3) To examine the influence of the introduction of AMV, we used the parameter
k ($k = 0, 0.1, 0.2, 0.3, 0.4, 0.5$). A larger k value indicate that the width of the

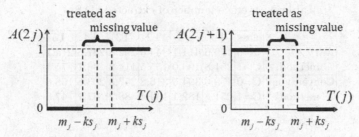

(a) Continuous variable transformation for the proposed method

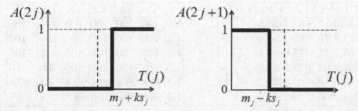

(b) Continuous variable transformation for the conventional method

Fig. 3. Continuous variable transformation in the experiments.

artificial missing values is larger. We denoted training data as T and test data as E. For example, $T2$ denotes T using $k = 0.2$, and $E5$ denotes E using $k = 0.5$. We evaluated 36 set combinations of the training and test data ($T0$-$E0$, $T0$-$E1$, ..., $T5$-$E5$).

4) For comparison, Fig. 3(b) shows the settings for transformation using the conventional rule-based method. It was not a simple matter to include comparative studies using well-known classifiers such as the Decision Tree, Random Forest, or artificial neural networks, because these methods do not handle missing values. In the experiment, we focused on the impact and potential for introducing AMVs in rule-based classification problems.

We extracted CARs for each class using the rule mining method described in [10]. We used $sup_{min} = 0.03$ in (1) and $\chi^2_{min} = 6.63$ in (2) and set the maximum number of attributes in the antecedent part of the rule to 8. The termination condition for the evolutionary process was 200 generations [10]. We coded all algorithms in C and performed the experiments on a 2.80GHz Intel(R) i7 860 with 4 GB RAM.

4.2 Classification Using AMV

Table 3 shows the average number of extracted CARs from the training data. The number of extracted rules decreased with an increasing number of AMVs, which can be due to the effect of the AMVs. However, the proposed method

Table 3. Average number of extracted CARs.

	class	T0	T1	T2	T3	T4	T5
AMV	C=0	479.6	411.1	344.6	292.8	247.6	214.1
method	C=1	251.8	196.0	155.5	114.2	96.7	75.1
Conventional	C=0	479.6	394.5	316.8	257.3	203.6	168.8
method	C=1	251.8	182.1	134.4	88.2	67.1	47.5

Table 4. Average cover rate (%) of test instances in classification.

(a) Using top 100 rules

	T0	T1	T2	T3	T4	T5
E0	82.3	87.8	93.6	98.8	–	–
E1	73.4	78.9	85.7	94.4	–	–
E2	65.1	71.8	78.8	90.1	–	–
E3	57.4	64.8	72.5	84.8	–	–
E4	50.0	56.8	65.2	77.4	–	–
E5	42.2	50.2	58.2	71.7	–	–

(b) Using all the extracted rules

	T0	T1	T2	T3	T4	T5
E0	100	100	100	100	100	100
E1	100	100	100	100	100	100
E2	100	100	100	100	100	100
E3	100	100	100	100	100	100
E4	99.8	99.8	99.8	99.8	99.8	99.8
E5	99.8	99.8	99.8	99.8	99.8	99.8

obtained more CARs than the conventional crisp set. In the proposed method, we calculate the *support* of rules by considering the number of available instances on a rule-by-rule basis. Equation (1) can work to some degree with the conventional method.

Table 4 shows the average cover rate of the test instances in the classification. We define the cover rate (%) for the classification as follows:

$$\frac{\text{Number of classified instances}}{\text{Number of instances for the test}} * 100.$$

Table 4(a) shows the results using the top 100 rules for $C = 1$ and $C = 0$ in building each classifier. In this experiment, we sorted the CARs extracted from the training data by *confidence* and used the top 100 rules for each class to build a classifier. We found that the proposed method could reduce the number of rules required to build a classifier. Table 4(a) does not show the results of $T4$ and $T5$, because the number of extracted rules was less than 100. Table 4(b) shows the results using all the extracted rules for $C = 1$ and $C = 0$ in building each classifier. Almost all of the instances were classified by each classifier.

Table 5 shows the averaged accuracy of the test instances in the classification. We defined the accuracy (%) of each classification as follows:

$$\frac{\text{Number of correctly classified instances}}{\text{Number of classified instances}} * 100.$$

The results indicate that the proposed AMV method did not reduce the classification accuracy. In contrast, conventional methods reduced the accuracy depending on their treatment of the discretization boundary. In this experiment, we

Table 5. Average accuracy (%) of test instances in classification.

(a) AMV method

	T0	T1	T2	T3	T4	T5
E0	69.7	69.7	69.5	68.9	69.1	69.0
E1	70.0	69.8	70.1	69.5	70.0	70.0
E2	69.8	69.3	69.5	68.8	69.0	68.2
E3	71.1	70.9	70.6	70.2	70.2	69.5
E4	71.2	71.1	70.6	70.7	70.6	69.9
E5	72.0	71.9	71.5	71.5	71.1	70.1

(b) Conventional method

	T0	T1	T2	T3	T4	T5
E0	69.7	69.4	69.0	68.8	68.9	68.3
E1	70.0	69.7	69.7	70.0	69.7	69.0
E2	70.1	69.7	69.8	68.8	68.3	67.6
E3	71.2	71.1	70.9	69.5	69.0	67.8
E4	71.2	71.1	70.6	70.0	69.0	67.5
E5	72.1	71.7	71.5	70.9	69.4	67.8

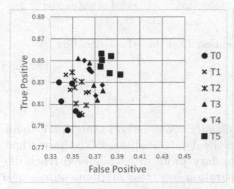

(a) Attention to the kind of training data set

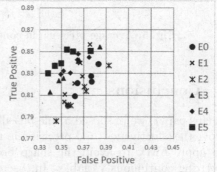

(b) Attention to the kind of test data set

Fig. 4. Average precision in classification using artificial missing values.

uniformly introduced AMVs for continuous attributes. In future work, we plan to optimize the AVM settings by considering the characteristics of the attributes.

Figures 4(a) and 4(b) show scatter diagrams of the average precision of the classification results using AMVs in the 36 combinations. In this experiment, we used all the extracted CARs to build a classifier. Figure 4(a) shows the precision based on the kind of training data set and Fig. 4(b) shows same results as Fig. 4(a) based on the kind of test data set. We found the precision trends to be $E0 < E3 < E5$ and $T0 < T3 < T5$, respectively. We also found that the proposed AMV method can potentially control the relation between false positive and true positive results for rule-based classifiers. Figures 5(a) and 5(b) show the same experimental results as Figs. 4(a) and 4(b) but using the conventional method. In these figures, we see that the degree of dispersion was greater than that in the AMV-based method and the *false positives* tended to have higher values. Our experimental results using the *Pima Indians Diabetes Database* indicated that the proposed method has the potential to improve the effectiveness and precision of CARs-based classification.

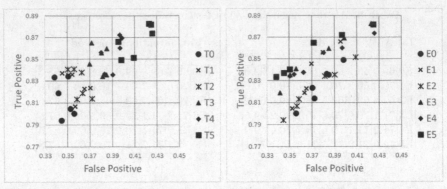

(a) Attention to the kind of training data set   (b) Attention to the kind of test data set

Fig. 5. Average precision in conventional rule-based classification.

5 Conclusions

In this paper, we proposed a rule-based classification method using AMV to improve the effectiveness and precision in medical data analysis. With this method, we apply AMV to avoid the sharp boundary problem encountered when discretizing continuous variables. In discretization, we treat attribute values near the boundary as missing values. we evaluated the performance of the proposed artificial missing value-based classification method and the results show that the proposed method is effective and has the potential to improve classification precision in the medical field. In future work, we will develop an extended method to determine the optimal combination of settings for introducing AMV to attributes in medical datasets.

Acknowledgment. This work was partly supported by JSPS KAKENHI Grant Number 16K00316.

References

1. R. Agrawal and R. Srikant, "Fast Algorithms for Mining Association Rules," in *Proc. of the 20th VLDB Conf.*, pp. 487–499, 1994.
2. J. Han, J. Pei, Y. Yin and R. Mao, "Mining Frequent Patterns without Candidate Generation: A Frequent-Pattern Tree Approach," *Data Mining and Knowledge Discovery*, Vol.8, pp.53–87, 2004.
3. V. Ivancevic, I. Tusek, J. Tusek, M. Knezevic, S. Elheshk, I. Lukovic, "Using association rule mining to identify risk factors for early childhood caries," *Computer Methods and Programs in Biomedicine*, Vol.122, pp.175–181, 2015.
4. F. Held, et al., "Polypharmacy in older adults: Association Rule and Frequent-Set Analysis to evaluate concomitant medication use," *Pharmacological Research*, 2016, http://dx.doi.org/10.1016/j.phrs.2016.12.018

5. G. Totia, R. Vilalta, P. Lindnerd, B. Leferb, C. Macias, D. Priceda, "Analysis of correlation between pediatric asthma exacerbation and exposure to pollutant mixtures with association rule mining," *Artificial Intelligence in Medicine*, Vol.74, pp.44–52, 2016.
6. B. Liu, W. Hsu and Y. Ma, "Integrating Classification and Association Rule Mining," in *Proc. of the ACM Int'l Conf. on Knowledge Discovery and Data Mining*, pp. 80–86, 1998.
7. W. Li, J. Han and J. Pei, "CMAR: Accurate and efficient classification based on multiple class-association rules," in *Proc. of the 2001 IEEE Int'l Conf. on Data Mining*, pp. 369–376. 2001.
8. J. W. Grzymala-Busse and W. J. Grzymala-Busse, Handling Missing Attribute Values Data Mining and Knowledge Discovery Handbook, 2nd ed., O. Maimon, L. Rockach (eds.), pp.33–51, Springer, 2010.
9. M. Saar-Tsechansky and F. Provost, Handling Missing Values when Applying Classification Models, *Journal of Machine Learning Research*, Vol.8, pp.1625-1657, 2007.
10. K. Shimada, "An Evolving Associative Classifier for Incomplete Database," Springer LNAI 7377: Advances in Data Mining, Perner P.(Ed.). pp.136–150, 2012.
11. K. Shimada and T. Hanioka, "An Evolutionary Method for Exceptional Association Rule Set Discovery from Incomplete Database," in *Proc. of Int'l Conf. on Information Technology in Bio- and Medical Informatics (ITBAM 2014)*, M. Bursa et al. (Eds.), Springer LNCS 8649, pp.133–147, 2014.
12. S. Mabu, C. Chen, N. Lu, K. Shimada and K. Hirasawa, "An Intrusion-Detection Model Based on Fuzzy Class-Association-Rule Mining Using Genetic Network Programming," IEEE Trans. on Syst., Man, and Cyber. -Part C-, Vol.41, pp.130–139, 2011.
13. K. Shimada, T. Arahira and T.Hanioka, "An Evolutionary Rule Mining Method for Continuous Value Prediction from Incomplete Database and Its Application Utilizing Artificial Missing Values," in *Proc. of the First IEEE Int'l Conf. on Big Data Computing Service and Applications*, pp.392–399, 2015.
14. J. W. Smith, J. E. Everhart, W. C. Dickson, W. C. Knowler and R. S. Johannes, "Using the ADAP learning algorithm to forecast the onset of diabetes mellitus," in *Proc. of the Symposium on Computer Applications and Medical Care*, pp. 261–265. IEEE Computer Society Press. 1988.
15. C. Blake and C. Merz, UCI repository of machine learning databases http://www.ics.uci.edu/ mlearn/MLRepository.html.

Mining Location-based Service Data for Feature Construction in Retail Store Recommendation

Tsung-Yi Chen[1], Lyu-Cian Chen[2], Yuh-Min Chen[3]

[1]Department of Electronic Commerce Management and Department of Information Management, Nanhua University, Chiayi County, Taiwan (R.O.C.)
tsungyi@mail.nhu.edu.tw
[2]Institute of Manufacturing Information and Systems, National Cheng Kung University, Tainan City, Taiwan (R.O.C.)
lcchen1993@gmail.com
[3]Institute of Manufacturing Information and Systems, National Cheng Kung University, Tainan City, Taiwan (R.O.C.)
ymchen@mail.ncku.edu.tw

Abstract. In recent years, with the popularization of mobile network, the location-based service (LBS) has made great strides, becoming an efficient marketing instrument for enterprises. For the retail business, good selections of store and appropriate marketing techniques are critical to increasing the profit. However, it is difficult to select the retail store because there are numerous considerations and the analysis was short of metadata in the past. Therefore, this study uses LBS, and provides a recommendation method for retail store selection by analyzing the relationship between the user track and point-of-interest (POI).

This study uses regional relevance analysis and human mobility construction to establish the feature values of retail store recommendation. This study proposes (1) architecture of the data model available for retail store recommendation by influential layers of LBS; (2) System-based solution for recommendation of retail stores, adopts the influential factors with specified data in LBS and filtered by industrial types; (3) Industry density, area categories and region/industry clustering methods of POIs. Uses KDE and KMeans to calculate the effect of regional functionality on the retail store selection, similarity is used to calculate the industry category relation, and consumption capacity is considered to state saturation feature.

Keywords: Urban mining, Spatial and temporal data mining, Location-based Service, Retail store recommendation

1 Introduction

Based on the well-developed LBS, many data are available for user behavior analysis. For example, the POI recommendation system uses the user track in location-based social network (LBSN) to analyze the user preference for recommendation. The POI recommendation system has been discussed extensively in previous studies [1-2], most

© Springer International Publishing AG 2017
P. Perner (Ed.): ICDM 2017, LNAI 10357, pp. 68–77, 2017.
DOI: 10.1007/978-3-319-62701-4_6

of which concerning the influential factors. The adaptive or personalized recommendation system uses data mining to obtain the users' personal data, so as to estimate their behaviors, characters and values [3].

These online user behavior evaluation methods based on internet mining are tolerable analysis materials for enterprises. In order to gain commercial profit, the customer segments and purchasing behavior are obtained by user behavior analysis. There are diversified analysis tools and purposes, the decision of retail location plays an important role in the purposes [4]. The data analysis technique derived from advanced technology takes a share gradually. The analysis of customer segments provides effective assistance for enterprises in deciding retail store, the complicated market survey is no longer a good scheme. Using big data to analyze the relationship of consumers with retail stores or other competitors by cluster, gravity model and neural networks has become an important reference for large enterprises to expand retail stores [5].

However, the traditional cluster analysis and gravity model methods rely on extensive calculation and data, while requiring the use of geography information system (GIS) [5]. The data collection process is very difficult, and the technical threshold is relatively high. As the present data is rapidly and frequently recorded via internet, this study adopts LBS to solve the problem of insufficient data sources in the decision-making of retail store by the location information and implicit user behavior.

Since 2010, there are many studies on the LBSN data analysis, most of which analyzed the relationship between mass user data and POIs according to the check-in data of users, so as to know the implicit information of preferences, sequential behaviors and social circle generated when the user checks in [6-7]. In addition, a few studies explored the relationship between locations and the popularity in LBS, and used machine learning to predict the possible optimum retail store location [4].

Most studies concerning LBSN intended to analyze the user behavior, while considering the factors of community, individual preference, regional composition and geographic property, but not the influence of all factors on the decision-making of retail store [1-3, 7, 9]. Therefore, this study discusses the data properties of LBS, point of view of marketing and the relationship inside the regional relationship of retail store and the human mobility, in order to develop a recommendation model for retail store expansion.

To sum up, although with well-developed LBSN and convenient techniques of data collection, there still lacks a recommendation method for retail stores based on LBS data mining. Therefore, this study develops the following purposes:

1. To build an influential factors model for retail stores recommendation system by POI recommendation, so as to identify the data types beneficial to retail store recommendation in LBSN.
2. To establish the retail store recommendation system architecture with LBS data as source.
3. To establish reasonable corresponding recommendation features according to the established influential factors by point of view of marketing.

2   Related work

In the LBSN, the sequential series connection of user's location behavior is called sequential behavior. A lot of useful information can be explored from this behavior, which may contain the information of user's daily routines, preferences and life circle. The sequential behavior is often discussed together with users' social network, that is social influence study. The social network of target user is brought into analysis to obtain the relationships of following between users and their friends. Moreover, multiple sequential behaviors are connected in series to estimate the diffusion relationship by algorithm [1]. Yang et al. proposed how to establish the users' activity characteristics in geo-spatial, so as to reduce the complexity in multi-dimensional condition [8].

The POIs, also known as venues, refers to the stores or attractions in the LBS graphical information. The POI recommendation system is a very important part of the urban mining research. The unique influence model of POI recommendation system is based on the factors in LBSN, such as social factors, individual preference, popularity of POI and geographic position. The target groups with close properties are identified by collaborative filtering, and the probable mobility is predicted for recommendation [3]. Ye also used collaborative filtering technique to calculate the similarity between social influence and geographical influence, and combined two influence aspects to build a fusion framework. With the user preferences for ranking, to achieve the best POI recommendation accuracy [9].

The factors that may be considered in POI recommendation model include social influence, individual preference, popularity of POI and geographic position. The individual preference uses classification of POI properties as main basis, the historical records of recommended target in LBS are used as recommendation reference for the recommended target. The popularity of POIs represents the recommendation value of the locations by the human clustering and following suit that is an important basis of recommendation. The geographic position also has significant effect on the recommended target. If it is too far from the user's relative position, or outside the life circle, this recommendation would not make sense [1-2].

In urban mining, the human mobility must be caught via internet for crowd monitoring. The best data collection method is to let the target feedback position via GPS in time, so as to implement real-time monitoring. The GPS monitoring has many advantages, such as real-time feedback of the location information, and the coherence for data gathering. Such mass continuous data can be used for crowd monitoring in large-scale activities to avoid dangers [10].

Considering the privacy problem, most studies based on crowd mobility do not use GPS as data source, while some use location-based network for research. Despite of the limited public data, the crowd behavior in location-based network still provide diversified research directions. Yuan et al. used Latent Dirichlet Allocation (LDA) to guess the division and function of regions in city according to the distribution of POIs and human mobility pattern, and presented the information that may be difficult to explore in an automated approach [11].

Developing the human mobility model is a challenge due to the computation complexity. Without any corresponding method to reduce the complexity, it is difficult to obtain the result. Wei et al. proposed the concept of grid that solved this. The adjacent POIs is classified into the same grid, and every frequent trajectory is displayed in region. This method greatly reduces the time complexity for calculating the frequent mobility route [12].

When planning for retail store, enterprises adopt data analysis methods of regression analysis, clustering analysis and gravity models. GIS is also used to identify the segmentation of target customer. However, even with technological assistance, decision-making needs to consider past experiences and common sense in commercial activity. If experience is considered in the decision-making process, the decision will be more accurate [5].

There are numerous influential factors in determining an appropriate location of the store. For the enterprises, the decision-making process is like looking for an appropriate portfolio, determining a correct site can bring enormous profit. The decision factors of retail store can be divided into two categories. One is consumer demand, the other one is appropriate location position. Huff proposed considering customer requirement appropriately, to predict if there was adequate demand at the location for sales estimation [13].

Ghosh & Craig indicated that the factors to be considered in retail store decision-making include the attractiveness of location for consumers. In addition, the competitive relation of the location in the region, the probable preference of customers and the demand for the industry should be considered [14]. Exploring user preference and behavior in urban mining is conducive to business operation. Upon LBS, the consumers' habits can be known by data mining. The analyses of visitors flowing rate and competitors distribution are greatly helpful to commercial activity [15].

3 Influential Factors and System Design

This study plans to design a retail store recommendation system based on LBSN. First, relational concept is imported based on the POI recommendation system in LBS to establish the influential factors related to retail store. The relationship between human mobility and the profit of related industry is observed in the POI recommendation system. The user preference is closely related to the types of industry, the recommended crowd is the target customers. Thus, the retail store recommendation system is built by using POI influential factors for analysis.

3.1 Construct Influential Factors with POI Recommendation System

Most of previous studies on POI recommendation use location influence and social influence to divide the influential factors in POI recommendation, and then use fusion rule, such as sum rule and product rule [1-2]. In addition, there are three major types based on feature extraction [3]. This study adopts in a similar concept, the following three types are described as follows:

- Individual preference: the user's personal preference ratio is defined according to the POI types the target user visited. Or the user's activity sequence is established by spatial temporal activity in the form of time correlation, and then the activity related preference types is established [8].
- Social influence: the social influence considers the user's friends, evaluates if there is stronger influence among the friends. Or disregards the intensity of influence, only aims at the behavioral aspects of friends, and measures the influence model (mobility or behavior sequence) for recommended target.
- Location influence: the most decisive factor in the POI recommendation system is the effect of geographic location. In the measurement of location factor, the location is usually represented by coordinates on a two-dimensional plane. The recommendations are ranked by the distance relation between users and stores. However, the users' individual preference shall be considered first when considering related candidate stores, so as to reduce the computational complexity preliminarily.

To sum up, in the analysis of social influence or individual preference, the influential factors may be correlated with geographic position. Based on the three major influential factors, corresponding to the influence layers of retail store recommendation, relevant information extraction model is built. Three data hierarchies are established by the original LBS data, including primitive layer, mobility and social layer, location layer (Fig.1.).

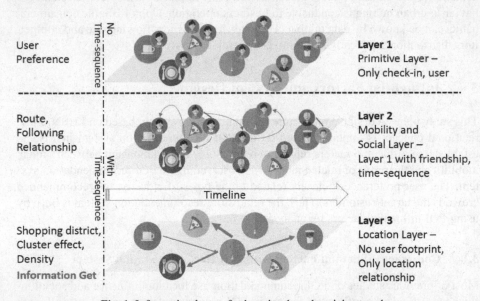

Fig. 1. Information layers for location-based social network

As shown in Fig. 1, the information of Layer 1 can be obtained easily, but the information of Layer 2 and Layer 3 can be obtained after extensive analysis. Moreover, in the study of social influence, it is difficult to obtain social network under the privacy policy. Therefore, removing the effect of social criteria, the features which influence the retail store can be simply divided into two types, one is regional relevance (RR) features, the other one is human mobility (HM) features.

3.2   System Development and Design

In order to recommend appropriate retail store, the enterprise can use the classification mechanism provided by real estate websites for screening. The recommendation system analyzes the implicit influential factors in the filtered candidate store, especially the customer segment and geographic relationships that the real estate websites are difficult to provide. Afterwards, other POIs within a radius of 200 meters from the candidate store and check-in data are collected to analyze the human mobility and industrial relations [4] (Fig. 2.).

As the check-in data is enormous, a filtering mechanism must be established to select the type of industry correlated with enterprise. This mechanism can be relevant information provided by enterprise, or the information derived from web text mining. The more correlated dataset can be filtered out before calculation by the industrial information provided by enterprise and the user preference in LBS, so as to increase the effectiveness of feature values.

Afterwards, the check-in data around each candidate retail store are analyzed, and the computing method of feature value extraction is proposed. The HM features are extracted from the mobility route formed of check-in sequence. In order to establish the mobility route, this study uses the route construction method proposed by Wei [12]. The computational complexity is reduced by grid division, and the industry filtering mechanism is used to select appropriate types to enhance the effectiveness; On the other hand, the RR features are extracted, considering the complex regional relation when the enterprise is selecting retail store location, probably including intensity, saturation, clustering degree, etc.

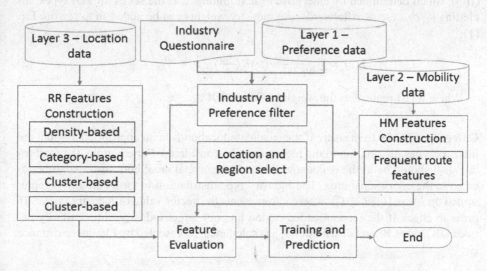

Fig. 2. Framework for feature discovering in retail store recommendation

4   Regional Relevance Features Construction

First we develop features with regional relevance. This paper provide four influence factors for regional relevance, and integrate these methods with industry-preference filter:

Density. The main concern of retailers is if there are competitive stores and crowds nearby the retail store, which are usually reflected in the store rent. The rent at busy areas is extremely high, thus the renters often doubt about the sales revenue and profit. These concerns are often tightly related to the store density. In order to measure the store density in the region, the Kernel Density Estimation (KDE) is used for calculating the density distribution. The KDE is a nonparametric entropy estimation. If the quantity of stores in the region is considered only, there may be fatal loss of some key factors. Only if the density nearby candidate point is calculated, the kernel of trading area is approached in deed. The KDE value of location l is calculated by using two-dimensional KDE, defined as follows:

$$\lambda(l) = \sum \frac{1}{|V_{l,r}|h^2} K(\frac{d_{v,l}}{h}) \, , \, v \in V_{l,r} \tag{1}$$

Where $d_{v,l}$ represents the distance between POI l and other POIs v in the range, $v \in V_{l,r}$, $V_{l,r}$ represents the set of all POIs in radius r of candidate point l, h is the bandwidth value for KDE calculation. Scott rule is used for estimation, K is kernel function, Gaussian function is used as kernel function in our experiment [11]. With density distribution generated by calculating KDE value may not be strong enough, so the industry impact filter is added to KDE function, the POI of different types of industry is given different weights [16]. $W = \{w_c \mid c \in C\}$, represents industry impact quality (IIQ), which determined by enterprise or text mining. C is the set of all POI types, including $c_1, c_2, \ldots, c_{|C|}$. Afterwards, the industry weights can be added in to rewrite Eq. (1):

$$\lambda(l) = \sum \frac{1}{|V_{l,r}|h^2} w(v) K(\frac{d_{v,l}}{h}) \, , \, v \in V_{l,r} \tag{2}$$

Where $w(v)$ represents the weights of each POI v.

Category. In order to measure if the candidate location is in an appropriate region, the industry types distribution in the region must be considered. For example, the food retail shall should be in the residential area or commercial area; bookstores shall be located in the educational area. The industry type distribution in target area l_r is represented by $I_{l,r} = \{i_c \mid c \in C\}$, where i_c represents the vector value of industry type c. In order to check if the recommended region hits our target industry, the similarity between IIQ value W and distribution $I_{l,r}$ is calculated, represented by Euclidean distance φ:

$$\varphi(W, I_{l,r}) = \sqrt{\sum_{x=1}^{|C|}(w_x - i_x)^2} \tag{3}$$

Clustering. Clustering is also very important for the store position. The distance between the candidate location and the business district will directly affect the customer's willingness to visit. Therefore, we use the clustering method to select the business district within the region, and after the clustering, analyze the cluster's industrial quality. First, we cluster the POI $V_{l,r}$ in the candidate region l_r by the K-Means clustering algorithm, and get the cluster label for each location. Second, the geometric center of each cluster is obtained, and the distance between the center and the candidate point is calculated. Finally, divide the distance by the industry correlation score φ(Eq. (3)), and then accumulate the results of all the clusters to get the feature score (Algorithm 1).

```
Algorithm 1
Cluster_feature(l)
    Initialize V_{l,r} and feature_score=0
    clusters=KMeans(V_{l,r}, ε, MinPts)
    for each cluster in clusters do
        avg = average coordinates of cluster
        φ = φ(W,I_cluster) according Equation (3)
        accumulate feature_score by φ/dis-
tance(avg , l.coordinate)
    end for
return feature_score
```

Saturation. Market saturation is a very important concept in economics, and its actual degree of saturation depends on the purchasing power of consumers [17]. We can simply denote the saturation degree $\theta_{l,r}$ of a target region l_r as:

$$\theta_{l,r} = -|A_{l,r}|/|V_{l,r}| \tag{4}$$

Where $A_{l,r}$ denotes all check-in activity in region l_r, which represents the purchasing power of consumers in this region. This purchasing power divided by the number of businesses in the region. And a negative number indicates that, the greater the number is, the stronger the negative impact on the recommendation. However, this does not fully explain the saturation degree of a single industrial category in the region. In addition, we do not define the saturation point, but only provide a relative comparison. Therefore, we modify the weight distribution of $I_{l,r}$ to be:

$$\Psi(I_{l,r}) = \begin{cases} 1, i_c \geq m \\ 0, i_c < m \end{cases}, c \in C \tag{5}$$

Where m is the relative limit decided by category set C, 0<m<1. And rewrite Eq. (4) based on the concept of industrial distribution:

$$\theta_{l,r,\Psi} = -\frac{|A_{l,r,\Psi}|/|V_{l,r,\Psi}|}{\theta_{max}} \tag{6}$$

θ_{max} represents the max value of saturation degrees for a specified region in the dataset.

5   Conclusion

In order to establish a suitable way for enterprises to find retail stores, we start with LBS and provide a model for mastering geography relation and crowd behavior. There are many implicit relationships that are not easily found through the geo-network. And we use information layered approach, the obvious expression of which can be obtained by data mining technology. In the research, an analytical framework which can take the LBS as the input is designed as the target of system design. In the future, we will use statistical tests to filter the selected features and use machine learning to analyze the best retail locations in the system.

In the selection of features, we take the retailer's point of view, carefully assess the influential factors when choosing retail stores. And then quantify these considerations into the feature values we need, e.g., density distribution, similarity, clustering are all retailers need to consider in detail the characteristics of the region. Based on these analysis of LBS, enterprises can save a considerable amount of cost in data collection. Not only more technical use of data mining methods, but also can improve the accuracy of the forecast.

Reference

1. Zhang, J. D., Chow, C. Y., & Li, Y. (2014, November). LORE: exploiting sequential influence for location recommendations. In *Proceedings of the 22nd ACM SIGSPATIAL International Conference on Advances in Geographic Information Systems* (pp. 103-112). ACM.
2. Zhang, J. D., & Chow, C. Y. (2015). CoRe: Exploiting the personalized influence of two-dimensional geographic coordinates for location recommendations. *Information Sciences, 293*, 163-181.
3. Ying, J. J. C., Lu, E. H. C., Kuo, W. N., & Tseng, V. S. (2012, August). Urban point-of-interest recommendation by mining user check-in behaviors. In*Proceedings of the ACM SIGKDD International Workshop on Urban Computing*(pp. 63-70). ACM.
4. Karamshuk, D., Noulas, A., Scellato, S., Nicosia, V., & Mascolo, C. (2013, August). Geospotting: mining online location-based services for optimal retail store placement. In *Proceedings of the 19th ACM SIGKDD international conference on Knowledge discovery and data mining* (pp. 793-801). ACM.
5. Hernandez, T., & Bennison, D. (2000). The art and science of retail location decisions. *International Journal of Retail & Distribution Management, 28*(8), 357-367.
6. Li, Y. M., Chou, C. L., & Lin, L. F. (2014). A social recommender mechanism for location-based group commerce. *Information Sciences, 274*, 125-142.

7. Wen, Y. T., Lei, P. R., Peng, W. C., & Zhou, X. F. (2014, December). Exploring social influence on location-based social networks. In *2014 IEEE International Conference on Data Mining* (pp. 1043-1048). IEEE.
8. Yang, D., Zhang, D., Zheng, V. W., & Yu, Z. (2015). Modeling user activity preference by leveraging user spatial temporal characteristics in LBSNs. *IEEE Transactions on Systems, Man, and Cybernetics: Systems, 45*(1), 129-142.
9. Ye, M., Yin, P., Lee, W. C., & Lee, D. L. (2011, July). Exploiting geographical influence for collaborative point-of-interest recommendation. In *Proceedings of the 34th international ACM SIGIR conference on Research and development in Information Retrieval* (pp. 325-334). ACM.
10. Blanke, U., Tröster, G., Franke, T., & Lukowicz, P. (2014, April). Capturing crowd dynamics at large scale events using participatory gps-localization. In *Intelligent Sensors, Sensor Networks and Information Processing (ISSNIP), 2014 IEEE Ninth International Conference on* (pp. 1-7). IEEE.
11. Yuan, J., Zheng, Y., & Xie, X. (2012, August). Discovering regions of different functions in a city using human mobility and POIs. In *Proceedings of the 18th ACM SIGKDD international conference on Knowledge discovery and data mining* (pp. 186-194). ACM.
12. Wei, L. Y., Zheng, Y., & Peng, W. C. (2012, August). Constructing popular routes from uncertain trajectories. In *Proceedings of the 18th ACM SIGKDD international conference on Knowledge discovery and data mining* (pp. 195-203). ACM.
13. Huff, D. L. (1966). A programmed solution for approximating an optimum retail location. *Land Economics, 42*(3), 293-303.
14. Ghosh, A., & Craig, C. S. (1983). Formulating retail location strategy in a changing environment. *The Journal of Marketing*, 56-68.
15. Wang, L., Gopal, R., Shankar, R., & Pancras, J. (2015). On the brink: Predicting business failure with mobile location-based checkins. *Decision Support Systems, 76*, 3-13.
16. Wang, B., & Wang, X. (2007). Bandwidth selection for weighted kernel density estimation. *arXiv preprint arXiv:0709.1616*.
17. O'Kelly, M. (2001). Retail market share and saturation. *Journal of Retailing and Consumer Services, 8*(1), 37-45.

Constraint-based Clustering Algorithm for Multi-Density Data and Arbitrary Shapes

Walid Atwa[1] and Kan Li[2]

[1]Faculty of Computer and Information, Menoufia University, Egypt
[2]School of Computer Science and Technology, Beijing Institute of Technology, China
Walid.atwa@ci.menofia.edu.eg, likan@bit.edu.cn

Abstract. The purpose of data clustering is to identify useful patterns in the underlying dataset. However, finding clusters in data is a challenging problem especially when the clusters are being of widely varied shapes, sizes, and densities. Density-based clustering methods are the most important due to their high ability to detect arbitrary shaped clusters. Moreover these methods often show good noise-handling capabilities. Existing methods are based on DBSCAN which depends on two specified parameters (*Eps* and *Minpts*) that define a single density. Moreover, most of these methods are unsupervised, which cannot improve the clustering quality by utilizing a small number of prior knowledge. In this paper we show how background knowledge can be used to bias a density-based clustering algorithm for multi-density data. First we divide the dataset into different density levels and detect suitable density parameters for each density level. Then we describe how pairwise constraints can be used to help the algorithm expanding the clustering process based on the computed density parameters. Experimental results on both synthetic and real datasets confirm that the proposed algorithm gives better results than other semi-supervised and unsupervised clustering algorithms.

Keywords: Semi-supervised clustering, pairwise constraint, multi-density data.

1   Introduction

Semi-supervised clustering algorithms have been received a significant amount of attention in data mining and machine learning fields [1, 2, 3, 4]. Unlike traditional clustering algorithms, semi-supervised clustering (also known as constrained clustering) is a category of techniques that tries to incorporate prior information like pairwise constraints into the clustering algorithms. Pairwise constraints provide the supervision information like must-link (*ML*) and cannot-link (*CL*), where must-link constraint specifies that the pair of instances should be assigned to the same cluster, and cannot-link constraint specifies that the pair of instances should be placed into different clusters.

With exponential growth of data scale and the enrichment of data types, some problems have been put forward on clustering algorithms as: dealing with multi-

© Springer International Publishing AG 2017
P. Perner (Ed.): ICDM 2017, LNAI 10357, pp. 78–92, 2017.
DOI: 10.1007/978-3-319-62701-4_7

density dataset, discovering clusters with arbitrary shapes, dealing with multi-dimensional data, and handling noise and outliers.

Recently, many clustering algorithms have been proposed in this area of research. Among all these methods, density based methods can detect arbitrary shaped clusters and show good noise-handling capabilities. But most existing methods cannot detect all the meaningful clusters for datasets with varied densities due to using global density parameters. DBSCAN is a density based clustering algorithm and its effectiveness for spatial datasets has been demonstrated in the existing literature [5, 13]. However, there are two distinct drawbacks for DBSCAN: (1) the performances of clustering depend on two specified parameters. One is the maximum radius of a neighborhood (*Eps*) and the other is the minimum number of the data points contained in this neighborhood (*Minpts*). In fact these two specified parameters define a single density. Nevertheless, without enough prior knowledge, these two parameters are difficult to be determined; (2) with these two parameters for a single density, DBSCAN does not perform well to datasets with varying densities [8].

For example, in Figure 1(a), DBSCAN fails to find the four clusters, because this dataset has four different densities and the clusters are not totally separated by sparse regions. In Figure 1(b), DBSCAN discovers only the three small clusters and considers the other two large clusters as noises, or merges the three small clusters in one cluster to be able to find the other two large clusters. These problems occur due to using global values of the parameters (*Eps*, *Minpts*).

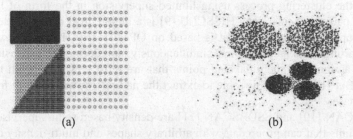

(a)   (b)

Fig. 1. Clusters with varying densities

In this paper, we propose a semi-supervised clustering (called SemiDen) algorithm that discovers clusters of different densities and arbitrary shapes. The idea of the proposed algorithm is to partition the dataset into different density levels and compute the density parameters for each density level set. Then, use the pairwise constraints for expanding the clustering process based on the computed density parameters. Evaluating SemiDen algorithm on both synthetic and real datasets confirms that the proposed algorithm gives better results than other semi-supervised and unsupervised density based approaches. In summary, our contribution in this paper is clustering multi-density datasets and arbitrary shapes using pairwise constraints.

The rest of the paper is organized as follows. Section 2 presents a brief review of the related work. Section 3 introduces our proposed algorithm for clustering multi-density datasets. Experimental results are presented in Section 4. Finally, we present the conclusions and future work in Section 5.

2   Related Work

Density-based methods consider that clusters are dense sets of data items separated by less dense regions; clusters detected by these may have arbitrary shape and data items can be arbitrarily distributed. There are a large number of density-based clustering algorithms, such as DBSCAN [5], OPTICS [6], and DENCLUE [7] and so on.

DBSCAN is a classic density-based clustering algorithm, it groups data points which are sufficiently dense into clusters, and the discovery process is based on the fact that a cluster can be expanded by any of its core objects. In DBSCAN, the density associated with a point is obtained by counting the number of points in a region of specified radius called *Eps* around this point. Points with a density above a specified threshold called *MinPts* are identified as core points. DBSCAN is able to handle clusters of different sizes and shapes, and also separate noise and outliers. But it fails to identify clusters with varied densities unless the clusters are totally separated by sparse regions.

Chen et al. proposed an enhancement of the DBSCAN algorithm that can deal with multi-density datasets called APSCAN [8]. APSCAN utilizes the Affinity Propagation (AP) algorithm to detect local densities for a dataset and generate a normalized density list, and then extends DBSCAN to generate final results. But APSCAN cannot deal efficiently with high dimensional data.

Semi-supervised clustering algorithms have been proposed to improve the performance of the clustering process using limited supervision in the form of labeled instances or pairwise constraints. HISSCLU [9] is a hierarchical density based method for semi-supervised clustering that is based on OPTICS [6]. HISSCLU expands the clusters starting at all labeled points simultaneously, and during the expansion, cluster labels are assigned to the unlabeled points that are most consistent with the cluster structure. But HISSCLU is not able to extract the natural cluster structure from multi-density dataset.

C-DBSCAN [10] and SSDBSCAN [11] are density based semi-supervised clustering algorithms that can group data with arbitrary shapes and multi-density dataset. C-DBSCAN partition the dataset into neighborhoods with a minimum number of data points and then builds local clusters, in which cannot-link constraints are enforced. Finally it uses must-link constraints to merge local clusters. C-DBSCAN ensures that all input constraints are satisfied. However, this has the side-effect of generating many singleton clusters, which correspond to cannot-link constraints. SSDBSCAN describes how labeled points can be used to help the algorithm detecting suitable density parameters to extract density-based clusters. However, SSDBSCAN decreases the clustering performance when applied on large dataset containing severely overlapping among the samples from different clusters.

3   Clustering Multi-Density Data

In this section, we propose a semi-supervised density-based clustering (SemiDen) algorithm that can find clusters of varying densities, shapes and sizes, even in the presence of noise and outliers. The proposed algorithm is divided into two main parts: (1) partitioning the dataset into different density levels; (2) using pairwise constraints for expanding the clustering process for each density level. We summarize our semi-supervised clustering (SemiDen) algorithm in Algorithm 1.

3.1 Partitioning Dataset

In this section, we describe the details of partitioning the dataset into different density levels. First our algorithm finds the k-nearest neighbors for each point in the given dataset. Based on the k-nearest neighbors, a local density function is used to find the density at each point. Where the local density function at point x is defined as the sum of the distances among the point x and its k-nearest neighbors, as shown in Eq. (1).

$$DEN(x) = \sum_{i=1}^{k} D(x, y_i) \tag{1}$$

where $D(x, y_i)$ is the Euclidean distance between point x and its k-nearest neighbors y_i.

$$D(x, y) = \sqrt{\sum_{j=1}^{n}(x_j - y_j)^2} \tag{2}$$

After computing the local density function for each data point, we sort them in ascending order and compute the density variation between each two adjacent points p_i and p_{i+1} denoted by $DENVAR(p_i, p_{i+1})$. Then, we get $DENVAR$ list (denoted by $DVList$) in which each element in $DVList$ is a density variation between two points in the dataset.

$$DENVAR(p_i, p_{i+1}) = \frac{|DEN(p_{i+1}) - DEN(p_i)|}{DEN(p_i)} \tag{3}$$

For datasets with widely varied densities, there will be some distinct variation depending on the densities of the data points. But for points in the same density level, the range of variation is small. Thus, we can acquire all density level sets by detecting these distinct variations of density.

For example, Figure 2 shows the density variation ($DENVAR$) values for the dataset in Figure 1(a) which has four density levels. From Figure 2, there are some sharp waves between two relatively smooth lines, i.e. a smooth line represents a density level set and a sharp wave indicates a sharp change between two density levels. In order to get these density level sets, each smooth line should be separated out.

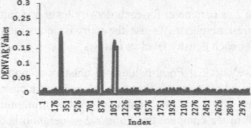

Fig. 2. Density Variation Values for Figure 1(a) Dataset

Definition 1: (Density Level Set). Density level set (*DLS*) consists of points whose densities are approximately the same. In other words, the density variations of the data points within the same *DLS* should be relatively small. Points p_i and p_j belong to the same *DLS* if they satisfy the following condition:

$$p_i, p_j \in DLS_k \quad if \; DENVAR(p_i, p_j) \le \tau$$

where τ is a density variation threshold which divides a multi-density dataset into several density level sets.

We implement partitioning method on *DVList*. Given a density variation threshold τ (Definition 1), remove *DENVAR* values which are bigger than τ out of *DVList*, then the points of remaining separated segments are considered as different density level sets. Here, we compute τ according to the statistical characteristics of the *DVList* as follows:

$$\tau = E(DVList) + \sigma(DVList) \tag{4}$$

where E is mathematical expectation and σ is standard deviation of *DVList*. According to the *DVList* values, there are only a small number of points with large *DENVAR* values which are used to divide the dataset into different sets according to the threshold τ.

After partitioning the dataset into different density level sets, we need to find representative value of the parameters (*Eps* and *Minpts*). We initialize the parameter *Minpts* as k-nearest neighbor and try to identify the value of parameter *Eps* for each density level. For a certain density level set (*DLS*), its corresponding *Eps* will be magnified by simply choosing the maximum *DEN* value. As we know, there are some points may correspond to border objects or noise and these points may have some influence on the *Eps* value. To deal with this problem, we compute Eps_i for DLS_i as follows:

$$Eps_i = maxDEN(DLS_i) \cdot \sqrt{\frac{medianDEN(DLS_i)}{meanDEN(DLS_i)}} \tag{5}$$

where *maxDEN*, *meanDEN* and *medianDEN* are the maximum, mean and the median density of DLS_i respectively.

3.2   Expanding Clusters

After computing the *Eps* parameter for each density level and initializing the parameter *MinPts* as k-nearest neighbor, we use the pairwise constraints for expanding the clustering process for each density level as follows:

- **In Step 11(a):** we check if *Point* belongs to clusters or noise set. Where the key idea of density-based clustering is that for each point of a cluster the neighborhood of a given radius (*Eps*) has to contain at least a minimum number of points (*MinPts*). Therefore we compute the *Point's Eps*-neighborhood. If the number of points in *Eps*-neighborhood less than *MinPts*, adding *Point* to noise set.

- **In Step 11(b):** satisfy Must-link constraints. If *Point* belongs to a transitive closure in *TCS*, all the points contained in the transitive closure are assigned to the current cluster, so as to satisfy the must-link constraints.
- **In Step 11(c):** satisfy Cannot-link constraints. Before adding point p into the current cluster, we should ensure that the adding operation does not violate cannot-link constraints. If there is a point q in the current cluster and a pair $\{p, q\} \in CL$, adding p into the cluster will violate the cannot-link constraints, therefore the points p should not be assigned to the current cluster.

Algorithm 1. SemiDen

Input: A set of data points X; set of must-link constraints ML; the set of cannot-link constraints CL

Output: Several clusters and a set of noises;

Begin

1. Compute local density function for each *Point* in X;
2. Sort the density values in ascending order;
3. Compute the density variation values between each two adjacent points;
4. Partition the dataset into different density level set (*DLS*) according to Definition 1;
5. **For** each density level set (DLS_i)
6.   Initialize *MinPts* as k-nearest neighbor;
7.   Initialize all points in DLS_i as *UNCLUSTERED*;
8.   Estimate the parameter Eps_i;
9.   *ClusterId* = 0;
10. **For** each *Point* in DLS_i
11.   **If** *Point.Id* = *UNCLUSTERED*
     Compute *Point's Eps*-neighborhood *neighborhood*;
     a. **If** the number of points in *neighborhood* < *MinPts*
       Point.Id = *NOISE*;
     Else
       Point.Id = *ClusterId*;
     b. **If** *Point* has must-link constraints
       For each point o in $ML(Point, o)$
         o.Id = *ClusterId*;
     c. **For** each point p in neighborhood **do**
       If *p.Id* = *UNCLUSTERED* and does not violated cannot-link constraints
       p.Id = *ClusterId*;
12.   *ClusterId* = *ClusterId* +1;
13. Return all clusters and a set of noises

End

---

4   Experimental Results

In this section we present two experimental results of SemiDen algorithm on a variety of datasets, including synthetic datasets and several real world datasets. We implement our algorithm in Java and work on a 2.4 GHz Intel Core 2 PC running windows XP with 2 GB main memory.

4.1 Competing Algorithms

Besides the proposed algorithm, we also implemented some competing counterparts as well as the baseline methods listed below for comparison.

1. **C-DBSCAN**: A density based semi-supervised clustering algorithm that is based on DBSCAN for clustering datasets with arbitrary structures. C-DBSCAN depends on two specified parameters (*Eps* and *MinPts*). We set the parameters *Eps*=0.5 and *MinPts* = 4 (default values in DBSCAN) [10].
2. **SSDBSCAN**: Semi-supervised density based clustering algorithm that automatically finds density parameters for each natural cluster in a dataset [11].
3. **HISSCUL:** A hierarchical semi-supervised density based clustering algorithm. HISSCLU use the parameters ρ and ξ to establish borders between clusters when there are no clear cluster boundaries. In order to maximally preserve the original cluster structure HISSCLU is recommended to set up with $\rho = 1.0$, $\xi = 0.5$ [9].
4. **APSCAN:** An unsupervised clustering algorithm that uses affinity propagation for clustering datasets with varying densities [8].

For each data set, we evaluated the performance with different numbers of pairwise constraints. We generated a varying number of pairwise constraints randomly for each data set, where each constraint was generated by randomly selecting a pair of samples. If the samples belong to the same class, a must-link constraint was formed. Otherwise, a cannot-link constraint was formed. The results are averaged over 100 independent runs.

To evaluate the performance of the algorithms, we used the normalized mutual information (NMI). NMI is an external validation metric, which is used to estimate the quality of clustering with respect to the given true labels of the datasets. NMI measures how closely the clustering algorithm could reconstruct the underlying label distribution in the data. If X is the random variable representing the cluster assignments of the instances and Y is the random variable representing the cluster labels of the instances, then NMI is defined as follows:

$$NMI = \frac{I(X;Y)}{(H(X)+H(Y))/2}$$

where $I(X; Y) = H(Y)-H(Y|X)$ is the mutual information between the random variables X and Y, H(Y) is the Shannon entropy of Y, and H($Y|X$) is the conditional entropy of Y given X. The range of NMI values is 0–1. In general, the larger the NMI value, the better the clustering quality.

4.2   Synthetic Datasets

We used different synthetic datasets to evaluate the performance of the proposed algorithm. The experiments were performed on six different datasets containing points in two dimensions of different densities, shapes, and sizes, and containing random noises as shown in Figure 3. The size of these datasets ranges from 3147 to 24000 points, and their exact sizes are indicated in Figure 3.

Figure 4 shows the clustering results of applying SemiDen algorithm on the six datasets in Figure 3. Different colors are used to indicate the clusters. The black color points are discarded as noises. It can be seen from Figure 4 that the proposed algorithm discovers the correct clusters in varying density datasets as in DS1, DS2, and DS3. Also it discovers datasets with arbitrary shapes and detects the noises as in DS4, DS5, and DS6.

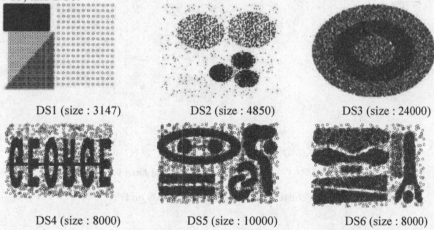

| DS1 (size : 3147) | DS2 (size : 4850) | DS3 (size : 24000) |
| DS4 (size : 8000) | DS5 (size : 10000) | DS6 (size : 8000) |

Fig. 3. Datasets used to evaluate the algorithm and their sizes

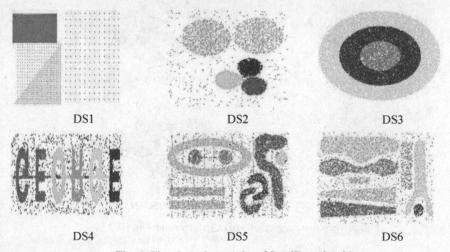

| DS1 | DS2 | DS3 |
| DS4 | DS5 | DS6 |

Fig. 4. The clustering results of SemiDen algorithm

Figure 5 shows the empirical results of SSDBSCAN algorithm on DS1 using different values for the parameters *Eps* and *MinPts*. We can see clearly that a little change of *Eps* or *MinPts* may lead to different number of clusters and different number of noise data. We set *MinPts* = 4 (default value in DBSCAN). If the density is too low, i.e. *Eps* is too little or *MinPts* is too large, SSDBSCAN ignore some clusters and select them as noise (noise is symbolized by black points), as shown as the result with *MinPts* = 4 and *Eps* = 0.2 On the other hand, if the density is too high, i.e. *Eps* is too large or *MinPts* is too little, SSDBSCAN merge several disjoint clusters into a single cluster as the result with *MinPts* = 4 and *Eps* = 0.8.

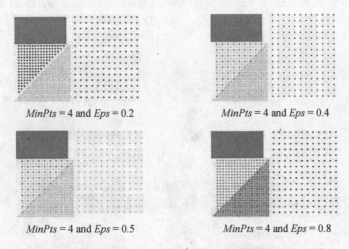

    MinPts = 4 and *Eps* = 0.2   *MinPts* = 4 and *Eps* = 0.4

    MinPts = 4 and *Eps* = 0.5   *MinPts* = 4 and *Eps* = 0.8

Fig. 5. The clustering results of SSDBSCAN on DS1

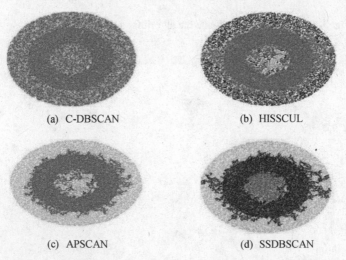

    (a) C-DBSCAN   (b) HISSCUL

    (c) APSCAN   (d) SSDBSCAN

Fig. 6. The clustering results of other algorithms on DS2

Also, we present the results of other semi-supervised and unsupervised clustering algorithms as shown in Figure 6. We can see that these algorithms cannot find out all the meaningful clusters for the dataset DS2. Different colors are used to indicate the clusters and the black color points are discarded as noises. We set the parameters *Eps*=0.5 and *MinPts* = 4 (default values in DBSCAN). HISSCLU use additional parameters ρ and ξ to establish borders between clusters when there are no clear cluster boundaries. In order to maximally preserve the original cluster structure HISSCLU is recommended to set up with $\rho = 1.0$, $\xi = 0.5$ [9].

We also evaluate the performance for the synthetic data (DS1, DS2, DS3, and DS4) with different number of pairwise constraints as shown in Figure 7. The horizontal axis indicates the number of pairwise constraints, and vertical axis indicates the clustering performance (as measured by *NMI*). As mentioned previously, each curve shows the average performance of a method across 50 independent random runs. From Figure 7 we can see that the clustering results of SemiDen are more efficient than other baseline methods. It is interesting to note that the performance of APSCAN algorithm is constant value as it is unsupervised clustering algorithm and not affected with constraints.

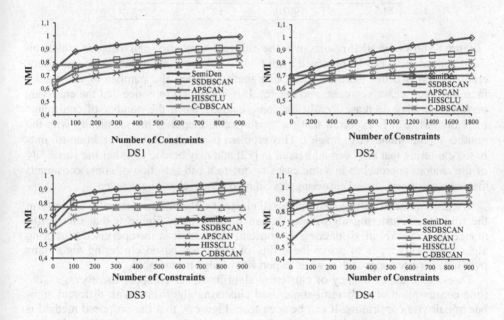

Fig. 7. Performance results over the different number of pairwise constraints on synthetic datasets.

4.3   Real Datasets

The experiments were performed on datasets from UCI repository[1] (ecoli, breast, yeast, segment, digits-389 and magic). These datasets provide a good representation of different characteristics: numbers of samples are ranges from 336 to 19,020, dimensionalities from 8 to 19, and number of clusters from 2 to 10. Some of these datasets (segment and magic) are severe overlapping among the samples from different clusters that will be appropriate for evaluating the robustness of the semi-supervised clustering algorithms. A summary of all the datasets used in this paper is shown in Table 1.

Table 1. Datasets used in the experiments

Dataset	#Samples	#Dimensions	#Clusters
Ecoli	336	8	8
Breast	683	9	2
Yeast	1484	8	10
Segment	2310	19	7
Digits-389	3165	16	3
Magic	19020	10	2

Figure 8 shows the NMI results over the different number of pairwise constraints on the real datasets. It can be observed from Figure 8 that our algorithm "SemiDen" generally performs better than the four other methods when the number of constraints increased (e.g. yeast, segment, and magic). It is interesting to notice that the clustering performance may decrease locally for some datasets when the number of constraints increases, while it is expected that the performance increases monotonically with the number of constraints (e.g. magic). This problem is a well known issue in constraint-based clustering that has been addressed in [12] and may be due to either the variability of the random approaches in some cases or due to a bad selection of some constraints that leads constraint -based clustering algorithm to poorer clustering results.

We also notice that the constraint based clustering algorithms generally outperform the traditional clustering algorithms. It can be seen from Figure 8 that the performance of APSCAN in all datasets is constant value as it is unsupervised clustering algorithms. This tends to prove the utility of constraint based clustering algorithms over unsupervised approaches when expert knowledge is available.

To evaluate the efficiency of clustering algorithms, we compare the average CPU time consumption of each semi-supervised clustering algorithm, with different number of pairwise constraints. It can be seen from Figure 9, that the proposed method is generally time efficient (about 6 seconds on the ecoli dataset with 336 samples processing 180 pairwise constraints, and about 1.5 minute on the magic dataset with 19020 samples possessing 1800 constraints). Although it is not always the fastest algorithm on all the data sets, it is much more advantageous than its counterparts in terms of the performance. It is also observed that the proposed method worked stably on all the datasets we have tried so far, as indicated by the small deviations of the

---

[1] http://www.ics.uci.edu/~mlearn/MLRepository.html.

execution time. By contrast, SSDBSCAN often showed large variations of the CPU-time, especially on the large-size data sets (e.g., magic). Further studies should be conducted on larger datasets to validate our approach. Furthermore, it could be interesting to reduce the complexity of our algorithm with approximate nearest neighbors search algorithms.

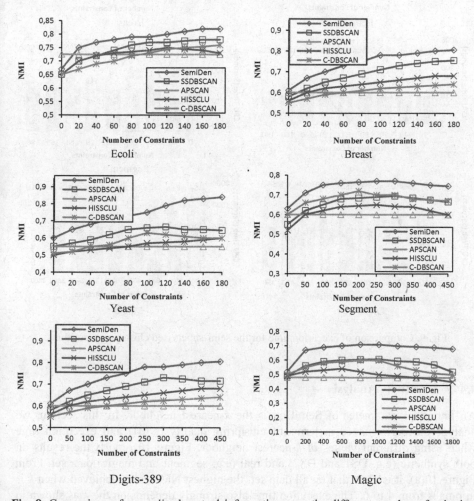

Fig. 8. Comparison of normalized mutual information over the different number of pairwise constraints

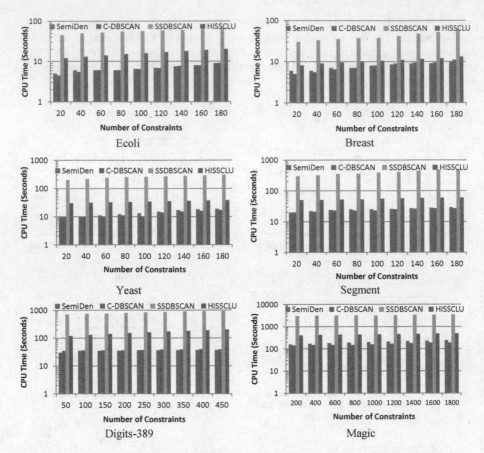

Fig. 9. Comparison of execution time for the semi-supervised clustering algorithms.

4.4   Sensitivity Analysis

An important parameter of SemiDen is the k-nearest neighbor. In this section, we analyze the performance in terms of clustering quality (NMI) and execution time when using different values of k-nearest neighbor. Figure 10 reports the results on both synthetic (e.g. DS1 and DS2) and real (e.g. segment and magic) data set. From Figure 10(a), it is clear that on all data set, the highest NMI value achieved when k is in ranges from 4 to 6. The execution time also strongly depends on the k as shown in Figure 10(b). When setting k to the range [4 - 6] SemiDen has obtained small execution time. Thus, setting $k = 5$ in our previous experiments can generate stable results.

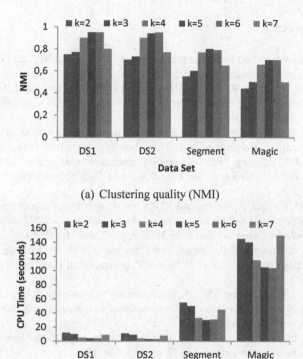

(a) Clustering quality (NMI)

(b) Execution Time

Fig. 10. Clustering quality and execution time vs. k-nearest neighbor.

5   Conclusion and Future Work

This paper presents a novel semi-supervised density based clustering algorithm that takes advantage of background knowledge in the form of pairwise constraints to find clusters of varying densities, shapes and sizes, even in the presence of noise and outliers. Our algorithm is able to extract meaningful clusters via partitioning the dataset into different density levels, and then using pairwise constraints for expanding clusters in each density level. Our experiments on both synthetic and real world datasets show that the proposed algorithm outperforms other algorithms especially for large scale datasets. In the experiments, the pairwise constraints are provided beforehand. In the future work, we plan to actively identify the most informative pairwise constraints for the clustering process.

6   References

1. X. Zhu, 'Semi-supervised learning literature survey'. Technical Report, Computer Sciences, University of Wisconsin-Madison, (2007).
2. M. Bilenko, S. Basu, and R. Mooney, 'Integrating constraints and metric learning in semi-supervised clustering'. In *Proceedings of the 21st International Conference on Machine Learning*, pp. 81–88, (2004).
3. K. Wagstaff, and C. Cardie, 'Clustering with instance-level constraints'. In *Proceedings of the 17th International Conference on Machine Learning*, pp. 1103–1110, (2000).
4. H. Zeng, and Y. Cheung, 'Semi-supervised maximum margin clustering with pairwise constraints'. *IEEE Transactions on Knowledge and Data Engineering*, vol 24, pp. 926–939, (2012).
5. M. Ester, H.P. Kriegel, J. Sander, and X. Xu, 'A density based algorithm for discovering clusters in large spatial databases with noise'. In *Proceedings of 2nd International Conference on Knowledge Discovery and Data Mining*, pp. 226–231, (1996).
6. M. Ankerst, M. Breunig, H.P. Kriegel, and J. Sander, 'OPTICS: Ordering points to identify the clustering structure'. In *ACM SIGMOD International Conference on the Management of Data,* (1999).
7. A. Hinneburg, and D. Keim, 'An efficient approach to clustering in large multimedia data sets with noise'. In *Proceedings of 4th International Conference on Knowledge Discovery and Data Mining*, pp. 58–65, (1998).
8. X. Chen, W. Liu, K. Qiu, and J. Lai, 'APSCAN: A Parameter Free Algorithm for Clustering'. *Pattern Recognition Letters*, vol 32, pp. 973–986, (2011).
9. C. Bohm, and C. Plant, 'HISSCLU: a hierarchical density-based method for semi-supervised clustering'. In *Proceedings of 11th International Conference on Extending Database Technology*, (2008).
10. C. Ruiz, M. Spiliopoulou, and E. Menasalvas, 'Density-based semi-supervised clustering'. *Data Mining and Knowledge Discovery*, vol 21, pp. 345–370, (2010).
11. L. Lelis, and J. Sander, 'Semi-Supervised Density-Based Clustering'. In *Proceedings of 8th IEEE International Conference on Data Mining*, pp. 842–847, (2009).
12. I. Davidson, K.L. Wagstaff,, and S. Basu, 'Measuring constraints-set utility for partitional clustering algorithms'. In Proceedings of European Conference on Machine Learning and Principles and Practice of Knowledge Discovery in Databases ECML, PKDD, pp. 115–126, (2006).
13. Fraley, Chris, and Adrian E. Raftery. "Model-based clustering, discriminant analysis, and density estimation." Journal of the American statistical Association 97.458, pp. 611-631, (2002).

Towards a Large Scale Practical Churn Model for Prepaid Mobile Markets

Amit Kumar Meher*, Jobin Wilson, R. Prashanth

R&D Department
Flytxt
Trivandrum-695581, India
{amit.meher, jobin.wilson, prashanth.ravindran}@flytxt.com

Abstract. Communication Service Providers (CSPs) are increasingly focusing on effective churn management strategies due to fierce competition, price wars and high subscriber acquisition cost. Though there are many existing studies utilizing data mining techniques for building churn prediction models, addressing aspects such as large data volumes and high dimensionality of subscriber data is challenging in practice. In this paper, we present a large scale churn management system, utilizing dimensionality reduction and a decision-tree ensemble learning. We consider subscriber behavior trends as well as aggregated key performance indicators (KPIs) as features and use a combination of feature selection strategies for dimensionality reduction. We report favorable results from evaluating our model on a real-world dataset consisting of approximately 5 million active subscribers from a popular Asian CSP. The model output is presented as an actionable lift chart which will help marketers in deciding the target subscriber base for retention campaigns. We also present the practical aspects involved in productionizing our model using the Apache Oozie workflow engine.

Keywords: Prepaid churn management, Large scale prediction, Feature engineering, Dimensionality reduction, Machine learning, Actionability, Operationalization

1 Introduction

Competition in the wireless telecommunication industry has become rampant due to its dynamic nature with new products, services, technologies, and carriers emerging. New rates and incentives are being frequently exercised by Communication Service Providers (CSPs) to attract new subscribers. With the availability of similar services with different price levels, subscribers prefer switching to different CSPs, resulting in significant revenue loss. Other factors such as inadequate connectivity, intermittent call drops and poor customer service also compel subscribers to switch to different CSPs. Wireless carriers across the world report churn rate varying from 1.5% to 5% per month [1, 2]. As subscriber acquisition costs are considerably higher as compared to retention costs [3, 4, 5],

© Springer International Publishing AG 2017
P. Perner (Ed.): ICDM 2017, LNAI 10357, pp. 93–106, 2017.
DOI: 10.1007/978-3-319-62701-4_8

CSPs are increasingly focusing on effective churn management to plan their retention strategies. Data mining and machine learning techniques are increasingly
applied for churn prediction. There have been many studies for developing churn
prediction models [3, 4],[6, 7, 8]. Existing works generally focus on comparing effectiveness of various machine learning algorithms in predicting churn. However,
these studies have the limitation that their test sets were small. In this paper, we
present a novel approach towards large scale churn modelling and report results
on a real world dataset pertaining to approximately 5 million prepaid subscribers
from a popular Asian CSP. We cover practical issues in churn modelling such as
high dimensionality, class imbalance, productionization of the churn model and
actionability in terms of campaign design based on churn scores.

Lack of contractual obligations between subscribers and CSPs in prepaid
mobile markets make predictive churn modelling a difficult problem [9]. Several subscribers sporadically stop consuming the network services and switch to
other CSPs, without prior notification. Many CSPs consider a continuous and
prolonged period of inactivity in terms of incoming and outgoing service usages
to be indicative of churn. In general, inactivity periods of 2 months, 3 months, 6
months or 1 year are used depending upon the CSP, geography and the government regulations, to mark the expiration of a Subscriber Identification Module
(SIM). After SIM expiration, all network services to the corresponding subscriber
would be terminated. In our opinion, using such long inactivity periods in predictive churn modelling is detrimental, as the CSPs loose actionability in terms
of providing counter offers and promotions to proactively retain their customers
and to encourage service consumption before they switch to other CSPs. Commonly used definitions of prepaid churn in the literature include two months
[10, 11] or three months [8, 10] of continuous inactivity. [12] uses an inactivity
period of 6 weeks to model prepaid churn. In this paper, we consider subscribers
with no outgoing activity in terms of voice usage, data usage and short messaging
service (SMS) usage for a period of 15 consecutive days as churners. Though our
definition may not be indicative of network churn, as subscribers may resume
service consumption after 15 days of inactivity, we believe that it allows CSPs
to engage with subscribers effectively by recommending personalized offers and
thereby influencing them to restart their service consumption.

Application of data mining techniques on telecom churn prediction is not
new. Most of the studies model churn prediction as a binary classification task.
Decision Trees [13], Neural Networks (NN) [14] and Support Vector Machines
(SVM) [15] have been extensively used as classifiers for building churn models. Many of the existing studies compare multiple classifiers in terms of their
predictive quality [16]. Decision Tree and NN classifiers are compared in terms
of their prediction quality on a dataset of 6000 instances and information gain
is used for feature selection in [17]. In [4], a binomial logit model was used on
a sample of 973 mobile users. [6] compares multiple approaches based on decision trees and Back Propagation Network (BPN) model to predict churn for
approximately 160,000 subscribers of a wireless telecom company. The study
also provided results for varying sizes of training sets created using different

sampling techniques. Data Mining by Evolutionary Learning (DMEL) to mine classification rules in large databases is presented in [18], with specific focus on churn modelling. DMEL searches through the huge rule spaces using an evolutionary approach. Its performance was observed to be better compared to NN and decision trees when applied to a database of 1 million subscribers. Genetic Programming (GP) and AdaBoost based approaches have been attempted in [19] to predict telecom churn. This work explores the evolution process in GP by integrating an AdaBoost style boosting to evolve multiple programs per class. Weighted sum of outputs from all GP programs is used to generate the final prediction. This approach provided better results as compared to KNN and Random Forest in terms of sensitivity, specificity and AUC on Orange Telecom (80,000 subscribers) and Cell2Cell (40,000 subscribers) datasets. However, these studies have limitations since the results are reported on small datasets, in a controlled experimental setting. In this work, we primarily focus on practical aspects involved in large-scale churn modelling such as feature selection, model evaluation, operationalization and providing actionabilty to marketers.

The outline of the paper is as follows. The modelling strategy is discussed in Section 2 and results in Section 3. Actionability and operationalization aspects of the developed system are presented in Sections 4 and 5 respectively. Section 6 concludes the paper.

2 Model Building

The data for the study is from a well known Asian CSP with a subscriber base of approximately 7 million. Uncompressed data in the form of Call Data Record (CDR) files, with datasize of approximately 80 GB is streamed on a daily basis to our data warehouse. We randomly sampled 0.5 million subscribers for developing the prediction model. Service usage pertaining to voice (incoming and outgoing), data and SMS along with their recharge information for a period of 90 days (3 months), with datasize of approximately 7200 GB, is used for the analysis. Subsequent subsections elaborate our modelling procedure.

2.1 Data Preparation and Preprocessing

Identifying the key KPIs is an important aspect of any modelling task. With the help of domain expertise and preliminary data analysis, we selected 36 candidate KPIs for building our model. Details of these KPIs are provided in Table 1. KPIs were extracted from the raw CDR data using custom PIG scripts [20] on a Hadoop cluster. Dormant users who were inactive for a period of 30 consecutive days were filtered out as some of them may have already churned out of the network. This reduced the sample size from 0.5 million to 0.34 million. Most of the KPIs were derived KPIs, which were obtained by aggregating subscribers' usage behavior over a period of one month or three months, as appropriate. KPIs were further categorized into snapshot or trending KPI. Snapshot KPIs contain a single value whereas trending KPIs comprise of more than one KPI, representing an usage trend. For example, *Days_Since_Last_Usage* is a snapshot KPI whereas *OG_MOU* is a trending KPI that represents total outgoing minutes of usage, for 3 consecutive months separately (represented by $M_1/M_2/M_3$ in Table 1) representing an usage trend. In other words, *OG_MOU*, which is a trending KPI, actually comprises of 3 snapshot KPIs namely $OG\_MOU.M_1$, $OG\_MOU.M_2$ and $OG\_MOU.M_3$, each of them corresponds to the usage for one month respectively. Description of all the KPIs is provided below.

1. **Days_Since_Last_Usage:** Number of days elapsed since the last service usage (data/voice/SMS) has been made by the subscriber
2. **Days_Since_Last_Recharge:** Number of days elapsed since the time the subscriber has made the last recharge
3. **Average_Recharge_Count:** Average number of recharges done per month by a subscriber within a time period
4. **Last_Recharge_MRP:** Monetary value of the last recharge done
5. **Average_Recharge_MRP:** Average monetary value of recharge done per month within a time period
6. **AON:** This is known as Age On Network, which represents the total number of days since the subscriber became active on the network
7. **Voice_OG_Call_Days:** Total number of days within a time period where at least one outgoing call has been made per day by the subscriber
8. **Voice_IC_Call_Days:** Total number of days within a time period where at least one incoming call has been made per day to the subscriber
9. **OG_MOU:** Total outgoing usage in minutes
10. **IN_DEC:** This is known as Intelligent Network Decrement, representing the total amount deducted from balance within a time period. Intelligent Network refers to the prepaid billing system.
11. **On_Net_Share:** Service usage in home network as a proportion of total service usage
12. **STD_Contribution:** STD call usages as a proportion of Total Outgoing call usages
13. **Max_Days_Between_Recharge:** Maximum number days between any two successive recharges done by the subscriber over a time period
14. **Median_Delay_Between_Recharge:** Median number days between any two successive recharges done by the subscriber over a time period
15. **Max_Consistent_Non_Usage_Days:** Maximum number of days at a stretch when no usage has been made by the subscriber over a time period

Churn rate, based on the proposed inactivity-based definition of churn was found to be 19.12%. The extracted dataset was split into a training set and a test set. The training set was balanced by under sampling of the majority class (non churners). In the test set, ratio of churners to non churners was kept same as in the original dataset. Figure 1 depicts the class distribution in the training and test sets.

2.2   Feature Selection

All 36 KPIs identified initially may not be equally important in characterizing the churn phenomena. Hence, feature selection was performed to identify most

Table 1: List of 36 KPIs, Their Types and Duration

KPI	Snapshot (S) or Trending (T)	Raw (R) or Derived (D)	Period To consider (In Months)
Days_Since_Last_Usage	S	D	1
Days_Since_Last_Recharge	S	D	3
Average_Recharge_Count	S	D	3
Last_Recharge_MRP	S	R	3
Average_Recharge_MRP	S	D	3
AON	S	D	NA
Voice_OG_Call_Days.$M_1/M_2/M_3$	T	D	1 x 3
Voice_IC_Call_Days.$M_1/M_2/M_3$	T	D	1 x 3
OG_MOU.$M_1/M_2/M_3$	T	D	1 x 3
IN_DEC.$M_1/M_2/M_3$	T	D	1 x 3
On_Net_Share.$M_1/M_2/M_3$	T	D	1 x 3
STD_Contribution.$M_1/M_2/M_3$	T	D	1 x 3
Max_Days_Between_Recharge.$M_1/M_2/M_3$	T	D	1 x 3
Max_Days_Between_Recharge.3_Months	S	D	3
Median_Delay_Between_Recharge.$M_1/M_2/M_3$	T	D	1 x 3
Median_Delay_Between_Recharge.3_Months	S	D	3
Max_Consistent_Non_Usage_Days.$M_1/M_2/M_3$	T	D	1 x 3
Max_Consistent_Non_Usage_Days.3_Months	S	D	3

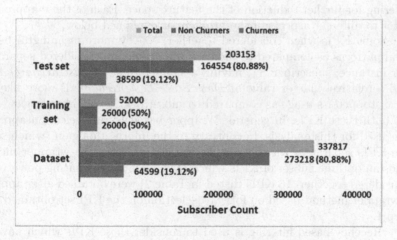

Fig. 1: Data Distribution

prominent KPIs from the initial set. We follow a 2-step strategy in selecting the prominent KPIs.

1. **Step-1 :** Information Gain [21] was used to rank the KPIs based on their information gain scores. Based on empirical experiments, a threshold of 0.12 was chosen and accordingly 19 KPIs were selected. Figure 2 shows the information gain scores of these 19 KPIs.

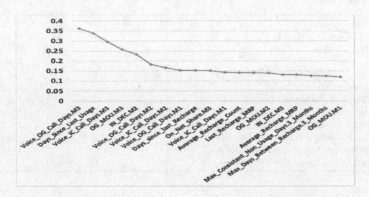

Fig. 2: Information Gain Scores of Top 19 KPIs

2. **Step-2 :** We use a combination of approaches such as wrapper method based on forward selection [22], recency-based filtering and computation-intensive filtering for further reduction of the feature space. Each of these approaches cater to different aspects of the problem as explained below.

Some KPIs when considered together, convey more meaningful behavioral patterns as compared to the case when they are considered separately. For instance, subscriber S_A, having a smaller value of *Last_Recharge_MRP* and a relatively larger value of *Days_Since_Last_Recharge* is more likely to have stopped its usage as compared to subscriber S_B, who has larger values for both these KPIs, in general. Wrapper based feature selection approach was used for this analysis. In contrary to the information gain (which scores the KPIs separately with respect to the class labels), the wrapper method finds an optimal subset of KPIs which have high discriminating power when considered together. 16 KPIs (listed in Table 2) were obtained after applying a wrapper method based on forward selection on the KPI set obtained from step 1.

Recency-based filtering is used to consider those KPIs which have the most recent values. This is applicable only to the trending KPIs. For instance, out of the KPIs namely $OG\_MOU.M_1$, $OG\_MOU.M_2$ and $OG\_MOU.M_3$, only $OG\_MOU.M_3$ will be chosen as it represents the most recent outgoing minutes of usage of a subscriber. This is based on the hypothesis that, most recent behavior better explains the churn phenomena as compared to the least recent behavior. In addition to this, we observed significant correlation among the trending KPIs. For instance, $corr(OG\_MOU.M_1, OG\_MOU.M_2)$ ≈ 0.78 and $corr(OG\_MOU.M_2, OG\_MOU.M_3) \approx 0.65$, where *corr* represents

Table 2: 16 KPIs using Wrapper method

Days_Since_Last_Recharge	*Voice_IC_Call_Days.M_2/M_3*
Last_Recharge_MRP	*IN_DEC.M_2/M_3*
Average_Recharge_MRP	*Days_Since_Last_Usage*
Average_Recharge_Count	*Max_Consistent_Non_Usage_Days.3_Months*
Voice_OG_Call_Days.M_1/M_2/M_3	*Max_Days_Between_Recharge.3_Months*
OG_MOU.M_3	*On_Net_Share.M_3*

the correlation between two KPIs. This motivated us to retain only the recent month's value of the trending KPIs instead of values corresponding to all 3 months.

Apart from this, few KPIs such as *Max_Consistent_Non_Usage_Days.3_Months*, *Max_Days_Between_Recharge.3_Months* pose significant overhead in terms of their computation time. We call these KPIs computationally intensive KPIs and can be ignored if doing so does not the affect the model accuracy significantly. Various combination of above techniques were tried and a Random Forest (RF) classifier (trained on the training set, using an ensemble of 100 Decision Trees) was used to make predictions on the test set, in order to understand the impact of the features on the model quality.

Table 3: Impact of Feature Selection Techniques on Model Quality using RF(Here IG: 'Information Gain')

Feature Selection Combination	Number of KPIs	Precision	Recall	F-measure
All KPIs (No Feature Selection)	36	0.552	0.882	0.679
IG	19	0.551	0.881	0.678
IG + Wrapper	16	0.552	0.88	0.678
IG + Recency Based Filtering	12	0.55	0.879	0.676
IG + Computation Intensive Filtering	17	0.55	0.88	0.677
IG + Wrapper + Computation Intensive Filtering	14	0.551	0.88	0.678
IG + Wrapper + Recency Based Filtering + Computation Intensive Filtering	10	0.547	0.877	0.674

As observed from Table 3, precision, recall and F-measure values are almost similar in all cases. Hence, it can be inferred that the trending KPIs are not contributing much to the discrimination between churners and non churners. So, for a real world deployment, we selected 10 most discriminating KPIs (as listed down in Table 4), without a significant loss in accuracy.

2.3 Machine Learning Classifiers

Several linear and non linear classifiers namely Logistic Regression (LR) [23], Random Forest (RF) [24], Support Vector Machine (SVM) [15] along with ensemble based Random Suspace (RS) [25] were used for this analysis. Many

Table 4: Final 10 KPIs

Days_Since_Last_Recharge	Days_Since_Last_Usage
Last_Recharge_MRP	On_Net_Share.M_3
Average_Recharge_MRP	Voice_IC_Call_Days.M_3
Average_Recharge_Count	OG_MOU.M_3
Voice_OG_Call_Days.M_3	IN_DEC.M_3

existing studies related to churn prediction have made use of standard classifiers, but prior works have not utilized Random Subspace (RS) classifier in churn prediction to the best of our knowledge. Here, we compare the performances of these alternate classifiers. RS is based on the following principle.

1. Given a training set $X = (X_1, X_2, ..., X_n)$, where $X_i = (x_{i1}, x_{i2}, ..., x_{id})$ is a d-dimensional feature vector, $p < d$ features are randomly selected. Modified bootstrapped training sets $\widetilde{X}^b = (\widetilde{X}_1^b, \widetilde{X}_2^b, ..., \widetilde{X}_n^b)$ are created, each containing p-dimensional training objects $\widetilde{X}_i^b = (x_{i1}^b, x_{i2}^b, ..., x_{ip}^b)(i = 1, ..., n)$.

2. Classifiers $C^b(x)$ are constructed in the random supspaces $\widetilde{X}^b$ and the prediction from each of the classifiers are combined by taking a simple majority voting (the most often predicted label)

$$\beta(x) = argmax \sum_b \delta_{sgn(C^b(x)),y} \tag{1}$$

where

$$\delta_{i,j} = \begin{cases} 1, & if \ i = j \\ 0, & if \ i \neq j \end{cases} \tag{2}$$

is the Kronecker symbol and $y \in \{-1, 1\}$ is a decision (class label) of the classifier.

2.4 Decile Wise Cumulative Coverage as a Model Evaluation Metric

We define a metric called Decile Wise Cumulative Coverage (DWCC) to assess the overall model quality along with other standard accuracy metrics namely precision, recall and F-measure. Here we describe the procedure to form deciles from the whole subscriber base. We first estimate the probability of churn for each subscriber, using a classfication algorithm as mentioned above. Then subscribers are ranked according to the decreasing order of their churn probabilities. The whole subscriber base is then divided into 10 equal bins, where each bin contains almost 10% of the total observations. Each of these 10 bins is called a decile. Decile wise cumulative coverage metric can be represented as follows:

$$DWCC_N = \sum_{d=1}^{N} \frac{TCC_d \times 100}{TCC}, 1 \leq N \leq 10 \tag{3}$$

where $DWCC_N$ denotes the decile wise cumulative coverage of churners (in percentage) upto N^{th} decile, TCC_d denotes the true churner count in d^{th} decile and TCC denotes the total churner count in the whole subscriber base (all 10 deciles combined). $DWCC_N$ can also be interpreted as the recall of churners in the top N deciles. We choose this evaluation metric as it describes the model efficacy in terms of proportion of actual churners covered if subscribers falling into top K deciles are targeted as part of retention campaigns. More details on subscriber targeting mechanism for retention campaigns is mentioned in Section 4.

Table 5: Comparision among classifiers using DWCC

Classifier	$DWCC_1$	$DWCC_2$	$DWCC_3$	$DWCC_4$	$DWCC_5$	$DWCC_6$	$DWCC_7$	$DWCC_8$	$DWCC_9$	$DWCC_{10}$
LR	38.26	68.13	83.34	91.4	95.63	97.82	99.04	99.61	99.89	100
SVM	32.7	68.13	86.11	93.74	95.86	96.33	96.82	97.66	98.79	100
RF	40.21	69.62	86.59	93.97	97.06	98.34	99.21	99.68	99.91	100
RS	40.31	70.05	86.53	93.86	97.07	98.53	99.33	99.8	100	100

Table 6: Comparison among classifiers using Precision, Recall and F-Measure

Classifier	Precision	Recall	F-Measure
LR	0.521	0.84	0.643
SVM	0.529	0.876	0.66
RF	0.548	0.877	0.674
RS	0.541	0.877	0.669

A grid search technique [26] was used to obtain the best model with respect to each of the classifiers. In this case, the balanced training set was used to train each of the candidate models and predictions were made on the test set. Best results obtained with respect to each classifier using the grid search technique are reported in Table 5 and 6. It is observed that the non-linear classifiers (SVM, RF, RS) performed slightly better than the linear counterpart (LR) in terms of $DWCC$ in top 3 deciles (83.34% for LR, 86.11% for SVM, 86.59% for RF and 86.53% for RS), precision, recall and F-Measure. However, SVM, RF, RS gave comparable performances except that, SVM covered only 32.7% of the actual churners in the 1st decile (which is the lowest among all classifiers used in this analysis). Among the non-linear classifiers, Random Forest (RF) and Random Subspace (RS) performed almost identically. Between RF and RS, RF was chosen, as its distributed implementation is readily available, which would be beneficial in case of a real world production deployment, where model needs to be trained on large-scale datasets, incurring minimal training time.

In our analysis, we have also considered the trending KPIs in terms of their relative values. A relative value indicates the incremental change (positive or negative) in percentage, observed in a KPI value in a month, with respect to the previous month. To this end, KPIs such as $OG\_MOU.M_1$, $OG\_MOU.M_2$, $OG\_MOU.M_3$ were replaced with $OG\_MOU.M_1$, ΔOG_MOU_1 and ΔOG_MOU_2 respectively, where Δ denotes the percentage change in outgoing minutes of usage in a month with respect to the previous month. We repeated similar experiments using the relative representation of the trending KPIs. However, it did not improve the results. Including KPIs such as number of dropped calls and number of calls made to customer care may further improve the model performance as they are known to be good indicators of churn. However, due to data unavailability, we could not use them in our analysis.

3   Large Scale Training and Prediction

It would be inappropriate to assume that the subscribers' behavior remain consistent over time. For example, let us assume that the frequency of recharge of a large number of subscribers decline suddenly because of the increased use of free messaging applications like WhatsApp. In this scenario, the same training set which was used to train the model may no longer be suitable for learning the newly emerging behavioral patterns and hence the generalization capability of the model will degrade. Hence, to capture latest trends in subscribers' usage, we use the most recent window to construct the training set. Once our model is used for prediction, we wait for next 15 days to collect the true labels of the predicted subscribers. A new training set is constructed by utilizing these true labels along with the required KPIs. The model is retrained using this newly created training set to make predictions for the next cycle.

Table 7: Decile wise Coverage of Churners and Non Churners

Decile	Total Subscriber count	Actual Churners Count	Actual Non Churners Count	Cumulative Population in %	% of Churners Covered per Decile	% of Non Churners Covered per Decile	Cumulative % of Churners Covered	Cumulative % of Non Churners Covered
1	456906	348356	108550	10	36.76	3.00	36.76	3.00
2	456906	267441	189465	20	28.22	5.23	64.98	8.23
3	456906	176573	280333	30	18.63	7.74	83.61	15.97
4	456906	85312	371594	40	9.00	10.26	92.61	26.23
5	456906	36382	420524	50	3.84	11.61	96.45	37.84
6	456906	16474	440432	60	1.74	12.16	98.19	50.01
7	456906	8458	448448	70	0.89	12.38	99.08	62.39
8	456906	4691	452215	80	0.50	12.49	99.58	74.88
9	456906	2681	454225	90	0.28	12.54	99.86	87.42
10	456907	1295	455612	100	0.14	12.58	100.00	100.00
TOTAL	4569061	947663	3621398		100.00%	100.00%		

Table 8: Confusion Matrix

Actual ↓	Predicted →	
	C	NC
C	818884	128779
NC	657184	2964214

Table 9: Precision, Recall and F-Measure

Precision	Recall	F-Measure
0.5548	0.8641	0.6757

Table 7,8 and 9 represent the real world prediction results when the model was applied to a full circle subscriber base [27] containing approximately 5 million active subscribers from a popular Asian operator. A balanced training set, consisting of approximately 0.9 million churners and an equal number of non churners was used for model training.

4 Actionability with Lift Chart

Campaign management is a crucial part of retention programs. To have effective campaigns, CSPs need actionability. We use lift chart [28] to provide actionability to marketers. These charts aid marketers in deciding the set of subscribers to be targeted for retention. Lift is a measure of effectiveness of a predictive model, calculated as the ratio between the results obtained with and without the predictive model which can be represented as follows:

$$Lift_{D(PM)} = \frac{DWCC_{D(PM)}}{DWCC_{D(RM)}} \tag{4}$$

where $Lift_{D(PM)}$ denotes lift obtained in top D deciles using our proposed model, $DWCC_{D(PM)}$ and $DWCC_{D(RM)}$ denotes the decile wise cumulative coverage of churners obtained in top D deciles using our proposed model and a random model respectively. In this context, a random model is equivalent to randomly guessing the likely churners. As evident from the decile formation described in the previous section, decile 1 contains subscribers who are most likely to churn and decile 10 contains subscribers who are least likely to churn, making it convenient for marketers to consider top K deciles as the target base for retention campaigns.

Choosing an optimal value of K depends mostly upon a marketer's subjectivity. One ideal way is to choose a value of K, beyond which the model's lift does not fall rapidly. Considering this case, optimal value of K that would have been chosen is 4, as seen from the lift chart in Figure 3. However, marketer can also set a threshold value of model's lift Th_L and select a value of K for which $Lift_{D(PM)} > Th_L$. For instance, if marketer chooses $Th_L = 2.5$, then value of K chosen would be 3. Moreover, marketer can also consider other factors such as cumulative decile wise coverage intended, cost of targeting a subscriber base and historically observed churn rate in choosing a target base for retention campaigns.

Table 10: Proposed Model's Lift Calculation

Decile	$DWCC_{PM}$	$DWCC_{RM}$	$Lift_{PM}$
1	36.74	10	3.674
2	64.98	20	3.249
3	83.61	30	2.787
4	92.61	40	2.315
5	96.45	50	1.929
6	98.19	60	1.636
7	99.08	70	1.415
8	99.58	80	1.244
9	99.86	90	1.109
10	100	100	1

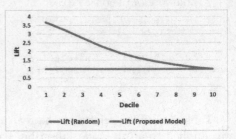

Fig. 3: Lift Chart

5   Operationalization

For the scaled implementation, the preprocessing and data preparation steps were carried out in a MapReduce framework [29] with the help of Apache spark [30] scripts. Apache Spark is a fast and general-purpose cluster computing system. In the production environment, the model was deployed as an Apache Oozie workflow [31] and was scheduled to run in every 15 days. We used MLlib library [32] for the training and prediction tasks. MLlib is Apache Spark's scalable machine learning library which contains several machine learning algorithms and utilities, including classification, regression, clustering, collaborative filtering, dimensionality reduction, as well as underlying optimization primitives. Our model utilities the distributed implementation of Random Forest, that can train multiple decision trees in parallel thereby reducing the training time significantly, especially when the size of the training set is huge. Figure 4 shows the sequence of actions carried out for productionizing the churn model.

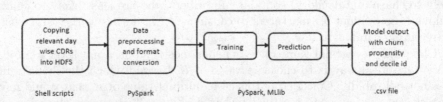

Fig. 4: DAG of Actions in Oozie Workflow

6 Conclusion

In this paper, we have proposed a practical approach for developing a usage based churn management framework for prepaid mobile markets. We have discussed about the KPI selection methodology using a combination of approaches. We also reported results from evaluating our model on a real-world dataset consisting of approximately 5 million active subscribers from a popular asian CSP. We further discussed about the actionability that the model provides in the form of lift charts. Finally, we described the operationalization part of the model in a real world scenario.

As a part of the future work, it would be interesting to apply PCA [33] for further dimensionality reduction. Incremental learning could also be incorporated, where model incrementally learns from the new set of instances, every time the model is trained, which will keep the model up-to-date and robust in case of any change in subscribers' usage patterns.

References

[1] Kang, C., Pei-ji, S.: Customer churn prediction based on svm-rfe. In: Business and Information Management, 2008. ISBIM'08. International Seminar on. Volume 1., IEEE (2008) 306–309

[2] Hadden, J., Tiwari, A., Roy, R., Ruta, D.: Computer assisted customer churn management: State-of-the-art and future trends. Computers & Operations Research **34** (2007) 2902–2917

[3] Wei, C.P., Chiu, I.T.: Turning telecommunications call details to churn prediction: a data mining approach. Expert systems with applications **23** (2002) 103–112

[4] Kim, H.S., Yoon, C.H.: Determinants of subscriber churn and customer loyalty in the korean mobile telephony market. Telecommunications Policy **28** (2004) 751–765

[5] Commission, F.C., et al.: Annual report and analysis of competitive market conditions with respect to mobile wireless, including commercial mobile services. Washington, DC (2010)

[6] Hung, S.Y., Yen, D.C., Wang, H.Y.: Applying data mining to telecom churn management. Expert Systems with Applications **31** (2006) 515–524

[7] Huang, B., Kechadi, M.T., Buckley, B.: Customer churn prediction in telecommunications. Expert Systems with Applications **39** (2012) 1414–1425

[8] Han, Q., Ferreira, P.: Determinants of subscriber churn in wireless networks: Role of peer influence. (2015)

[9] Kraljević, G., Gotovac, S.: Modeling data mining applications for prediction of prepaid churn in telecommunication services. AUTOMATIKA: časopis za automatiku, mjerenje, elektroniku, računarstvo i komunikacije **51** (2010) 275–283

[10] Radosavljevik, D., Van Der Putten, P., Larsen, K.K.: The impact of experimental setup on prepaid churn modeling: Data, population and outcome definition. In: Industrial Conference on Data Mining-Workshops, Citeseer (2010) 14–27

[11] Kusuma, P.D., Radosavljevik, D., Takes, F.W., van der Putten, P.: Combining customer attribute and social network mining for prepaid mobile churn prediction. In: Proceedings of the 23rd Annual Belgian Dutch Conference on Machine Learning (Benelearn). (2013) 50–58

[12] Owczarczuk, M.: Churn models for prepaid customers in the cellular telecommunication industry using large data marts. Expert Systems with Applications **37** (2010) 4710–4712

[13] Quinlan, J.R.: Simplifying decision trees. International journal of man-machine studies **27** (1987) 221–234

[14] Bishop, C.M., et al.: Neural networks for pattern recognition. (1995)

[15] Cortes, C., Vapnik, V.: Support-vector networks. Machine learning **20** (1995) 273–297

[16] Powers, D.M.: Evaluation: from precision, recall and f-measure to roc, informedness, markedness and correlation. (2011)

[17] Umayaparvathi, V., Iyakutti, K.: Applications of data mining techniques in telecom churn prediction. International Journal of Computer Applications **42** (2012) 5–9

[18] Au, W.H., Chan, K.C., Yao, X.: A novel evolutionary data mining algorithm with applications to churn prediction. Evolutionary Computation, IEEE Transactions on **7** (2003) 532–545

[19] Idris, A., Khan, A., Lee, Y.S.: Genetic programming and adaboosting based churn prediction for telecom. In: Systems, Man, and Cybernetics (SMC), 2012 IEEE International Conference on, IEEE (2012) 1328–1332

[20] Foundation, T.A.S.: Apache pig: https://pig.apache.org/ (2015)

[21] Kent, J.T.: Information gain and a general measure of correlation. Biometrika **70** (1983) 163–173

[22] Kohavi, R., John, G.H.: Wrappers for feature subset selection. Artificial intelligence **97** (1997) 273–324

[23] McCullagh, P., Nelder, J.A.: Generalized linear models. Volume 37. CRC press (1989)

[24] Breiman, L.: Random forests. Machine learning **45** (2001) 5–32

[25] Skurichina, M., Duin, R.P.: Bagging, boosting and the random subspace method for linear classifiers. Pattern Analysis & Applications **5** (2002) 121–135

[26] Wikipedia: Hyperparameter optimization — wikipedia, the free encyclopedia (2017) [Online; accessed 7-April-2017].

[27] Wikipedia: Mobile telephone numbering in india — wikipedia, the free encyclopedia (2017) [Online; accessed 7-April-2017].

[28] Uregina: Cumulative gains and lift charts (2015) [Online; accessed 7-April-2017].

[29] Chu, C., Kim, S.K., Lin, Y.A., Yu, Y., Bradski, G., Ng, A.Y., Olukotun, K.: Map-reduce for machine learning on multicore. Advances in neural information processing systems **19** (2007) 281

[30] Spark, A.: Apache spark-lightning-fast cluster computing (2014)

[31] Islam, M., Huang, A.K., Battisha, M., Chiang, M., Srinivasan, S., Peters, C., Neumann, A., Abdelnur, A.: Oozie: towards a scalable workflow management system for hadoop. In: Proceedings of the 1st ACM SIGMOD Workshop on Scalable Workflow Execution Engines and Technologies, ACM (2012) 4

[32] Meng, X., Bradley, J., Yuvaz, B., Sparks, E., Venkataraman, S., Liu, D., Freeman, J., Tsai, D., Amde, M., Owen, S., et al.: Mllib: Machine learning in apache spark. JMLR **17** (2016) 1–7

[33] Abdi, H., Williams, L.J.: Principal component analysis. Wiley Interdisciplinary Reviews: Computational Statistics **2** (2010) 433–459

Smart Stores: A scalable foot traffic collection and prediction system

Soheila Abrishami[1], Piyush Kumar[1], and Wickus Nienaber[2]

[1] Florida State University, Tallahassee, FL 32306, USA
abrisham,piyush@cs.fsu.edu
[2] Bloom Intelligence, Tallahassee, FL 32303, USA
wickus@bloomintelligence.com

Abstract. An accurate foot traffic prediction system can help retail businesses, physical stores, and restaurants optimize their labor schedule and costs, and reduce food wastage. In this paper, we design a large scale data collection and prediction system for store foot traffic. Our data has been collected from wireless access points deployed at over 100 businesses across the United States for a period of more than one year. This data is centrally processed and analyzed to predict the foot traffic for the next 168 hours (a week). Our current predictor is based on Support Vector Regression (SVR). There are a few other predictors that we have found that are similar in accuracy to SVR. For our collected data the average foot traffic per hour is 35 per store. Our prediction result is on average within 22% of the actual result for a 168 hour (a week) period.

Keywords: Time series forecasting, Regression algorithms, Foot traffic, Wireless access point, Bloom Intelligence

1 Introduction

Smart stores are an integral part of smart cities, which improve prosperity, save costs, reduce wastage and resource usage, and improve quality of life [15]. Demand and sales forecasting are one of the important inputs for smart stores. Based on foot traffic predictions, businesses can adjust staffing and product stock levels [14]. The further a business can accurately forecast foot traffic into the future, the more it can optimize operations management (example: labor scheduling), product management (example: stock levels) and in consequence grow profits [13,17]. Therefore, an accurate forecast of foot traffic, by a store manager (for example a restaurant), can help increase customers' satisfaction, increase sales, and reduce food waste. Foot traffic forecasting is valuable for independent retail stores, franchises, and corporate chains alike. Accurate foot traffic information and predictions can also help reduce energy usage and improve safety in offices and buildings [8].

Compared to stock forecasting, food sales forecasting, and similar problems, the subject of foot traffic prediction has received limited research focus. The main reason for insufficient research in foot traffic forecasting is lack of real

© Springer International Publishing AG 2017
P. Perner (Ed.): ICDM 2017, LNAI 10357, pp. 107–121, 2017.
DOI: 10.1007/978-3-319-62701-4_9

world data. To our knowledge there does not exist a public dataset that has foot traffic per minute, hour, or day, for a large number of stores. Due to lack of data, restaurants and physical stores have not had a reliable way of predicting foot traffic. To collect large scale and reliable foot traffic data, a comprehensive system with sensors is required. In this paper, we describe such a scalable data collection system that we designed and used for our experiments.

A study by Veeling [18] worked on improving foot traffic forecasting in 11 retail stores with neural networks. In the study, security gates are used to count people entering or exiting the physical stores. Beams placed at the gates are broken when people walk through them. To keep complexity low, only daily foot traffic is used in the prediction model. The study uses the Non-linear AutoRegressive with Exogenous inputs (NARX) model to predict one-step-ahead. In a different study Cortez et al. [10] used digital cameras to detect foot traffic at a sports store. Data mining methods are applied to build foot traffic forecasting models. The camera is linked with a human facial recognition system which counts the foot traffic and groups the traffic into three categories: all faces, female faces and male faces. The daily foot traffic combined with other factors like weather and special events are used to build a prediction model. They compare six forecasting methods, and their model predicts daily up to 7 days in advance. Further, we found a patent application for predicting traffic at a retail store [2], which indicates that our problem is of practical interest but not well researched.

A problem closely related to foot traffic prediction is retail sales forecasting [1]. Accurate demand forecasting is used to help retail businesses to organize and plan production. Time series forecasting models are often used for retail sales forecasting, but improving quality of forecasts still remains a challenge. Forecasting accuracy can be impacted by different factors such as time, weather conditions, economic factors, random cases, etc [15]. Several methods such as Winters exponential smoothing, the Autoregressive Integrated Moving Average (ARIMA) model [3], multiple regression, and artificial neural networks (ANNs) have been the most widely used approaches to time series forecasting because these models have the ability to capture trend and seasonal fluctuations present in aggregate retail sales. However, all these methods have shown difficulties and limitations. Therefore, it is necessary to investigate further how to improve the quality of forecasts. For sales predictions, data is usually collected when an order is placed, the transaction is automatically processed through the point of sale (POS) system, and then stored in a database.

In this research, we aim to predict hourly foot traffic based on historical data that contain foot traffic of stores for every hour. Data for our work was gathered by Cisco Meraki wireless access points that are installed in different stores, such as gyms, restaurants and bars. WiFi access points installed in businesses detect smartphones with WiFi turned on, whether or not a user connects his or her device to the wireless network. However, the device's presence can only be detected while the device is within range of the network [7]. After preprocessing the raw data, we built a model for prediction. We choose traditional machine learning approaches rather than time series analysis such as ARIMA. Time series models with multi-step-ahead forecast have high error rate [6], and would reduce the effectiveness of the prediction model beyond one step-ahead (one hour-ahead). Our goal is to predict foot traffic each hour of the next 7 days; we are predicting $h = 168$ steps ahead (hour).

We are able to predict hourly foot traffic for the next hour, day and week by using historical data and regression algorithms such as Random Forest Regression [4] and Support Vector Regression [9]. The variation in foot traffic can be caused by different factors such as weather and holidays. Hence, we consider these influencing factors in the forecasting model to improve forecast accuracy. The prediction models learn from the collected data for different stores to see how the prediction models work for each different type of business and location.

The remainder of this paper is organized as follows. Section 2 presents the description of data collection system in the study. Section 3 explains the prediction model and implementation of the developed model. Experimental design and results are presented in Section 4. Finally, conclusions and future work are described in Section 5.

2 Data Collection System

For this research raw data is collected from Cisco Meraki Wireless Access Points (AP) by Bloom Intelligence. The AP's are installed at customers of Bloom Intelligence such as gyms, bars, restaurants, etc. The raw data is processed and classified to extract foot traffic and other engagement metrics for the location. This data is then made available to Bloom Intelligence customers through an analytics dashboard.

2.1 Data Collection

The IEEE 802.11 specification for wireless communication provide a mechanism for devices to discover other compatible 802.11 devices. An 802.11 enabled WiFi device, a mobile phone for instance, would broadcast a probing requests that can be received by any other compatible device, a Cisco Meraki Wireless Access Points. The AP captures these probing request for each device when they are in range and is able to collect information about the device. Some of the data collected and utilized in this research is highlighted in table 1.

Table 1. Information collected from access points

AP MAC	MAC address of the observing AP
Client MAC	MAC address of the probing device
Seen Epoch	Observation time in seconds since the UNIX epoch
RSSI	Device receiver signal strength indication (RSSI) as seen by the AP. This determines the proximity of the device to the AP.

Devices create probing events at different intervals and can be affected by many factors such as the operating system in use, applications that are installed, etc. Cisco Meraki with their analytics partners have found that the request interval can vary greatly. From our collected data we have seen similar results [7].

The data collected by the AP are relayed to the Bloom Intelligence (BI) big data processing servers in an aggregated JSON format that defines each probe as an observation. At the BI servers the data is validated and normalized followed by classified and analytics generation. In the final analytics step the foot traffic per hour is aggregated and made available for the prediction model.

2.2   Preprocessing

The observations, collected by AP's, are analyzed using a proprietary heuristic to discover patterns. The heuristic looks for two types of patterns that classify the foot traffic as either a visit or passerby for a given location. The visit patterns are then analyzed to aggregate hourly foot traffic per hour and per location.

We will now briefly describe the preprocessing (classifying heuristic) shown in algorithm 1. Using the data defined in table 1, a passerby is defined as any device that is detected by the AP at least once. A visitor is a device that is considered present at the location for 5 minutes or more within a 20-minute window. A 20-minute window is initiated by an observation with an RSSI of at least 15. The window is maintained (device is present) when the observations has an RSSI of 10 or more. Once a window is classified as a visit the duration of the visit can be determined. The 20- minute window period is increased by checking for continued activity at the AP. If there is no activity for 20 minutes from the last observations time, the visit window is deemed terminated. Any new observations that occur after this 20-minute period would result in a new pattern analysis. The implication of this procedure is that a device can have multiple visits to a location within a time period like a day or an hour and, visits can also span multiple consecutive hours or days. Figure 1 shows seen devices that are classified as visitors.

If there are n visitors (client MAC address) and m stores (access points), time complexity for calculation of hourly visitor numbers will be $\mathcal{O}(nm)$. Since the number of observations of a particular client MAC address is small, we assume it to be $\mathcal{O}(1)$ in this analysis. In the DAM and cache-oblivious model, our algorithm runs in $\mathcal{O}(nm/B)$ I/Os [11].

This study uses hourly foot traffic of different retails locations that are collected from August 2015 to October 2016. The preprocessed statistics have precomputed the foot traffic for every hour for each calendar day in this period.

3   Our Prediction System

In this section we present our prediction model and the metrics for measuring the forecast error.

Algorithm 1: Calculate visitor numbers per hour for one access point

---

Input : observations time O and rssi values corresponding to observations
      time R
Output: # of visitors per hour

    function CALCULATE_VISITOR_NUMBERS(O, R)
        total number of visitors $= 0$

        foreach *client mac address* **do**

        create a list of sorted observations time (already sorted in DB)
        create a list of rssi values corresponding with sorted observations

          while *list of observations time and rssi are not empty* **do**

          slides window till first observation with rssi ≥ 15

          ANALYSIS_THE_WINDOW_FOR_THE_PRESENCE($window$)

            if *window is visit* **then**

            extend the window till to 20 minutes gap
            total number of visitors $++$
            reset the window

            end

          end

        end

        return *total number of visitors*
    end function

    function ANALYSIS_THE_WINDOW_FOR_THE_PRESENCE($window$)

      if *presence in window* $\geq$ *5 minutes* **then**

        return window is visit

      end
    end function

3.1 Forecast error measures

For forecasting models there is no general applicable accuracy measure because
there are a variety of forecasting objectives, and also data scales and patterns
are different [12]. Thus, in order to reduce possible bias generated by one single accuracy measure, in this study we use three measures, including root mean
square error (RMSE), mean absolute percentage error (MAPE), and mean absolute error (MAE). If $\bar{Y}$ is a vector of n predictions, and Y is the vector of actual
observation then RMSE, MAPE and MAE are defined as follows:

$$RMSE = \sqrt{\frac{1}{n}\sum_{i=1}^{n}\left(\bar{Y}_i - Y_i\right)^2} \tag{1}$$

$$MAE = \sum_{i=1}^{n}\left|\bar{Y}_i - Y_i\right| \tag{2}$$

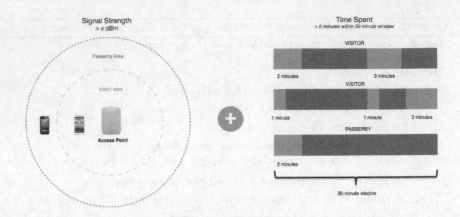

Fig. 1. Computing visitor state [7].

$$MAPE = \sum_{i=1}^{n} \left| \frac{\bar{Y}_i - Y_i}{Y_i} \right| * 100 \qquad (3)$$

These measure have some advantages and disadvantages. For example the MAE is easy to understand, but it is not appropriate for comparing forecast performance between different data sets. MAPE is scale-independent and suitable for comparison of prediction accuracy of multiple data sets, however it has large value when any y_i is close to zero. The MAPE will be infinite if $y_i = 0$.

3.2 Prediction Model

After collecting and preprocessing data we have datasets which contain foot traffic per hour. In other words, for each retail store we have a dataset that shows us how many people visit the store every hour. Now we want to build a prediction model to forecast foot traffic of each store for every hour of the next seven days.

For building the prediction model we choose traditional machine learning approaches instead of using time series analysis such as ARIMA. We prefer using regression algorithms such as Random Forest Regression and Support Vector Regression because time series models have high error for multi-step-ahead forecast [6]. This work aims to predict foot traffic per hour for next week which means $h = 168$ hour-ahead (24 hours × 7 days = 168).

Function Learning. In order to use the regression algorithms for building the prediction model we block up the data into overlapping windows of time, then use it to predict the next days (or weeks). Moreover, we believe that future traffic will resemble past traffic and also foot traffic will be different depending on the

Table 2. Factors that may impact the foot traffic

Factor	Range or an example of the factor
Weather	Temperature, rainfall level, snowfall level
Holidays	Public holidays, school holidays
Special events	Happy hour, sport games, local concerts, conferences, other events
Location	Close to schools, tourist cities

time of year, day of week, and time of day. For instance, in a restaurant, traffic may be higher on weekends, or at specific times such as during periods of lunch or dinner. Therefore, we can conclude foot traffic on hour h and day d, where $0 \leq h \leq 23$ and $0 \leq d \leq 6$, is close to foot traffic on hour h and day d of last weeks. For example, if we want to predict foot traffic on Monday at 8 pm, we should look at the foot traffic on the previously observed Mondays at 8 pm.

Thus, based on this logic, we can define a function learning problem. Let foot traffic at timestamp t be equal to v_t. Our function f takes N inputs, and will output the foot traffic for an hour. The N inputs to f correspond to the foot traffic from the same hour and day in previous N weeks. Formally:

$$f(v_{t-w}, v_{t-2w}, v_{t-3w}, \ldots, v_{t-Nw}) = v_t$$

where $w = 24 \times 7 = 168$. Now we can easily use our data to solve for f using regressions of different types.

Building a Prediction Model Considering Factors. Once function learning is created for the datasets, we can use regression algorithms and build a model that can predict how traffic moved in our historic datasets. Historical data alone is not sufficient in producing accurate prediction values. The foot traffic can be affected by factors such as weather, special events and public holidays. In table 2 we list the factors that we have considered for our prediction model.

Weather: Before adding this feature to our prediction model, we performed some analysis to explore the potential relationships between the weather features and foot traffic, and no significant correlation observed. Moreover, because weather forecasts are inaccurate and they can cause error propagation, we do not use these features in our prediction model.

Holiday: We classify holidays into two categories: regular holidays and festival holidays. The regular holidays include the official four holidays in Florida (Martin Luther King Day, Memorial Day, Independence Day, and Labor Day). The festival holidays include New Year's day, Thanksgiving day, and Christmas day which the stores are usually closed. We believe based on the business type of stores the holiday effect needs to be considered for the day before and after the holiday as well. For example, for a bar, foot traffic is impacted before, on, and after the holiday.

In order to incorporate the effect of holidays into the model, for festival holidays we only need to return zeros for all 24 hours because the stores are

---

Algorithm 2: Prediction with holiday consideration

1: **function** PREDICTION_HOLIDAY(*Data*)
2:     PREDICTION_WITH_RANDOM_FOREST_REGRESSION(*Data*)
3:     DETECT_HOLIDAY_BEHAVIOR(*Data*)
4:     **for** *the days that are holiday* **do**
⌊   *rate* × (*prediction results for the holiday*)
5:     **return** *updated prediction results for a week*
6: **end function**

7: **function** PREDICTION_WITH_RANDOM_FOREST_REGRESSION(*Data*)
8:     **return** prediction results for a week
9: **end function**

10: **function** DETECT_HOLIDAY_BEHAVIOR(*Data*)
11:     **if** *holiday behavior is close to Sundays* **then**
| $rate = Average(\frac{foot\ traffic\ on\ last\ Sundays}{foot\ traffic\ on\ last\ days\ that\ have\ same\ weekday\ as\ holiday})$
12:     **else if** *holiday behavior is close to last holidays* **then**
| $rate = \frac{foot\ traffic\ on\ the\ last\ holiday}{foot\ traffic\ on\ last\ day\ that\ has\ same\ weekday\ as\ holiday}$
13:     **else**
| $rate = 1$
14:     **return** *rate*
15: **end function**

---

usually closed. However, for regular holidays we need an approach to consider holiday effect. The simplest way for including holidays into the model is using dummy variables. For instance, normal days are coded as 0, regular holidays as 1 and festival days as 2. This simple way is effective with large data sets, but we have only around a year of data, and there are only a few holidays in one year. Instead, we use an alternative approach to improve the forecast accuracy for holidays.

Our alternative method for prediction with holiday consideration is using a rate, in a way that the prediction results are updated with a rate. Once, the prediction model returns results for holidays, the results are updated with a rate. Details regarding the rate calculation and prediction model with holiday consideration are shown in algorithm 2.

Special events: The impact of special events depends on the type of event. Events like a happy hour occur on a regular, predictable time, like a Friday from 5pm to 6pm, that can be captured by the prediction model. However, events like sport games, local concerts, and conferences can have irregular or one-off occurrences need to be considered as special days and their behavior could be considered similar to holidays. To incorporate their effects into the prediction model the same as holidays, dummy variables can be used.

Location: The location of some stores induces specific irregularities in foot traffic. For instance, those stores that are close to universities or located in tourist cities show complex behavior. See experiment 5 for more details.

Table 3. Data collected from different stores

Group #	Business type	Avg. foot traffic per hour	# of stores
1	Gym	72	4
2	Coffee shop	26	6
3	Restaurant	28	26
4	Bar	40	19
5	Barbershop	16	1

4 Experimental Results

This section presents several experiments that were conducted to evaluate the forecasting performance of the multiple regression algorithms using real-world store foot traffic data. In order to validate the general forecasting performance of our proposed prediction models, multiple experiments were performed to explore foot traffic prediction in different types of businesses and locations with a variety of foot traffic. Further, we do some experiments to show how holidays can affect foot traffic, and how our prediction model with holiday considerations can manage this irregular behavior.

We have access to data from over 100 different stores. However, in some cases data collection has only been active for 6 months or less. At the results we have chosen 56 different stores that their data collection has been performed for more than one year. The stores are categorized as: gym, restaurant, coffee shop, bar and barbershop. The collected data is from August 2015 to October 2016, and contain the aggregated foot traffic counts for each hour of the day in this period.

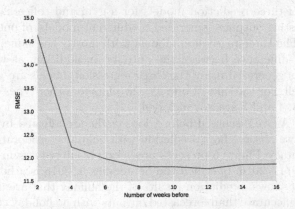

Fig. 2. Correlation between number of weeks before (N) and RMSE.

As mentioned in section 3.2, we had to compute a constant value of N (number of weeks before) for our function learning problem. There are 9504 (396 days × 24 hours) rows in our dataset. However, after creating the learning function described in section 3.2, number of rows in our dataset will be reduced. For choosing N we run an experiment to see the quality of the prediction models with different values of N. As figure 2 shows, after $N = 8$ there is no significant improvement in error. There is a trade-off between increasing N and computation time for function learning. Increasing N decreases our total dataset size for learning. Hence we chose $N = 8$ for our experiments. To create the learning function, we process each row, for which we need data from the 8 previous weeks. For the first 8 weeks there is no data. Therefore, the number of rows available for function learning, will be 9504 - (8 weeks × 7 days × 24 hours) = 8160. Details regarding the datasets are shown in table 3.

For all the experiments the datasets are partitioned into training and test datasets. We use Random Forest Regression from Python package sklearn [16], Support Vector Regression from Libsvm [5], and Google cloud predictor [1].

Now we explain 5 experiments that show the quality of our prediction model:

Experiment 1: In this experiment we compare the forecast measures for three different prediction models of the five random stores: a gym, restaurant, coffee shop, bar and barbershop. The values of RMSE, MAE and MAPE of the one-week forecasts are presented in table 4. Although support vector regression has the least error compared to other models for most stores, this difference is not significant and can be ignored.

Experiment 2: In order to compare error for different types of businesses, MAPE is compared for stores described in table 3. As table 5 illustrates, the gyms have around 13% errors, while coffee shops, restaurants, bars and barbershops have 20% , 24%, 22%, and 17% errors respectively. This is due to the types of stores (datasets); we believe the reason is that gyms have members that go to gyms routinely, therefore foot traffic patterns are less random.

Experiment 3: This experiment describes the comparisons of actual and predicted outputs of three prediction models for a gym and coffee shop. Figure 3 shows the forecasting outputs, generated by different models, of one-week in advance for gym. The three regression models work mostly the same, and there is not a significant difference between the outputs. In addition, as we can see in Figure 4, outputs generated using the three regression models are very close for coffee shop as well. For the sake of brevity, the plots for other types of businesses in experiments 3, 4 and 5 are not included.

Experiment 4: As we discussed before, foot traffic are affected by holidays. In this experiment we observe how prediction with holiday consideration improves the forecast accuracy. Figure 5 presents prediction results of a gym for the week of 9/4/2016 to 9/11/2016, of which the September 5, 2016 is a holiday (Labor day). As the plot shows, prediction results of the holiday that are updated with the rate are more accurate than forecasted outputs with no holiday consideration.

---

[1] https://cloud.google.com/prediction/

Table 4. Comparison of different prediction model

Business Type	Model	RMSE	MAE	MAPE
	Random Forest Regression	16.622	12.22	0.122
Gym	Support Vector Regression	14.856	10.5	0.101
	Google cloud predictor	14.929	10.910	0.114
	Random Forest Regression	6.80	4.66	0.178
Coffee shop	Support Vector Regression	6.988	4.75	0.178
	Google cloud predictor	6.892	4.821	0.179
	Random Forest Regression	7.341	5.059	0.1965
Restaurant	Support Vector Regression	7.057	4.744	0.173
	Google cloud predictor	6.879	4.654	0.163
	Random Forest Regression	10.133	6.863	0.197
Bar	Support Vector Regression	8.344	5.684	0.1657
	Google cloud predictor	8.593	6.702	0.202
	Random Forest Regression	4.237	2.869	0.175
Barbershop	Support Vector Regression	4.182	2.934	0.179
	Google cloud predictor	4.624	3.13	0.175

Table 5. Error comparison for different type of business

Group #	Business type	MAPE
1	Gym	13%
2	Coffee shop	21%
3	Restaurant	24%
4	Bar	22%
5	Barbershop	17%

For the non-holidays, solid red and blue line are overlapping and only the red one can be seen. The comparison of predicted values with and without holiday consideration for a restaurant are shown in figure 6. The prediction results are for the week of 7/2/2016 to 7/9/2016, of which 4th of July is a holiday.

Experiment 5: In this experiment we see how proximity to universities and locating in tourist cities can affect the prediction results. For a bar in downtown

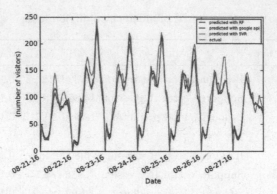

Fig. 3. Comparison of the predicted results using the three different models for a gym.

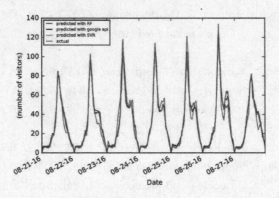

Fig. 4. Comparison of the predicted results using the three different models for a coffee shop.

Orlando (a tourist city) which has an average of 55 hourly foot traffic, MAPE by SVR is 30% which is more than some other bars that have MAPE of 22%. Also, for a bar close to Florida State University in Tallahassee with an average of 21 hourly foot traffic, MAPE by SVR is 31%. As figure 7 shows the traffic behavior is very random which causes poor prediction. Furthermore, for a coffee shop with an average of 21 hourly foot traffic which is close to FSU, the MAPE is 33%.

5   Conclusion and future work

In this paper, we presented a scalable data collection and prediction system to forecast hourly foot traffic, one-week in advance. Raw data for more than one year was gathered by using wireless access points installed in more than 100 stores. We preprocessed the raw data in order to calculate the foot traffic

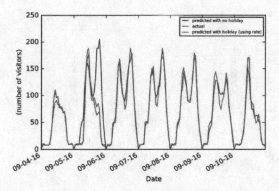

Fig. 5. Comparison of the predicted results with and without holiday consideration for a gym

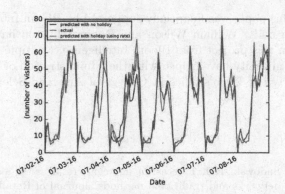

Fig. 6. Comparison of the predicted results with and without holiday consideration for a restaurant.

per hour. After preprocessing data, we used a regression algorithm to build a prediction model, predicting the foot traffic for the next 168 hours. Experiments show that the best results are for SVR, however a few other regression models exist that have accuracies close to SVR. Average error for hourly prediction for one-week in advance is 22%.

Future research will focus on three different aspects: handling the location impact on the forecast model, on top of the prediction model for the next 168 hours; separately build a forecast model only for the next day to improve the accuracy of the next 24 hours; and finally, develop the prediction model to a real-time system and integrate the model into the store environment.

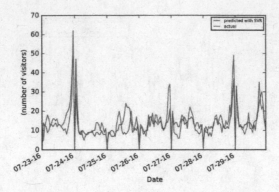

Fig. 7. Random traffic behavior which causes poor prediction.

6   Acknowledgments

The research for this paper was financially supported by Bloom Intelligence and Florida State University. William Wilson and Robin Johnston have been particularly helpful on this project from Bloom Intelligence. Multiple departments at Florida State University have helped with the administration of this research project, particularly the Dean's office at College of Arts and Sciences and the FSU Foundation.

References

1. Alon, I., Qi, M., Sadowski, R.J.: Forecasting aggregate retail sales:: a comparison of artificial neural networks and traditional methods. Journal of Retailing and Consumer Services 8(3), 147 – 156 (2001), http://www.sciencedirect.com/science/article/pii/S0969698900000114
2. Anand, J., Young, K., Choudhary, R., Howell, D.: Predicting shopper traffic at a retail store (Aug 9 2011), https://www.google.com/patents/US7996256, uS Patent 7,996,256
3. Box, G.E., Jenkins, G.M., Reinsel, G.C.: Time Series Analysis: Forecasting and Control. Wiley-Blackwell (2008)
4. Breiman, L.: Random forests. Mach. Learn. 45(1), 5–32 (Oct 2001), http://dx.doi.org/10.1023/A:1010933404324
5. Chang, C.C., Lin, C.J.: Libsvm: A library for support vector machines. ACM Trans. Intell. Syst. Technol. 2(3), 27:1–27:27 (May 2011), http://doi.acm.org/10.1145/1961189.1961199
6. Cheng, H., Tan, P.N., Gao, J., Scripps, J.: Multistep-ahead time series prediction. In: Ng, W.K., Kitsuregawa, M., Li, J., Chang, K. (eds.) Advances in Knowledge Discovery and Data Mining: 10th Pacific-Asia Conference, PAKDD 2006, Singapore, April 9-12, 2006. Proceedings, pp. 765–774. Springer Berlin Heidelberg (2006), http://dx.doi.org/10.1007/11731139_89
7. Cisco: Location Analytics (CMX). https://documentation.meraki.com (2015)

8. Clark, G., Mehta, P.: Artificial intelligence and networking in integrated building management systems. Automation in Construction 6(5), 481–498 (1997)
9. Cortes, C., Vapnik, V.: Support-vector networks. Mach. Learn. 20(3), 273–297 (Sep 1995), http://dx.doi.org/10.1023/A:1022627411411
10. Cortez, P., Matos, L.M., Pereira, P.J., Santos, N., Duque, D.: Forecasting store foot traffic using facial recognition, time series and support vector machines. In: International Joint Conference SOCO'16-CISIS'16-ICEUTE'16: San Sebastián, Spain, October 19th-21st, 2016 Proceedings, pp. 267–276. Springer International Publishing (2017), http://dx.doi.org/10.1007/978-3-319-47364-2_26
11. Frigo, M., Leiserson, C.E., Prokop, H., Ramachandran, S.: Cache-oblivious algorithms. In: Proceedings of the 40th Annual Symposium on Foundations of Computer Science. pp. 285–. FOCS '99, IEEE Computer Society, Washington, DC, USA (1999), http://dl.acm.org/citation.cfm?id=795665.796479
12. Hyndman, R.J., Koehler, A.B.: Another look at measures of forecast accuracy. International Journal of Forecasting 22(4), 679 – 688 (2006), http://www.sciencedirect.com/science/article/pii/S0169207006000239
13. Özgür Kabak, Ülengin, F., Aktaş, E., Şule Önsel, Topcu, Y.I.: Efficient shift scheduling in the retail sector through two-stage optimization. European Journal of Operational Research 184(1), 76 – 90 (2008), http://www.sciencedirect.com/science/article/pii/S0377221706010277
14. Lam, S., Vandenbosch, M., Pearce, M.: Retail sales force scheduling based on store traffic forecasting. Journal of Retailing 74(1), 61 – 88 (1998), http://www.sciencedirect.com/science/article/pii/S0022435999800888
15. Lasek, A., Cercone, N., Saunders, J.: Restaurant sales and customer demand forecasting: Literature survey and categorization of methods. In: Leon-Garcia, A., Lenort, R., Holman, D., Staš, D., Krutilova, V., Wicher, P., Cagáňová, D., Špirková, D., Golej, J., Nguyen, K. (eds.) Smart City 360°: First EAI International Summit, Smart City 360°, Bratislava, Slovakia and Toronto, Canada, October 13-16, 2015. Revised Selected Papers, pp. 479–491. Springer International Publishing (2016), http://dx.doi.org/10.1007/978-3-319-33681-7_40
16. Pedregosa, F., Varoquaux, G., Gramfort, A., Michel, V., Thirion, B., Grisel, O., Blondel, M., Prettenhofer, P., Weiss, R., Dubourg, V., Vanderplas, J., Passos, A., Cournapeau, D., Brucher, M., Perrot, M., Duchesnay, E.: Scikit-learn: Machine learning in python. J. Mach. Learn. Res. 12, 2825–2830 (Nov 2011), http://dl.acm.org/citation.cfm?id=1953048.2078195
17. Perdikaki, O., Kesavan, S., Swaminathan, J.M.: Effect of traffic on sales and conversion rates of retail stores. Manufacturing & Service Operations Management 14(1), 145–162 (2012)
18. Veeling, B.: Improving Visitor Traffic Forecasting in Brick-and-Mortar Retail Stores with Neural Networks. Bachelor thesis, Universiteit Twente (2014)

Multivariate Time Series Representation and Similarity Search Using PCA

Aminata Kane✉️ and Nematollaah Shiri

Computer Science and Software Engineering
Concordia University, Montreal, Canada
{am_kane,shiri}@cse.concordia.ca

Abstract. Multivariate time series (MTS) data mining has attracted much interest in recent years due to the increasing number of fields requiring the capability to manage and process large collections of MTS. In those frameworks, carrying out pattern recognition tasks such as similarity search, clustering or classification can be challenging due to the high dimensionality, noise, redundancy and feature correlated characteristics of the data. Dimensionality reduction is consequently often used as a preprocessing step to render the data more manageable. We propose in this paper a novel MTS similarity search approach that addresses these problems through dimensionality reduction and correlation analysis. An important contribution of the proposed technique is a representation allowing to transform the MTS with large number of variables to a univariate signal prior to seeking correlations within the set. The technique relies on unsupervised learning through Principal Component Analysis (PCA) to uncover and use, weights associated with the original input variables, in the univariate derivation. We conduct numerous experiments using various benchmark datasets to study the performance of the proposed technique. Compared to major existing techniques, our results indicate increased accuracy and efficiency. We also show that our technique yields improved similarity search accuracy.

Keywords: Multivariate time series · Dimensionality reduction · Similarity search

1 Introduction

Innovation and advances in technology have led to the growth of data at a phenomenal rate. Paradoxically, the existing MTS data reduction, analysis and mining techniques do not scale well to its current challenges. Among those challenges, the high dimensionality of the data both in terms of the number of variables and the length of the time series, but also the presence of noise and redundancies makes it difficult to uncover important patterns for many practical applications.

Hence, most pattern recognition tasks rely on dimensionality reduction as a crucial preprocessing step, for reasons of efficiency and interpretability, for a

© Springer International Publishing AG 2017
P. Perner (Ed.): ICDM 2017, LNAI 10357, pp. 122–136, 2017.
DOI: 10.1007/978-3-319-62701-4_10

better understanding of the underlying processes that generated the data, but also to afford a framework that allows downstream pattern recognition tasks to perform more efficiently.

The adequacy of the chosen reduction technique is very important as it will greatly affect the overall quality of search. In similarity search for instance, MTS reduction techniques often follow one of three approaches. In the first approach, each variable is considered independently as a time series [6]. While being easier to process, this approach often requires much more computation time. The second approach consists of concatenating all data contained within all variables as a long univariate time series (UTS) [13]. Like the first approach, this often overlooks the relationships that exist among the variables and cannot efficiently process a relatively large number of variables. The third approach transforms the MTS into a lower dimensional representation that still captures its main characteristics while rendering the data more manageable. Although this approach presents more complexity, it provides more accurate results for the similarity search.

In this paper we propose a similarity search technique based on dimensionality reduction and time series correlations analysis. An important aspect for this technique is the proposed representation based on PCA that allows to transform the MTS with large number of variables to a UTS prior to seeking correlations. This is particularly important because, on one hand, the representation takes into account the correlation between variables within each multivariate dataset, in addition to decreasing redundancy and noise and, reducing the intrinsic high dimensionality. Other proposed univariate representations are often not able to retain the correlation between variables within each multivariate dataset [6]. On the other hand substantial research and progress in making UTS pattern recognition tasks in general, and similarity search in particular, very efficient on large datasets has occurred in recent years [23, 26, 4, 20]. The proposed representation will allow efficient UTS techniques to be easily extended to MTS.

In what follows, we formulate the problem in Section 2, review the related work in Section 3, and provide preliminaries in Section 4. Our proposed technique is introduced in Section 5. Section 6 presents our experimental results. Concluding remarks and future directions are presented in Section 7.

2 Problem Formulation

A UTS $X = < x_1, x_2, ..., x_n >$, of dimension n is a sequence of real values for a variable measured at n different timestamps. A MTS $A_{n,m}$ of n instances for m variables can be represented as a $n \times m$ matrix A (shown below) in which $a_{i,j}$ is the value of variable $X_{*,j}$ measured at time-stamp i, for $1 \leq i \leq n$, $1 \leq j \leq m$.

$$A_{n,m} = \begin{bmatrix} a_{1,1} & a_{1,2} & \cdots & a_{1,m} \\ a_{2,1} & a_{2,2} & \cdots & a_{2,m} \\ \vdots & \vdots & \ddots & \vdots \\ a_{n,1} & a_{n,2} & \cdots & a_{n,m} \end{bmatrix}$$

We are interested in the problem of similarity search in MTS defined as follows:

Definition 1. *(MTS similarity search)*
Let $D = \{A_{n,m}^1, A_{n,m}^2 ..., A_{n,m}^q\}$ be a set of MTS, each of which containing n instances and m variables; and ϵ be a user specified threshold value. A MTS similarity search retrieves all pairs of times series A^i and A^j in D such that their correlation distance does not exceed ϵ, for $1 \leq i, j \leq q$.

Similarity search techniques in time series can be classified in two categories: subsequence search and whole sequence search. In this paper, our focus is on whole sequence search and we use Pearsons product-moment coefficient [21] as the measure to assess similarity between two time series.

3   Related Work

Transforming MTS into lower-dimensional time-series have had some interest and many dimensionality reduction methods have been proposed. Those broadly adopted in the literature include Independent Component Analysis (ICA) [5, 8], Random Projection (RP) [3, 7], and Principal Component Analysis(PCA) [27, 3, 7].

The Independent Component Analysis technique allows to find a new basis in which to represent the multivariate data. It can be considered a generalization of the PCA technique since the latter can be used as a preprocessing step in some ICA algorithms. However, while the goal in PCA in is to capture the maximum variance of data or minimize reconstruction error, the goal of ICA is to minimize the statistical dependence between the basis vectors. ICA however presents limitations that include the inability to determine the order of the independent components and the need for the input time-series data to have non-Gaussian distribution.

The Random Projection technique relies on projecting and embedding the multivariate data onto a lower dimensional subspace, randomly. It is based on the Johnson - Lindenstrauss lemma proposed in 1984 [10]. The lemma states that, given a set of points in a high-dimensional space, they can be projected and embedded into a lower dimensional subspace, such that, the distances between the points are nearly preserved. For random projections, the lower dimensional subspace is randomly chosen based on some distribution and, we can seek to have a probabilistic guaranty that the distance between two time series in the higher dimensional space will have some sort of correspondence with the distance between the same two time series in the lower dimensional space. This data reduction technique is efficient for frameworks with a relatively small number of very long time series due to the fact that, the data size k resulting from the reduction does not depend on the length of the time series but rather the number of time series [31]. It is however known to be less effective than PCA for severe dimensionality reduction [7].

The PCA technique is an orthogonal linear transformations in which one assumes all basis vectors to form an orthonormal matrix. It projects the original dataset in a new coordinate system where the directions are pairwise orthonormal. A main advantage of PCA in our work is that it guaranties the uncovering of an optimal new embedding with minimal approximation error, and hence retains the crucial underlying structure of the original data. In addition to reducing dimensionality, the transformation decreases redundancy and noise, highlights relationships between the variables and reveals patterns by compressing the data while expressing it in such a way that highlights their similarity and dissimilarity. In addition, if two MTS are similar, their PCA representations will also be similar [15]. Many similarity search techniques [28, 27, 2, 15] have relied on PCA for MTS processing as it is known to be one of the most efficiently computable techniques and a powerful tool of choice in high dimensional data environments for linear dimensionality reduction. PCA is however also limited by the fact that, as a new set of features is generated, the reduced form of the data is still a matrix. Retaining the first principal component in order to vectorize the data has been explored with some level of success in the literature [27]. However, since principal components carry in decreasing order portions of the explained variance from the data, in order to retain enough information in the new representation, one would need to retain at least a few principal components in most cases. Hence the reduced form of the data would remain in a matrix form.

4 Preliminaries

In this section we review some background, definitions and notions needed in later sections.

4.1 Notation

This section provides the notation used in this paper, if not specified otherwise.

- D denotes the set of multivariate time series(or, D' if normalized).
- D^U denotes the set of UTS resulting from the STEP1 reduction.
- $A_{n,m}$ is the multivariate time series with n instances and m variables.
- A is such that $A = [a_{ij}]$ is a matrix representing the multivariate time series.
- A^T is the matrix transpose of A.
- V is the right eigenvector matrix of size m×m, V_k is the right eigenvector matrix of size m×k.
- S denotes the diagonal matrix of the singular values of A, S_k is the diagonal matrix of the k largest singular values of A.
- $[a_i]$ is the column vector of the matrix.
- $[a]_{i,*}$ denotes the i-th row of the matrix.
- $[a]_{*,j}$ denotes the j-th column of the matrix.
- $[a]_{i,j}$ denotes the element entry at the i-th row and j-th column of the matrix.
- θ is the cumulative explained variance in the data that are represented within k retained principal components.
- ρ is the Pearson correlation coefficient.
- ϵ is the user specified correlation threshold.

4.2 Principal Component Analysis

Carrying out PCA on raw MTS data often requires some preprocessing such as mean-centering and scaling to adjust values measured on different scales to a relatively common scale, since PCA is a variance maximizing exercise. The technique is often explained through the original data matrix's covariance matrix $(A^T A)$ eigen-decomposition. It however can also be performed through the Singular Value Decomposition(SVD) of the data matrix. In this paper, we consider the latter, framed as per below.

Let $A_{n,m}$ be a matrix with n instances and m variables, and k be the dimension of the space in which we wish to embed the data. Using a Singular Value Decomposition of the matrix, PCA returns the top k left and top k right singular vectors of A. It subsequently projects the original data on the k-dimensional subspace spanned by the chosen column singular vectors.

Definition 2. *(Singular Value Decomposition) Let A be a n×m data matrix with r as its rank. The singular value decomposition (SVD) of A is the factorization $A = USV^T$, where:*

- *U is a column-orthonormal n×r matrix whose columns are the eigenvectors of AA^T,*
- *S is a diagonal r×r matrix of the singular values s_i for A, otherwise related to the eigenvalues λ_i of the covariance matrix $A^T A$ by $\lambda_i = s_i^2/(n-1)$, where $\lambda_1 \geq \ldots \geq \lambda_r \geq 0$, and,*
- *V is a column-orthonormal r×m matrix whose columns are the eigenvectors of $A^T A$.*

Often, a rank-k approximation of the dataset works well because many datasets occurring in practice present a structure that leads to only the first few principal components being non-negligible.

To identify the number of principal components to retain from each MTS, we use the relative percentage variance criterion [11] to translate the amount of variance we wish to retain in the data to the number of principal components. The number k of relevant principal components may vary for different MTS, consequently k_{max}, representing the largest of all identified ks, are to be retained. Algorithm 1 summarizes the steps in uncovering k_{max}.

5 The Proposed Technique

Given a set of MTS $D = \{A_{n,m}^1, A_{n,m}^2 \ldots, A_{n,m}^q\}$ and a user specified correlation threshold ϵ, our goal is to identify all pairs of time series whose Pearson correlation value is no less than ϵ. The proposed technique follows a two-steps resolution process. It first uses a novel transformation technique (M2U) to transform the MTS to a UTS, then seeks pairwise correlations within the set of newly generate univariate series, using the Pearson product moment correlation. An important aspect about the proposed representation resulting from the M2U transformation is that, it allows efficient UTS pattern recognition techniques to be easily extended to MTS.

Algorithm 1 - Find k_{max} the number of PCs to retain

Input: $D' = \{A_{n,m}^1, A_{n,m}^2 ..., A_{n,m}^q\}$ a set of normalized MTS, θ cumulative variance to retain from each MTS.

Output: The number k_{max} of principal components to retain from each MTS s.t. $k_{max} = max(k_1, k_2, ..., k_q)$

begin

1: $k_{max} \leftarrow 0$

2: **for** $i \leftarrow 1$ *to* q **do**

3:      Uncover fraction of total explained variance

4:      f(k) $\leftarrow \Sigma_{z=1}^k \lambda_z / \Sigma_{z=1}^r \lambda_z$ for all z = {1, ..., r}

5:      Choose the smallest k so that f(k) $\geqslant \theta$ and retain that
         number of k eigenvectors to keep explained variance θ
         in the new embedding.

6:      **if** $k > k_{max}$ **then** $k_{max} \leftarrow k$

7: **end for**

8: return k_{max}

end

5.1 M2U : Multivariate Time Series to a Univariate Time Series Transformation

In this section, we formally define the transformation process then describe its underlying intuition. Algorithm 2 provides the transformation steps from line 2 to 13.

Definition 3. *(M2U (Multivariate to Univariate Transformation)). Given the matrix $A \in R^{n \times m}$ with rank $r = rank(A)$ s.t. $r \leqslant min\{n, m\}$ and $k \leqslant r$. Let V_k be the matrix containing the top k right singular vectors of A, and S_k be the matrix containing the top k singular values of A. Then, the (rank-k) univariate representation of A is defined as*
$[U_{n,1}]_i^k = \Sigma_{v=1}^m a_{i,v} \hat{w}_v$, *for* $i = 1, 2, ..., m$ *where:*

- $a_{i,v}$ *is the element of matrix A at row i, and column v.*
- $\hat{w}_j = \Sigma_{z=1}^k w_z e_{j,z}$, *for* $j = \{1, 2, ..., m\}$ *is the weight of the column variable j within the given multivariate dataset, called weighted score and below defined.*
- $[U_{n,1}]_i^k = \Sigma_{v=1}^m a_{i,v} \hat{w}_v$ *is the i-th entry of the newly generated UTS* $U_{n,1}$.

We assume that each MTS $A_{n,m}^i$ of n instances for m variables within D can be represented as a $n \times m$ normalized matrix A (shown below).

$$X_{*,1} \quad X_{*,2} \quad \cdots \quad X_{*,m}$$

$$A_{n,m} = \begin{bmatrix} a_{1,1} & a_{1,2} & \cdots & a_{1,m} \\ a_{2,1} & a_{2,2} & \cdots & a_{2,m} \\ \vdots & \vdots & \ddots & \vdots \\ a_{n,1} & a_{n,2} & \cdots & a_{n,m} \end{bmatrix}$$

Each column variable $X_{*,j}$ holds a particular weight or importance $\hat{w}_j$ with respect to the whole data matrix $A_{n \times m}$ [14]. Let us consider $\hat{w}$ the weight vector containing all variable weights. Intuitively, if we seek to transform the MTS to a UTS in a new framework, there will be a need to uncover and take into account the variable's importance or weight in the reconstruction process.

Finding the Weighted Scores(Variable Weights) We rely on unsupervised learning through a principal component analysis of the input data to uncover the variable weights (weighted scores) within $\hat{w}$. We use information drawn from the diagonal of the matrix S and the rows of matrix V (from the factorization $A = USV^T$) to computed statistics that reveal influence on the columns of the original matrix A.

Let us first note that the entries in each column of $V = A^T US^\dagger$ (where $S^\dagger$ denotes the Moore pseudo-inverse of S) provide the regression coefficients of a corresponding principal component, which in turn is expressed as a linear combination of all variables from the original matrix. More precisely, the coefficient of the ith new feature component uncovered through PCA is expected to be the ith entry of the eigenvector. The first k principal components can be expressed as below illustrated if we consider $X_1, ..., X_m$ to be the original variables within the data matrix A.

$$e_{1,1}X_1 + e_{1,2}X_2 + e_{1,m}X_m = PC_1$$
$$e_{2,1}X_1 + e_{2,2}X_2 + e_{2,m}X_m = PC_2$$
$$...$$
$$e_{k,1}X_1 + e_{k,2}X_2 + e_{k,m}X_m = PC_k$$

Just as the principal components can be expressed as a linear combination of all variables from the original matrix, the original variables can also be defined as linear combinations of the principal components. The rows of V hence each concern a specific variable and are considered rescaled data projected onto the principal components; the data is indeed rescaled according to the singular values to ensure that the covariance is identity.

In the multivariate to univariate transformation process, we wish to uncover the influence of the original variables with respect to the input data, hence we will seek to retain coefficients that are "unscaled". Such coefficients will need to account for the relative portions of variance carried by the principal components.

Definition 4. *(Weighted Scores) Given the matrix $A \in R^{n \times m}$ with rank $r = rank(A)$ s.t. $r \leqslant min\{n, m\}$ and $k \leqslant r$. Let V_k be the matrix containing the top k right singular vectors of A, and S_k be the matrix containing the top k singular values of A. Then, the (rank-k) weighted score of the i-th column of A is defined as $\hat{w}_i^{(k)} = |\Sigma_{j=1}^k w_j e_{i,j}|$, for $i = 1, 2, ..., m$*
where:

- *$w_j = \lambda_j / \Sigma_{z=1}^r \lambda_z$, the fraction of variance carried by the j-th column in $[V_k]$, for $1 \leqslant j \leqslant k$ and,*

$- \lambda_j = \sigma_j^2/(n-1)$ *is the variance corresponding to the jth singular value(σ_j), consequently to the jth column of $[V_k]$, and $\lambda_1 \geq \ldots \geq \lambda_r \geq 0$*

Let us note that the weight w within the weighted score is reflected through the proportion of explained variance retained by the specific principal component. For instance if we consider the jth principal direction, its weight labeled w_j is $w_j = \lambda_j/\Sigma_{z=1}^r \lambda_z$.
A matrix $[wV_k]$ of weighted principal directions is then constructed by multiplying each component within the retained matrix of eigenvectors V_k by its corresponding weight w_j.

$$
\begin{array}{cccc} E_1 & E_2 & \cdots & E_k \end{array}
$$
$$
V_k = \begin{bmatrix} e_{1,1} & e_{1,2} & \cdots & e_{1,k} \\ e_{2,1} & e_{2,2} & \cdots & e_{2,k} \\ \vdots & \vdots & \ddots & \vdots \\ e_{m,1} & e_{m,2} & \cdots & e_{m,k} \end{bmatrix}
$$

$$
\begin{array}{cccc} w_1E_1 & w_2E_2 & \cdots & w_kE_k \end{array}
$$
$$
[wV_k] = \begin{bmatrix} w_1e_{1,1} & w_2e_{1,2} & \cdots & w_ke_{1,k} \\ w_1e_{2,1} & w_2e_{2,2} & \cdots & w_ke_{2,k} \\ \vdots & \vdots & \ddots & \vdots \\ w_1e_{m,1} & w_2e_{m,2} & \cdots & w_ke_{m,k} \end{bmatrix}
$$

Subsequently, the row entries of the weighted matrix $[wV_k]$ are aggregated as per line 9 of Algorithm 2 to provide the variable weights vector $\hat{w}$.

$$
\hat{w} = \begin{bmatrix} |w_1e_{1,1} + w_2e_{1,2} + \cdots + w_ke_{1,k}| \\ |w_1e_{2,1} + w_2e_{2,2} + \cdots + w_ke_{2,k}| \\ \vdots \qquad \vdots \qquad \ddots \qquad \vdots \\ |w_1e_{m,1} + w_2e_{m,2} + \cdots + w_ke_{m,k}| \end{bmatrix} = \begin{bmatrix} \hat{w}_1 \\ \hat{w}_2 \\ \vdots \\ \hat{w}_m \end{bmatrix}
$$

The variable weights vector $\hat{w}$ entries expressed as $\hat{w}_j = |\Sigma_{z=1}^k w_z e_{j,z}|$ for $j = \{1, 2, ..., m\}$, are the original weights for the column-variables within the given multivariate dataset.

Deriving the Univariate Signal. Once the variable weights are uncovered, the next step consists of building a weighted matrix $[\hat{w}A]$ by factoring the original data matrix $A_{n\times m}$ and the variable weights vector $\hat{w}$. More precisely, as shown on lines 10 and 11 of Algorithm-2, each column of $A_{n\times m}$ is factored by its corresponding weight and the row entries of the weighted matrix are subsequently aggregated to form the new univariate derivation.

$$
U_{n,1} = \begin{bmatrix} a_{1,1} & a_{1,2} & \cdots & a_{1,m} \\ a_{2,1} & a_{2,2} & \cdots & a_{2,m} \\ \vdots & \vdots & \ddots & \vdots \\ a_{n,1} & a_{n,2} & \cdots & a_{n,m} \end{bmatrix} * \begin{bmatrix} \hat{w}_1 \\ \hat{w}_2 \\ \vdots \\ \hat{w}_m \end{bmatrix} = \begin{bmatrix} \Sigma_{v=1}^m a_{1v}\hat{w}_v \\ \Sigma_{v=1}^m a_{2v}\hat{w}_v \\ \vdots \\ \Sigma_{v=1}^m a_{nv}\hat{w}_v \end{bmatrix}
$$

An important aspect for this representation technique is that, it uses statistics drawn from the PCA to leverage the relative importance of each variable and uncovers a univariate derivation of the time series. The new derivation takes into account the correlation between variables in the MTS dataset and, decreases redundancy and noise. The proposed representation will allow efficient UTS pattern recognition techniques to be easily extended to MTS.

5.2 Similarity Measure

We use Pearson's product-moment coefficient [21] as the measure to assess similarity between two time series. The Pearson correlation measure is known to be more robust against data that is not normalized and to respond better to baseline and scale shifts when compared to other measures [31].

Let X, Y be two normally distributed time series of equal dimension n. The Pearson correlation coefficient of X, Y denoted $\rho(X,Y)$ is a value in [-1,1] that measures of the linear dependency between X and Y, defined as follows:

$$\rho(X,Y) = \frac{\sum_{t=1}^{n} (x_t - \overline{x})(y_t - \overline{y})}{\sqrt{\sum_{t=1}^{n} (x_t - \overline{x})^2}\sqrt{\sum_{t=1}^{n} (y_t - \overline{y})^2}} \qquad (1)$$

where $\overline{x}_t$ is the mean of X over n and, $\overline{y}$ is the mean of Y over n. The Pearson correlation coefficient can be approximated to the Pearson product moment, expressed as follows:

$$\rho(X,Y) = \frac{1}{n-1} \sum_{t=1}^{n} \frac{xy}{S_x S_y} \qquad (2)$$

where $x = (x_t - \overline{x})$, $y = (y_t - \overline{y})$,
$S_x = [(1/n-1)\sum_{t=1}^{n} x^2]^{1/2}$, and $S_y = [(1/n-1)\sum_{t=1}^{n} y^2]^{1/2}$.

Given a user specified correlation threshold ϵ, our goal is to identify all pairs of time series whose Pearson correlation value is no less than ϵ. Algorithm 2 summarizes the steps for the pairwise correlation search from line 14 to 17.

6 Experimental Set Up and Results

To evaluate the effectiveness of our proposed technique, we implemented the code in Matlab and conducted numerous experiments on benchmark datasets, using a configured PC with Intel Quad core i7 2.00 GHz CPU, 8 GB RAM, running Windows 7.

6.1 Benchmark Datasets

The experiments were ran on benchmark datasets drawn from several widely used repositories [1, 16, 9] in the current literature. Experiments and results pertaining to three of the used benchmark datasets are reviewed in this section.

Algorithm 2 - M2U and Pairwise Correlation Search

Input: $D' = \{A^1_{n,m}, A^2_{n,m}..., A^q_{n,m}\}$ a set of normalized MTS, θ (cumulative variance explained), ϵ a user specified Pearson correlation threshold.

Output: A set C of all pairs (A^i, A^j) in D' whose correlation is not less than ϵ.

begin

1: Estimate k_{max} using Algorithm 1.
   $k \leftarrow k_{max}$

2: **for** $i \leftarrow 1$ *to* q **do**

3:     **STEP1: Reduce MTS $A^i \in D'$ to UTS U^i, add it to D^U**

4:     $A \leftarrow$ *the ith*MTS of rank r, in D', $A^i_{n,m}$

5:     Compute the Singular Value Decomposition
       $[U, S, V^T] \leftarrow SVD(A)$

6:     Retain a matrix of k eigenvectors

7:     $M \leftarrow V_k$

8:     Build the weighted matrix $[wV_k]$
       For $z \leftarrow 1$ to k
       $w_z \leftarrow \lambda_z / \Sigma^r_{z=1} \lambda_z$
       $[wV_k]_{*,z} \leftarrow w_z * [M]_{*,z}$
       end for

9:     Compute the weighted score for each variable
       $\hat{w}^{(k)}_j \leftarrow |\Sigma^k_{z=1} w_z e_{j,z}|$, for all j = {1, 2, ..., m}.

10:     Build the weighted matrix $[\hat{w}A]$
       For $v \leftarrow 1$ to m
       $\hat{w}_v \leftarrow \hat{w}^{(k)}_v$
       $[\hat{w}A]_{*,v} \leftarrow [A]_{*,v} * \hat{w}_v$
       end for

11:     Uncover row entries for the new univariate signal $U_{n,1}$
       $[U_{n,1}]_i \leftarrow \Sigma^m_{v=1} a_{i,v} \hat{w}_v$, for i = {1, 2, ..., n}

12:     add $[U_{n,1}]$ to D^U

13: **end for**

14: **STEP2: Uncover correlated pairs**

15:     For all $(U^i, U^j) \in D^U$

16:     Compute their pairwise Pearson correlations

17:     If ($|\rho(U^i, U^j)| \geqslant \epsilon$) then add (A^i, A^j) to C

end

The Australian language sign dataset(AUSLAN) [12] was gathered through two gloves, with 22 sensors while native AUSLAN speakers signed. The dataset contains 95 signs having 27 examples each, hence a total of 2565 of signs gathered. This dataset is well used in similarity search problems due to its complexity.

The INRIA Holidays images dataset (INRIA HID) [9] is a collection of images that have served in testing the robustness to various transformations: rotations, viewpoint and illumination changes, blurring, etc. The dataset contains 500 high resolution image groups representing a large variety of scene types to incorporate diversity in representation.

The Transient classification benchmark dataset (Trace) [25] was gathered for power plant diagnostics. The dataset has 5 variables (4 process variables and a class label) and 16 operating states. The class label is set to 0 until the transient occurs, at which time it is set to 1. The part of the data that is of interest for us is the subset where the transient occurs.

6.2   Evaluation and Results

We designed experiments to assess the performance of the proposed technique. In this section, we compare our performance against those from primarily five other techniques: the Correlation Based Dynamical Time Warping (CBDTW) [2], the 2-D correlation measure for matrices(see section 6.2)(Corr2), the Dynamical Time Warping(DTW), Eros [29], Euclidien Distance(ED).

The recall-precision ratios recorded for all techniques on the AUSLAN and TRACE datasets are shown on Fig. 1 and Fig. 2 respectively. On both datasets, we can see that the Euclidien Distance(ED), Dynamical Time Warping(DTW) perform worst than the remaining techniques. This may be due to the fact that, both techniques do not take into account the existing correlations between the variables of the MTS while the remaining four techniques do. Our technique outperforms the remaining techniques on both datasets. In another set of ex-

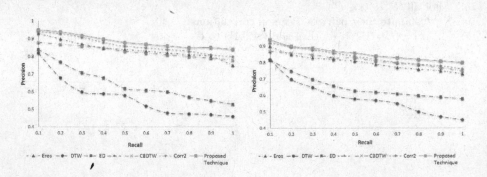

Fig. 1. Recall-Precision on AUSLAN   **Fig. 2.** Recall-Precision on TRACE

periments, we further assess how using the proposed univariate representations compares to the case where the original matrices are used to find pairwise correlations within a set of MTS. Our results confirm that our technique yields improved similarity search accuracy. To illustrate with an example from the INRIA Holidays images dataset, let us consider the six images on the left side of Fig. 3, with three scenes taken at different points in time.

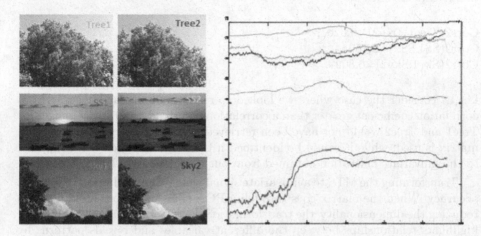

Fig. 3. Left - Six images from the INRIA HID of three scenes taken at different points in time, identified as closest matches. Right - Univariate signals of the six images after M2U transformation. The image name is color-coded with its corresponding univariate signal.

For the purpose of the experiment, the images were converted to the grayscale intensity images, then subsequently to double precision to transform the true-color image RGB to 2-dimensional matrices. Each image was then represented by a 2816×2112 matrix.

Using our proposed transformation technique M2U, each matrix is transformed into a univariate signal represented on the right of Fig. 3. The color of each univariate signal (Fig. 3 right) matches the color of the text on its corresponding image to the left side (Fig. 3 left). We can see that similar images generated similar univariate signals. Furthermore, using our technique, the Pearson correlation coefficients post transformation are:

$\rho(\text{Tree1},\text{Tree2})=0.9661,$
$\rho(\text{SS1},\text{SS2})=0.9413,$
$\rho(\text{Sky1},\text{Sky2})=0.9982.$

To uncover the correlation coefficient that would result from using the original matrices, without transformation, we use the 2-D correlation coefficient *Corr2*, framed as:

$$Corr2(A^i, A^j) = \frac{\Sigma_n \Sigma_m (A^i_{mn} - \bar{A}^i)(A^j_{mn} - \bar{A}^j)}{\sqrt{(\Sigma_n \Sigma_m (A^i_{mn} - \bar{A}^i)^2)(A^j_{mn} - \bar{A}^j)^2}}$$

where $\bar{A}^i = mean2(A^i)$ and $\bar{A}^j = mean2(A^j)$.

For this set of experiments on the full image matrices, the Pearson correlation coefficients are such as:

Corr2(Tree1,Tree2)=0.6261,
Corr2(SS1,SS2)=0.7594,
Corr2(Sky1,Sky2)=0.8027.

Let us consider the case where we looked to retrieve all similar images, with a correlation coefficient greater than a correlation threshold $\epsilon = 0.7$, the images for Tree1 and Tree2 would not have been retrieved as pairwise correlated if the full matrix is used, while it would be identified if the Pearson correlation is applied to the univariate derivation obtained from our technique (M2U).

Transforming the MTS to a univariate signal yields improved similarity search accuracy. When the matrix goes through a PCA transformation, in addition to reducing the dimensionality, the transformation decreases redundancy and noise, highlights relationships between the different variables and reveals patterns by compressing the data while expressing it in such a way that highlights their similarity and dissimilarity. In addition, since we are not discarding any of the relevant principal components, but rather re-combing them, we preserve much of the relevant and needed information from the data.

7   Conclusion

We propose a novel technique for multivariate time series representation, analysis and search. The technique relies on dimensionality reduction and correlation analysis to uncover similar multivariate time series. It uses statistics drawn from the Principal Component Analysis to find a unique derivation of the MTS into a univariate time series prior to seeking correlations. Our experiment results indicate increased accuracy and efficiency when compared to major existing techniques. The proposed representation will allow efficient techniques for univariate time series to be easily extended to multivariate time series. We are currently working on extending the proposed technique to application frameworks for streaming time series.

Acknowledgments: This work was partially supported by Natural Sciences and Engineering Research Council (NSERC) of Canada and Concordia University.

References

1. A. Asuncion and D. Newman. Uci machine learning repository, 2007.
2. Z. Bankó and J. Abonyi. Correlation based dynamic time warping of multivariate time series. *Expert Systems with Applications*, 39(17):12814–12823, 2012.
3. E. Bingham and H. Mannila. Random projection in dimensionality reduction: applications to image and text data. In *Proceedings of the seventh ACM SIGKDD international conference on Knowledge discovery and data mining*, pages 245–250. ACM,2001.
4. A. Camerra, T. Palpanas, J. Shieh, and E. Keogh. isax 2.0: Indexing and mining one billion time series. In *Data Mining (ICDM), 2010 IEEE 10th International Conference on*, pages 58–67. IEEE, 2010.
5. B. A. Draper, K. Baek, M. S. Bartlett, and J. R. Beveridge. Recognizing faces with pca and ica. *Computer vision and image understanding*, 91(1):115–137, 2003.
6. B. Esmael, A. Arnaout, R. K. Fruhwirth, and G. Thonhauser. Multivariate time series classification by combining trend-based and value-based approximations. In *Computational Science and Its Applications–ICCSA 2012*, pages 392–403. Springer, 2012.
7. D. Fradkin and D. Madigan. Experiments with random projections for machine learning. In *Proceedings of the ninth ACM SIGKDD international conference on Knowledge discovery and data mining*, pages 517–522. ACM, 2003.
8. A. Hyvärinen and E. Oja. Independent component analysis: algorithms and applications. *Neural networks*, 13(4):411–430, 2000.
9. H. Jegou, M. Douze, and C. Schmid. Inria holidays dataset, 2008.
10. W. B. Johnson and J. Lindenstrauss. Extensions of lipschitz mappings into a hilbert space. *Contemporary mathematics*, 26(189-206):1, 1984.
11. I. Jolliffe. *Principal component analysis*. Wiley Online Library.
12. M. W. Kadous. *Temporal classification: Extending the classification paradigm to multivariate time series*. PhD thesis, The University of New South Wales, 2002.
13. T. Kahveci, A. Singh, and A. Gurel. Similarity searching for multi-attribute sequences. In *Scientific and Statistical Database Management, 2002. Proceedings. 14th International Conference on*, pages 175–184. IEEE, 2002.
14. A. Kanc and N. Shiri. Selecting the top-k discriminative features using principal component analysis. In *Data Mining Workshops (ICDMW), 2016 IEEE 16th International Conference on*, pages 639–646. IEEE, 2016.
15. L. Karamitopoulos, G. Evangelidis, and D. Dervos. Pca-based time series similarity search. In *Data Mining*, pages 255–276. Springer, 2010.
16. E. Keogh. Ucr time series archive www. cs. ucr. edu/~ eamonn, 2006.
17. E. Keogh, K. Chakrabarti, M. Pazzani, and S. Mehrotra. Dimensionality reduction for fast similarity search in large time series databases. *Knowledge and information Systems*, 3(3):263–286, 2001.
18. J. Lin, E. Keogh, S. Lonardi, and B. Chiu. A symbolic representation of time series, with implications for streaming algorithms. In *Proceedings of the 8th ACM SIGMOD workshop on Research issues in data mining and knowledge discovery*, pages 2–11. ACM, 2003.
19. M. Moinester and R. Gottfriedb. Sample size estimation for correlations with pre-specified confidence interval.
20. A. Mueen, S. Nath, and J. Liu. Fast approximate correlation for massive time-series data. In *Proceedings of the 2010 ACM SIGMOD International Conference on Management of data*, pages 171–182. ACM, 2010.

21. K. Pearson. Mathematical contributions to the theory of evolution. xix. second supplement to a memoir on skew variation. *Philosophical Transactions of the Royal Society of London. Series A, Containing Papers of a Mathematical or Physical Character*, pages 429–457, 1916.

22. Quandl *url http://www.quandl.com/help/api*.

23. T. Rakthanmanon, B. Campana, A. Mueen, G. Batista, B. Westover, Q. Zhu, J. Zakaria, and E. Keogh. Searching and mining trillions of time series subsequences under dynamic time warping. pages 262–270, 2012.

24. C. Ratanamahatana, E. Keogh, A. J. Bagnall, and S. Lonardi. A novel bit level time series representation with implication of similarity search and clustering. In *Advances in knowledge discovery and data mining*, pages 771–777. Springer, 2005.

25. D. Roverso. Plant diagnostics by transient classification: The aladdin approach. *International Journal of Intelligent Systems*, 17(8):767–790, 2002.

26. J. Shieh and E. Keogh. isax: disk-aware mining and indexing of massive time series datasets. *Data Mining and Knowledge Discovery*, 19(1):24–57, 2009.

27. Y. Tanaka, K. Iwamoto, and K. Uehara. Discovery of time-series motif from multidimensional data based on mdl principle. *Machine Learning*, 58(2-3):269–300, 2005.

28. K. Yang and C. Shahabi. A pca-based similarity measure for multivariate time series. In *Proceedings of the 2nd ACM international workshop on Multimedia databases*, pages 65–74. ACM, 2004.

29. K. Yang, H. Yoon, and C. Shahabi. A supervised feature subset selection technique for multivariate time series.yang2004pca 2005.

30. B.-K. Yi and C. Faloutsos. Fast time sequence indexing for arbitrary lp norms. VLDB, 2000.

31. Y. Zhu. *High performance data mining in time series: techniques and case studies*. PhD thesis, New York University, 2004.

Message Passing on Factor Graph: A Novel Approach for Orphan Drug Physician Targeting

Yunlong Wang*, Yong Cai

Advanced Analytics Department, QuintilesIMS Corporation

Abstract. To successfully market an orphan drug requires a different business model than traditional blockbuster drugs. An orphan drug that treats a rare disease condition affects only a small patient population. Pharmaceutical companies rely on effective sales and marketing models under limited budget. The small sales field force has limited reach ability and relies on a well defined target list. But in practice it is often difficult to accurately identify physicians who are treating diagnosed or underdiagnosed rare disease patients. The challenges come from the extreme data imbalance and look-alike patient physician profiles between true and negative classes. Many classical targeting tools such as segmentation and profiling developed for mass market are unsuitable for orphan drug market. In addressing this task, the authors propose a graphical model approach to predict targets by jointly modeling physician and patient features from different data spaces and utilizing the extra relational information. Through an empirical example with medical claim and prescription data, the proposed approach demonstrates enhanced accuracy in identifying targets. The graph representation also provides visual interpretability of relationship among physicians and patients. The model can be extended to incorporate more complex dependency structures.

1 Introduction

Collectively, rare diseases affect up to 30 million Americans [1]. And many of them are serious and life threatening. To encourage the treatment development for rare conditions, United States 1983 Orphan Drug Act provides financial incentives to developers. Given these opportunities, pharmaceutical companies still face great challenges to successfully develop and launch orphan drugs into market. An orphan drug is a pharmaceutical product that treats rare medical conditions. The orphan drug market requires different business model in order to be financially successful. One major challenge comes from the small market size. According to the United States Rare Disease Act of 2002, a rare condition only affects fewer than 200,000 people. In some extreme cases, for example, Hutchinson-Gilford progeria syndrome only affects a few dozen children [2]. An orphan drug business model often has a small field force and limited marketing budget. To be financially successful, it calls for a rather precise targeting list in order to reach out to the right audience such as physician and patients.

---

* yunlong.Wang@us.imshealth.com

© Springer International Publishing AG 2017
P. Perner (Ed.): ICDM 2017, LNAI 10357, pp. 137–150, 2017.
DOI: 10.1007/978-3-319-62701-4_11

In practice, it is very difficult to identify physician targets from historical treatment database. Many patients are underdiagnosed for rare disease because physician rarely encounter such cases and the rare conditions can hide behind other common conditions for a long time. Also there can be no corresponding diagnosis codes for certain rare conditions in the database. Some predictive or classification models are needed to find the targets[3–5].

The current approaches for identifying targets in database marketing or target marketing are developed under assumption for large markets [6, 7]. They prioritize customers by defined value. For example, one method is to derive customer targets through segmentation or clustering framework [8, 9]. These approaches dont perform well in rare disease market where the class of interest is small and extremely imbalanced [10–12]. The physicians, especially primary care physicians, who treat rare disease patients have similar characteristics or patient profiles as other physicians. Segmentation and profiling methods group all look-alike physicians together and do not differentiate well for the true rare disease physicians. The researchers can also use supervised classification approaches. But the traditional classification models have difficulties to predict smaller classes well [13]. Moreover, in all the above literature, the authors address the targeting problem either from physician or patient perspective, without utilizing the physician and patient interactions. In pharmaceutical marketing, there exist complex relationships among various stakeholders. But to our knowledge, there is limited effort to take advantage of such information.

In this paper, We propose a novel method to structurally model physician patient features together and utilize the additional relational information to improve target identification accuracy. Our hope is that the information from dependencies among physicians, patients, and between physician-patient can contribute to the accuracy gain. Specifically, we first build a probabilistic model in the form of a joint distribution that include all the physician and patient features and labels. This joint distribution also depicts the connections between patients and physicians. Then we compute the posterior distribution of all the physician labels conditioned on known patients and physicians information. Finally, we use factor graph and sum-product message passing algorithm [14] to efficiently compute the marginalized physician label posterior for prediction of individual physician targeting.

We organized the remaining of the article in the following way: in the next section, we discuss background of rare disease and data acquisition. Then we mathematically formulate the problem in Section 3. In Section 4, we present the proposed model and the factor graph for target label prediction, and parameter estimation. Finally, experiments with real data as well as concluding remarks will be given in the last two sections.

2   Data Acquisition

To limit the scope, we focus on only one particular rare disease market. The selected rare disease is an inherited blood disorder caused by genetic defect. According to Genetic

Home Reference (GHR) from NIH[1], this disease is estimated to occur in about 1 in 10,000 to 1 in 50,000 people. The data available in this work have been extracted from IMS longitudinal prescription (Rx) and diagnosis (Dx) medical claims data. Each medical claim records the visiting time, physician and patient demographic, and the prescription and diagnosis results.

The Rx data is derived from electronic records collected from pharmacies, payers, software providers and transactional clearinghouses. This information represents activities that take place during the prescription transaction and contains information regarding the product, provider, payer and geography. The Rx data is longitudinally linked back to an anonymous patient token and can be linked to events within the data set itself and across other patient data assets. Common attributes and metrics within the Rx data include payer, payer types, product information, age, gender, 3-digit zip as well as the scripts relevant information including date of service, refill number, quantity dispensed and day supply. Additionally, prescription information can be linked to office based claims data to obtain patient diagnosis information. The Rx data covers up to 88% for the retail channel, 48% for traditional mail order, and 40% for specialty mail order.

The Dx data is electronic medical claims from office-based individual professionals, ambulatory, and general health care sites per year including patient level diagnosis and procedure information. The information represents nearly 65% of all electronically filed medical claims in the US. All data is anonymous at the patient level and HIPAA compliant to protect patient privacy.

2.1   Patient Data

For model development, we pull the diagnoses, procedures and prescriptions at transaction level using selection period from January 1, 2012 to July 31, 2015. From the extracted data, we positively identify 1,233 true rare disease patients and for modeling purpose, we match each rare disease patient with 200 randomly selected control patients. We carry out this match is although the selected rare disease occurs less than 1 in 10,000 among all people, its occurrence ratio among patients who have clinical records in IMS database is around 0.5%. In the rest of the article, we name a patient with the rare disease condition as "positive patient" and name the rest of them as "negative patient".

In creating features for patients, we select 58 types of diagnosis, 8 procedures, and 12 types of prescriptions, which are believed to be relevant with the studied rare disease according to the a priori medical knowledge. For each patient, we count the number of occurrence for those 78 clinical codes as this patient's clinical features. Namely, these features describe how many times the event of prescription, procedure or diagnosis this patient has had during the study period. Moreover, for each patient sample, we also have his/her demographic features including age, gender and region.

---

[1] The National Institutes of Health, https://www.nih.gov/

To demonstrate the quality of data, we use two matrices "missing ratio" and "sparse level", which are defined as the ratios of missing entries and value-zero entries over the total number of entries respectively. A larger value of missing ratio or sparse level implies larger proportion of missing or value-zero entries, and vice versa. We summarize the patient features into four categorizes and list the data missing ratio of each category in Table 1. We can see that the sparse level of diagnosis, procedures, and prescription features are high. This is because for most patients, during the studied period, they only have records in a few types of clinical treatment. Then most of the clinical features have value zero. The patient demographic features are categorical, so the sparse level is a not suitable metric for these features.

Table 1: Patient data overview

Feature category	Missing ratio	Sparse level
Diagnosis features (58 columns)	13.47%	75.52%
Procedure features (8 columns)	6.55%	63.74%
Prescription features (12 columns)	8.52%	70.85%
Demographic features (3 columns)	3.79%	N.A.

2.2 Physician Data

Based on the common anonymous IDs in the patient data, we further pull records for physicians who have treated those patients in the same redefined selection period from January 1, 2012 to July 31, 2015. Similar to the definition of positive patients, a physician is classified as "positive physician" if he or she has treated at least one positive patient, and "negative physician" otherwise. Then we end up with 68,898 unique physicians in total and among them 8,346 are positive.

Associated with each physician sample, the physician's demographic features are specialty, gender, total patient count, and the state where his or her office locates. The second part of physician features accounts for physicians overall office claims histories. For this part, we create maximum, minimum, sum and average of observed number of claims for each physician. The overview of physician features are listed in table 2.

Table 2: Physician data overview

Feature category	Missing ratio	Sparse level
Demographic features (4 columns)	4.34%	N.A.
Medical claim features (4 columns)	5.25%	7.93%

3 Problem Formulation

With the 68,898 physicians and 247,833 patients in the pulled data set, one can draw a bipartite graph shown in Figure 1, which contains 68,898 square nodes and 247,833 circle nodes representing the physicians and patients respectively. A node is shaded if the physician or patient is positive, and unshaded otherwise. In Figure 1, each pair of physician and patient are linked if they both show up in at least one medical claim record. There are 1,463,030 physician-patient links in the graph associated with the studied database. On average each patient has visits 5.9 physicians, and each physician has treated 21.23 patients in Figure 1.

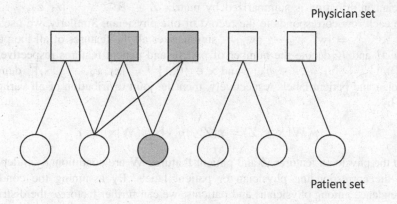

Fig. 1: This schematic depicts the physician-patient network

With these defined components, we can formulate the problem as follows. Given known patient features in Table 1 and physician features in Table 2, identify the positive physicians who is linked with at least one positive patient.

Here we remark that this is a supervised learning task because although when using the model, patient or physician labels are unknown, these labels are known when training the model. We also point out the challenges of this task from three aspects. First, the imbalance among positive and negative classes limits the performance of many common machine learning and statistical models like regression, support vector machine and decision trees. Second, the complicated relationships between patients and physicians make it difficult to directly generate meaningful features as model input from raw data. Third, the large amount of data calls for an efficient inference algorithm instead of naive marginalization. We'll propose our solutions in the next sections.

4 Methodology

In this section, we will first build a probabilistic model and introduce the parameter estimation method for the model. Second, we will present the factor graph that interprets the proposed model, and the inference algorithm associated with the factor graph. Finally, model performance evaluation metrics will be introduced in the last part of this section.

4.1 Proposed Model

To account for the dependencies among physicians and patients, we build a Bayesian model representing the joint distribution of all observed and latent variables[2]. Let N and L denote the number of physicians and features of each physician, then the physician information is summarized by matrix $\mathbf{Z} \in \mathbb{R}^{N \times L} = [\mathbf{z}_1, \mathbf{z}_2, \cdots, \mathbf{z}_N]^\top$, where each row corresponds to the record of one physician. Similarly, we use matrix $\mathbf{W} \in \mathbb{R}^{M \times K} = [\mathbf{w}_1, \mathbf{w}_2, \cdots, \mathbf{w}_M]^\top$ summarizes all the features of all the patients, where M and K denote the number of patient and patient features respectively. Let $\mathbf{y} \in \{0, 1\}^N = [y_1, y_2, \cdots, y_N]^\top$ and $\mathbf{x} \in \{0, 1\}^M = [x_1, x_2, \cdots, x_M]^\top$ denote the physician and patient labels respectively, then the joint distribution of all variables is given by

$$p(\mathbf{W}, \mathbf{x}, \mathbf{y}, \mathbf{Z}) = p(\mathbf{Z}|\mathbf{y})p(\mathbf{y}|\mathbf{x})p(\mathbf{W}|\mathbf{x})p(\mathbf{x}), \tag{1}$$

where the physician features $\mathbf{Z}$ and patient features $\mathbf{W}$ are conditionally independent given the corresponding physician or patient labels. By assuming the conditional independence among physicians and patients, we can further factorize the distribution by

$$p(\mathbf{x}) = \prod_{j=1}^{M} p(x_j), \tag{2}$$

$$p(\mathbf{Z}|\mathbf{y}) = \prod_{i=1}^{N} p(\mathbf{z}_i|y_i), \tag{3}$$

$$p(\mathbf{W}|\mathbf{x}) = \prod_{j=1}^{M} p(\mathbf{w}_j|x_j), \tag{4}$$

where $p(x_j) = 1/201, \forall j$ means that one in two hundred and one patient has the rare disease, and Eq. 3 and Eq. 4 are because one physician/patient's feature variables are independent with other physician/patient's feature variables given the physician/patient labels.

---

[2] Observed variables refer to the variables whose values are given. Oppositely, latent variables refer to the variables with unknown value.

Last but not least, we need to derive the the physician labels joint distribution $p(\mathbf{y}|\mathbf{x})$. Let $\mathbf{x}^i$ be the labels of the patients linked with physician A_i, then we have

$$p(\mathbf{y}|\mathbf{x}) = \prod_{i=1}^{N} p(y_i|\mathbf{x})$$

$$= \prod_{i=1}^{N} p(y_i|\mathbf{x}^i) \tag{5}$$

where the second equal sign is due to the fact that one physician's label is defined by her patients' labels only, i.e., $p(y_i|\mathbf{x}^i) = p(y_i|\mathbf{x}^i, \mathbf{x}^j), \forall j \neq i$. Accoding to the definition of positive physician in Subsection 2.2, the $p(y_i|\mathbf{x}^i)$ in Equation 5 is given by

$$p(y_i|\mathbf{x}^i) = \left(1 - \prod_{j \in \mathcal{N}_i} \mathbb{I}(x_j = 0)\right)^{y_i} \left(\prod_{j \in \mathcal{N}_i} \mathbb{I}(x_j = 0)\right)^{1-y_i} \tag{6}$$

where the indicator function $\mathbb{I}()$ equals to one if the condition inside the parentheses is true and equal to zero otherwise.

4.2   Parameter Estimation

From machine learning point of view, the model parameter estimation process can be understood as a training stage, during which the computer is trained to make meaningful predictions for the variables of interest, e.g., the physician label in our problem. In the remaining of this subsection, we assume that the values of the observed variables are from the training data set. In order to find the maximum likelihood estimate of the parameters, we aim at solving the following optimization problem:

$$\hat{\theta} = \underset{\theta}{\mathrm{argmax}} \; p(\mathbf{W}, \mathbf{x}, \mathbf{y}, \mathbf{Z}; \theta) \tag{7}$$

$$= \underset{\alpha, \beta}{\mathrm{argmax}} \; L(\mathbf{Z}, \mathbf{y}; \alpha) + L(\mathbf{W}, \mathbf{x}; \beta) \tag{8}$$

where $L(\mathbf{Z}, \mathbf{y}; \alpha) = \log(p(\mathbf{Z}|\mathbf{y}; \alpha))$, $L(\mathbf{W}, \mathbf{x}; \beta) = \log(p(\mathbf{W}|\mathbf{x}; \beta))$, and where α and β represent all parameters in physician and patient feature distributions, respectively, and the second equal sign is because of the joint distribution can be factorized as in 1. Here we remark that since there is no unknown parameters in $p(\mathbf{y}|\mathbf{x})$ and $p(\mathbf{x})$, both of these two distributions do not play a role in the objective function. Because the objective function is separable, we can estimate α and β by maximizing $L(\mathbf{Z}, \mathbf{y}; \alpha)$ and $L(\mathbf{W}, \mathbf{x}; \beta)$ respectively.

4.3   Predictive Inference

With the proposed model and the estimated parameters, one can predict the physician label by using graphical model and the message passing algorithm. The target of this subsection is to efficiently compute the following marginal distribution:

$$p(y_i|\mathbf{W}, \mathbf{Z}) = \sum_{y_{-i}=0}^{1} \sum_{\mathbf{x}} p(\mathbf{y}, \mathbf{x}|\mathbf{W}, \mathbf{Z}), \qquad (9)$$

where $\sum_{y_{-i}=0}^{1}$ denotes a summation over all random variables expect y_i in $\mathbf{y}$, and where $\sum_{\mathbf{x}}$ denotes a summation over all x_j variables. Clearly, if we use brutal force marginalization, the computation cost is $O(2^{M+N})$, which is technically impossible because of the large value of M and N.

An alternative way of computing the marginal distribution is by utilizing the conditional independences among those variables. Specifically, we first present the factor graph for the the proposed model in Figure 2, where each circle represents a variable node and each square represents a factor node.

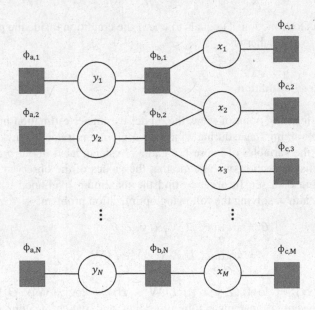

Fig. 2: Factor graph depicting the relationships between variables nodes and factor nodes.

For the variable nodes, they have the exact same definition as the variables in Subsection 4.1.For the factor nodes, the definitions and descriptions are given in Table 3.

As is shown by the factor graph topology in Figure 2, it has more than one components, whereas each components of it has a tree structure. According to [14], one can show that the sum-product algorithm yields an exact result of $p(y_i, \mathbf{W}, \mathbf{Z})$ by exact $2L$ message passings, where L is the number of edges in the factor graph. According to

Table 3: Physician data overview

Factor	Probability distribution	Description
$\phi_{a,i}(\mathbf{z}_i, y_i)$	$p(\mathbf{z}_i \vert y_i)$	Joint Poisson distribution
$\phi_{b,i}(y_i, \mathbf{x}^i)$	$p(y_i, \mathbf{x}^i)$	Indicator distribution in Eq. 5
$\phi_{c,j}(\mathbf{w}_j, x_j)$	$p(\mathbf{w}_j \vert x_j)$	Joint Poisson distribution

the algorithm, there are two types of messages. When the message is from a factor node s to a variable node v, the message is a probability distribution given by

$$\mu_{s \to v}(v) = \sum_{u \in \mathcal{N}_s \backslash v} \phi_s(\mathcal{N}_s) \prod_{u \in \mathcal{N}_s \backslash v} \mu_{u \to s}(u),$$

where $\mathcal{N}_s$ represents all variable nodes in the neighbor set of node s, and node u is a node in node s's neighbor set but not equal to v. On the other hand, the message from a variable node u to a factor node s is given by

$$\mu_{u \to s}(u) = \prod_{\omega \in \mathcal{N}_u \backslash s} \mu_{\omega \to s}(u),$$

where ω denotes a node in node u's neighbor set but not equal to s.

4.4 Model Performance Evaluation

In a binary decision problem, a classifier labels data sample as either positive or negative. The decision made by the classifier can be represented in a structure known as a confusion matrix (Table 4, "1" for positive class and "0" for negative class). The confusion matrix has four categories: True Positives (TP) are examples correctly labeled as positives; False Positives (FP) refer to negative examples incorrectly labeled as positive; True Negatives (TN) correspond to negatives correctly labeled as negative; finally, False Negatives (FN) refer to positive examples incorrectly labeled as negative.

Table 4: Confusion Matrix

	Actual=1	Actual=0
Predicted=1	True Positive (TP)	False Positive (FP)
Predicted=0	False Negative (FN)	True Negative (TN)

Based on the confusion matrix, we will be able to further define several metrics to evaluate model performance as listed in Table 5. The Precision denotes the number of correct positive results divided by the number of the predicted positive results, and

Recall is the number of correct positive results divided by the number of all positive results. Similarly, False Positive Rate (FPR) measures the fraction of negative examples that are misclassified as positive, and True Positive Rate (TPR) measures the fraction of positive examples that are correctly labeled. When comparing two binary classifiers, with same recall or FPR, the better one should have a larger precision value or TRP. Obviously, all the above four metrics have value between zero and one.

Table 5: Model Performance Metrics

Metric	Definition
Precision	TP/(TP+FP)
Recall	TP/(TP+FN)
True Positive Rate (TPR)	TP/(TP+FN)
False Positive Rate (FPR)	FP/(FP+TN)
AUC	Area under the ROC curve
AUPR	Area under the PR curve
MCC	$\dfrac{TP \times TN - FP \times FN}{\sqrt{(TP+FP)(TP+FN)(TN+FP)(TN+FN)}}$

Secondly, in ROC space, one plots the False Positive Rate (FPR) on the x-axis and the True Positive Rate (TPR) on the y-axis. In PR space, one plots Recall on the x-axis and Precision on the y-axis. Recall is the same as TPR, whereas Precision measures that fraction of examples classified as positive that are truly positive.

The third performance metric we use is the Matthews correlation coefficient(MCC)[15]. It is generally regarded as a balanced performance measure for binary classification which can be used even if the classes are of very different sizes. A coefficient equal to 1 means perfect prediction, 0 denotes random guess and -1 indicates total disagreement between prediction and observation. Noting that the data set used for validation is highly imbalanced, we use MCC as the third performance metric for its consistency in data balance.

5   Results

To validate the proposed model, in this section we provide experiments demonstrating the performance of our method, as well as the comparisons with three benchmark methods. Noting the imbalance of in our data set, the first benchmark we choose is weighted logistic regression (denoted by "Logit" hereafter), where the minority class weight as a hyper parameter is empirically set to be 10^3, and majority weight is 1.

---

[3] This weight is set according to the procedure of hyper-parameter tuning with respect to model prediction accuracy.

Also because of the data imbalance, the second and third benchmarks we use are the bootstrap aggregating LASSO and bootstrap aggregating random forest(hereby denoted as "Bagging LASSO" and "Bagging RF" respectively). Specifically, we firstly perform a random under sampling of the majority pool (positive) and combine it with all negative physicians to build an artificial balanced data set, then LASSO[17] or Random Forest[16] is applied to the balanced sampled data. This process is repeated for 1,000 iterations, with each implementation generating a predictive model. The final model is an aggregation of models over all iterations.

For the benchmark methods, unlike the proposed model, it cannot incorporate patient and physician dependency directly. To aggregate patient information to physician features, for each physician, we average all patients records that link to this physician and create similar patient features at physician level. The benchmark methods are implemented through the Python package scikit-learn [18].

Table 6: Model Performance (Ten-fold Cross-Validation)

Metric	Logit	Bagging LASSO	Bagging RF	Proposed Method
AUPR	23.07%	27.24%	28.57%	35.53%
Precision (Recall=45%)	6.33%	7.23%	3.65%	20.19%
Precision (Recall=40%)	7.86%	8.85%	4.49%	29.17%
Precision (Recall=35%)	10.67%	10.10%	5.89%	49.37%
Precision (Recall=30%)	43.61%	51.74%	48.25%	61.43%
Precision (Recall=25%)	55.11%	53.81%	59.46%	70.13%
Precision (Recall=20%)	78.46%	75.27%	78.26%	80.36%

Then we perform Ten-fold cross-validation with the training data. Specifically, we split all the physicians into ten folds, and for each given fold, we train a model with the remaining nine folds and calculate the performance metrics on the given fold, the final performance outputs are metrics averaged over the ten folds. Take the first fold as an example, The testing set has 6,000 (10%) physicians, where 713 physicians are positive and 5,287 are negative. The testing set contains the rest 62,898 physicians including 55,265 positive and 7,633 negative physicians respectively. To avoid information leakage from the patient label in the training data set, any patient connected with any physician in testing data set shall be regarded as a testing patient. As a result, according to the train-test split in the physicians, 161,681 patients are grouped into the training set, and 86,152 patients are grouped into the testing set. There are 736 positive and 160,945 negative patients in the training set, and there are 497 positive and 85,655 negative patients.

Summary of the results are listed in Table 6. The area under curve serves as a single variable summary of the PPV performance, which is 0.3553 for the proposed method, and 0.2857 for the best benchmark, Bagging RF, suggesting that the overall relative performance increase is 24.31%. In particular, when the recall is greater than 30%, the Precision of the benchmarks are less than 11%, suggesting that traditional classification method barely works in such scenarios. In comparison, our proposed method still provide an acceptable Precision even when recall is greater than 30%.

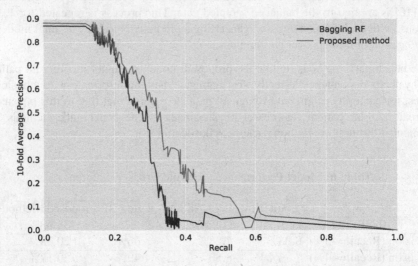

Fig. 3: Precision with respective to Recall curves for proposed and benchmark methods averaged by ten-fold validation

In Figure 3, we plot the precision-recall curve where y-axis denotes the average precision for results from ten data folds, and x-axis denotes the recall level. Each dot on the curve represents a sensitivity precision-recall pair. We can see almost for every value of recall, the proposed method has a higher precision value than the benchmark method.

The Final result we present is the MCC comparisons. In Table 7, we show the MCC of the proposed method and bagging RF with respect to the recall level. We can see that the proposed method outperforms the Bagging RF in each recall level. And particularly for Recall level greater than 30%, again we see the bagging RF hardly work whereas the proposed method shows good robustness.

As is shown in table 6, the predictive accuracy of Bagging LASSO is inferior to Logistic Regression in our case. The reason could be that the high dimensional features have much correlation and in this case less sparse model is preferred.

In searching for better performance, we also test cost-sensitive learning methods such as weighted SVM and weighted LASSO on the same training data. Both of the results don't show improvement over the standard or random under-sampling ensemble methods.

Table 7: Comparison results between proposed and benchmark methods in ten-fold validation

Metric	Bagging RF	Proposed Method
Average MCC (Recall=45%)	4.41%	12.95%
Average MCC (Recall=40%)	6.90%	23.32%
Average MCC (Recall=35%)	7.54%	42.61%
Average MCC (Recall=30%)	33.42%	57.32%
Average MCC (Recall=25%)	46.64%	68.39%
Average MCC (Recall=20%)	64.69%	80.46%

6 Conclusion and Discussion

In this article, we proposed a factor graph based learning algorithm for rare disease physician targeting problem. We presented a graphic representation and probability model to join physician and patient features together with their network relationship. Through the graphical structure, researchers can visualize the connectivity among physicians and patients. The graphic representation provides clear interpretability of data entities and correlation. The proposed model also has flexibility to specify additional dependencies, add features or extend to more complicated network structure. In the empirical example, we tested and compared the bootstrap aggregating and factor graph based learning algorithms on a real world physician-patient data case. The training data set contains only 0.5% positive patients and it is a good real world example to demonstrate imbalanced learning challenge. We observed that at high sensitivity level, the proposed method showed significant improvement over benchmark. In practice, this means when a smaller target is needed under tight marketing budget, the proposed method can yield superior results by identifying more real targets.

Although in our study the proposed model improves rare disease targeting accuracy to certain degree, the precision still leaves a lot to be desired. We hope to see more innovative methods to tackle this hard problem in the future research.

References

1. Field, Marilyn J., and Thomas F. Boat, eds. Rare diseases and orphan products: Accelerating research and development. National Academies Press, 2011.
2. Pollex, R. L., and R. A. Hegele. "HutchinsonGilford progeria syndrome." Clinical genetics 66.5 (2004): 375-381.
3. Narayanan, Sridhar, and Puneet Manchanda. "Heterogeneous learning and the targeting of marketing communication for new products." Marketing Science 28.3 (2009): 424-441.
4. Dong, Xiaojing, Puneet Manchanda, and Pradeep K. Chintagunta. "Quantifying the benefits of individual-level targeting in the presence of firm strategic behavior." Journal of Marketing Research 46.2 (2009): 207-221.
5. Manchanda, Puneet, Peter E. Rossi, and Pradeep K. Chintagunta. "Response modeling with nonrandom marketing-mix variables." Journal of Marketing Research 41.4 (2004): 467-478.

6. Hughes, Arthur Middleton. Strategic database marketing. McGraw-Hill Pub. Co., 2005.
7. Van den Poel, Dirk. Predicting mail-order repeat buying: which variables matter?. No. 03/191. Ghent University, Faculty of Economics and Business Administration, 2003.
8. DeSarbo, Wayne S., Rajdeep Grewal, and Crystal J. Scott. "A clusterwise bilinear multidimensional scaling methodology for simultaneous segmentation and positioning analyses." Journal of Marketing Research 45.3 (2008): 280-292.
9. Cameron, Marcia J., et al. "Evaluation of academic detailing for primary care physician dementia education." American journal of Alzheimer's disease and other dementias 25.4 (2010): 333-339.
10. Akbani, Rehan, Stephen Kwek, and Nathalie Japkowicz. "Applying support vector machines to imbalanced datasets." European conference on machine learning. Springer Berlin Heidelberg, 2004.
11. Fox, Chester H., et al. "Improving chronic kidney disease care in primary care practices: an upstate New York practice-based research network (UNYNET) study." The Journal of the American Board of Family Medicine 21.6 (2008): 522-530.
12. de Vrueh, R., E. R. Baekelandt, and J. M. de Hann. "Update on 2004 background paper: BP 6.19 rare diseases." Geneva: World Health Organization (2013).
13. Chawla, Nitesh V. "Data mining for imbalanced datasets: An overview." Data mining and knowledge discovery handbook. Springer US, 2005. 853-867.
14. Kschischang, Frank R., Brendan J. Frey, and H-A. Loeliger. "Factor graphs and the sum-product algorithm." IEEE Transactions on information theory 47.2 (2001): 498-519.
15. Matthews, Brian W. "Comparison of the predicted and observed secondary structure of T4 phage lysozyme." Biochimica et Biophysica Acta (BBA)-Protein Structure 405.2 (1975): 442-451.
16. Breiman, Leo. "Random forests." Machine learning 45.1 (2001): 5-32.
17. Tibshirani, Robert. "Regression shrinkage and selection via the lasso." Journal of the Royal Statistical Society. Series B (Methodological) (1996): 267-288.
18. Pedregosa, Fabian, et al. "Scikit-learn: Machine learning in Python." Journal of Machine Learning Research 12.Oct (2011): 2825-2830.

Fast GPU-based Influence Maximization within Finite Deadlines via Node-level Parallelism

Koushik Pal[1], Zissis Poulos[2], Edward Kim[1], and Andreas Veneris[2]

[1] Sysomos L.P., Toronto, ON M5J 2V5, Canada
{kpal, ekim} @sysomos.com
[2] University of Toronto, Toronto, ON M5S 3G4, Canada
{zpoulos, veneris} @eecg.toronto.edu

Abstract. Influence maximization in the continuous-time domain is a prevalent topic in social media analytics. It relates to the problem of identifying those individuals in a social network, whose endorsement of an opinion will maximize the number of expected follow-ups within a finite time window. This work presents a novel GPU-accelerated algorithm that enables node-parallel estimation of influence spread in the continuous-time domain. Given a finite time window, the method involves decomposing a social graph into multiple local regions within which influence spread can be estimated in parallel to allow for fast and low-cost computations. Experiments show that the proposed method achieves up to x85 speed-up vs. the state-of-the-art on real-world social graphs with up to 100K nodes and 2.5M edges. In addition, our optimization solutions are within 98.9% of the influence spread achieved by current state-of-the-art. The memory consumption of our method is also substantially lower. Indicatively, our method can achieve, on a single GPU, similar running time performance as the state-of-the-art, when the latter distributes execution across hundreds of CPU cores.

Keywords: Social Media Analytics, Influence Maximization, GPUs

1 Introduction

Influence maximization is one of the dominant topics in viral marketing. It pertains to the problem of identifying a subset of the population that, within a certain time window (deadline), can trigger the maximum number of expected follow-ups in a given network. Understanding the temporal dynamics of influence diffusion is of paramount importance to marketing departments, as it enables them to plan their campaigns operating within strict time-sensitive constraints.

The problem of influence maximization has been extensively studied in the discrete-time domain with infinite deadlines [15, 9, 1, 10, 7]. However, optimizing influence spread over infinitely long time horizons does not always reflect realistic scenarios. For example, a marketer often wishes for an opinion to become viral in a matter of minutes or days, not decades.

© Springer International Publishing AG 2017
P. Perner (Ed.): ICDM 2017, LNAI 10357, pp. 151–165, 2017.
DOI: 10.1007/978-3-319-62701-4_12

As such, maximizing influence spread within finite (and often short) time windows is a variant of the problem which is closer to real world needs. Enforcing such time-sensitive constraints requires influence diffusion models that accurately capture the temporal dynamics of the process to predict how future events unfold in time. A sequence of recent studies on real world data highlights the superiority of continuous-time models over discrete-time ones in expressing the temporal properties of influence diffusion [4, 5, 12, 16].

Motivated by these findings, recent work [6, 3] introduced continuous-time generative models to address influence maximization within finite time windows. The authors have modelled node-to-node influence propagation by transmission rates obeying densities over time, and designed methods for computing exact and approximate influence spread. The method that computes exact spread [6] is not scalable; influence spread from a particular node is computed over the whole network. Yet, in the continuous-time setting, influence decays rapidly towards the network regions that are further away from the source. As such, a big fraction of the computations is wasted analyzing regions of the graph where influence is minuscule or zero, especially when the influence deadline is relatively short.

Based on this observation, we propose a novel approximation method that uses deadline constraints to identify, for each node, a local graph region where the volume of its influence is restricted. The method entails an inexpensive preprocessing step that extracts a decomposition of the social graph into possibly overlapping trees, where the influence of each node is restricted within its own local tree region. This enables us to avoid exhaustive graph inference, thus speeding-up computations with minimal impact on accuracy. Further, it enables GPU-based parallelization, since the influence spread of each node can be computed independently within each local tree region. We build upon this node-level parallelism and harness the parallel processing capacity of commercial-level GPUs to achieve orders of magnitude faster computations than the current state-of-the-art.

Efforts to address the scalability issue have also been taken up by the authors in [3], where an approximation method is developed and shown to be orders of magnitude faster compared to exact inference [6]. This speed-up, however, comes at the cost of enormous memory consumption, which hampers GPU-based acceleration. Indicatively, for social graphs with millions of edges, the method necessitates the instrumentation of massive clusters consisting of 192 CPU cores. Consequently, it is not suitable for parallelization using inexpensive alternatives, such as (multi-) GPU systems. The application of GPUs to this problem has been previously explored [13], but is restricted solely to the discrete-time domain.

2   Preliminaries

Our work is based on the continuous-time generative model for network diffusion that has been introduced in [6]. Given a social network, modeled as a directed graph $\mathcal{G} = (\mathcal{V}, \mathcal{E})$, the influence propagation process begins from an initial set $\mathcal{S} \subset \mathcal{V}$ of source nodes, referred to as the *seed set*. The seed set is assumed to be influenced by means of adopting an opinion at time zero.

Influence propagates via directed edges from the seed nodes towards their out-neighbours. The newly influenced nodes influence their out-neighbours in turn, and this process continues. An influenced node is assumed to remain influenced for the entire duration of the diffusion process. Consequently, the node that influences a given node at the earliest time will be its parent in the induced influence propagation graph (also called the *cascade*), effectively imposing a Directed Acyclic Graph (DAG) cascade structure, even if $\mathcal{G}$ contains cycles.

The spread of influence from a node u to an out-neighbour v is assumed to consume *random* time, drawn from a conditional density function $f_{uv}(t_v|t_u)$. This models the time it takes for node u to influence node v at time-stamp t_v given that node u has been previously influenced at time-stamp t_u. These *transmission times* can be distributed differently across the edges, but they are assumed to be mutually independent. We further assume that the transmission function $f_{uv}(t_v|t_u)$ is *shift invariant*: $f_{uv}(t_v|t_u) = f_{uv}(\tau_{uv})$, where $\tau_{uv} := t_v - t_u$, and *nonnegative*: $f_{uv}(\tau_{uv}) = 0$ if $\tau_{uv} < 0$. Examples include exponential and Rayleigh distributions. Consequently, each directed edge $(u,v) \in \mathcal{E}$ is associated with a density function $f_{uv}(\tau_{uv})$, which models the time it takes for u to influence v (independent of the actual timestamps when u and v are influenced). Because of the mutual independence assumption, one obtains a fully factorized joint density of the set of transmission times $p(\{\tau_{uv}\}_{(u,v)\in\mathcal{E}}) = \prod_{(u,v)\in\mathcal{E}} f_{uv}(\tau_{uv})$.

An useful property of the above continuous-time *Independence Cascade* (IC) model is that, for a given sample of edge weights corresponding to their respective transmission times, the time t_u taken to influence a node u is the length of the shortest path in $\mathcal{G}$ from the seed set $\mathcal{S}$ to node u. This *shortest path property* is leveraged for influence spread estimation in [3] and is also utilized in the work presented here, as it reduces the problem of approximating influence spread to a well-studied graphical optimization problem, namely that of finding shortest paths. Because of this property, the infection times $\{t_u\}_{u\in\mathcal{V}}$ can be obtained from the transmission times $\{\tau_{uv}\}_{(u,v)\in\mathcal{E}}$ via the transformation $t_u = g_u(\{\tau_{vw}\}_{(v,w)\in\mathcal{E}}) := \min_{q\in\mathcal{Q}_u} \sum_{(v,w)\in q} \tau_{vw}$, where $\mathcal{Q}_u$ is the collection of all directed paths in $\mathcal{G}$ from each of the source nodes to u, and $g_u(\cdot)$ is the value of the shortest-path minimization. With this setup, one can then compute the probability of u being influenced within the deadline T as

$$\Pr\{t_u \leq T\} = \Pr\{g_u(\{\tau_{vw}\}_{(v,w)\in\mathcal{E}}) \leq T\}.$$

By standard definition [6], the *influence spread* $\iota(\mathcal{S},T)$ of the seed nodes $\mathcal{S}$ in the deadline T can then be computed as

$$\iota(\mathcal{S},T) = \mathbb{E}\left[\sum_{u\in\mathcal{V}} \mathbb{I}\{t_u \leq T\}\right] = \sum_{u\in\mathcal{V}} \mathbb{E}\left[\mathbb{I}\{t_u \leq T\}\right] = \sum_{u\in\mathcal{V}} \Pr\{t_u \leq T\}$$

$$= \sum_{u\in\mathcal{V}} \Pr\{g_u(\{\tau_{vw}\}_{(v,w)\in\mathcal{E}}) \leq T\} = \mathbb{E}\left[\sum_{u\in\mathcal{V}} \mathbb{I}\{g_u(\{\tau_{vw}\}_{(v,w)\in\mathcal{E}}) \leq T\}\right],$$

where $\mathbb{I}\{\cdot\}$ is the indicator function, $\mathbb{E}\{\cdot\}$ is the expectation function, and the expectation is taken over the set of independent variables $\{\tau_{vw}\}_{(v,w)\in\mathcal{E}}$. The sum

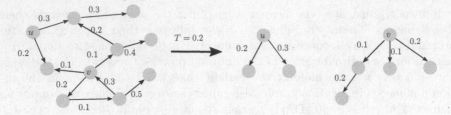

Fig. 1: Extracting local regions in the IC model

$\sum_{u\in\mathcal{V}}\mathbb{I}\{\cdot\}$ in the above formula is essentially the influence spread of the seed nodes for a given sample of transmission times $\{\tau_{vw}\}_{(v,w)\in\mathcal{E}}$.

Finally, *influence maximization* is the problem of finding an optimal set $\mathcal{S}$ of seeds of a fixed size $\mathcal{C}$ such that $\iota(\mathcal{S}, T)$ is maximized, i.e., we seek to solve

$$\mathcal{S}^{\star} = \arg\max_{\{\mathcal{S}\,:\,|\mathcal{S}|\,\leq\,\mathcal{C}\}} \iota(\mathcal{S}, T). \tag{1}$$

We take $\mathcal{C}$ (determined by budgetary constraints) as an input in this paper. It is noteworthy that the above optimization problem is NP-hard in general.

3   Methodology

A fundamental step in maximizing influence is to compute the influence spread of each node of the graph $\mathcal{G}$. The most basic way of doing that, as suggested by Equation 1, is via Naïve Sampling, where one generates a random sample of $\{\tau_{uv}\}_{(u,v)\in\mathcal{E}}$ from the corresponding edge distributions $\{f_{uv}(\tau_{uv})\}_{(u,v)\in\mathcal{E}}$, runs a *single source shortest path* (SSSP) algorithm from each node, and computes the influence of that node for that sample as the number of nodes whose shortest distance from the source node is less than the deadline T. The process repeats for several iterations (say, N_s times). On termination, the average spread of each node across all N_s samples is computed. Due to its exhaustive nature, Naïve Sampling is a costly process, as it runs a single source shortest path algorithm from each node for each sample. Assuming Dijkstra's algorithm at the core of the process, the overall time complexity is $O(N_s|\mathcal{V}|(|\mathcal{E}| + |\mathcal{V}|\log|\mathcal{V}|))$, which is prohibitively expensive. Space complexity is, however, minimal — $O(|\mathcal{V}| + |\mathcal{E}|)$.

Naïve Sampling is massively parallelizable across samples, but with shortcomings. All samples need to be generated at once, which requires $O(N_s(|\mathcal{V}| + |\mathcal{E}|))$ space. Moreover, it needs to run a SSSP algorithm from every node of the graph to every other reachable node, which is redundant for smaller deadlines. Thus, it is reasonable to identify, for each node, a "large enough, yet small" subgraph wherein its influence within the deadline is primarily restricted to, and search for its influence there (instead of the whole graph). We refer to this method as Localized Naïve Sampling (LNS) (cf. Algorithm 1 and Algorithm 2). We obtain such a subgraph for each node u by running Dijkstra's algorithm from u with the means of the edge distributions as the corresponding weights for each edge, and keeping those nodes in the subgraph of u whose shortest distance from u in $\mathcal{G}$ is "slightly greater" than T. Figure 1 depicts the process, and shows the subtrees extracted for nodes u and v in the graph. The precise condition for deciding whether to include a node v in the subgraph of u is mentioned in Algorithm 2 (Line 25). The explanation of the criterion and the choice of the free parameter σ are explained further in the Appendix. The quantity $Var(u, v)$ in Line 20 of Algorithm 2 is the variance of f_{uv} corresponding to the edge $(u, v) \in \mathcal{E}$.

Looking for the influence spread of each node in a smaller subgraph potentially reduces computations. The subgraph we choose is a subtree of the shortest path tree obtained when running Dijkstra's algorithm. Computing shortest paths in a tree for a given sample of weights is cost-effective, since there is exactly one path from the source node to any other node. For a tree of size $|L|$, the time complexity of computing the distance of each node from the root is $O(|L|)$. Once we extract a local subtree for each node, we generate multiple weight samples for that subtree and compute the average influence spread of that node in that subtree over all iterations. This process gives us an approximate spread for each node, as opposed to its true spread under the IC model. However, it does not largely affect the quality of the seeds obtained, as results in Section 4 show.

Assuming the size of each subgraph is bounded by $|L|$, the time complexity of the serial version of LNS is $O(|\mathcal{E}| + |\mathcal{V}| \log |\mathcal{V}| + N_s |V||L|)$, which in the worst case is $O(N_s |V|^2)$, as $|L| = O(|V|)$ in the worst case. But LNS is node-level parallelizable (cf. Parallel Block in Algorithm 1), which leads, for the parallel version, to a worst case runtime of $O(|\mathcal{E}| + |\mathcal{V}| \log |\mathcal{V}| + N_s |V|)$. This enables us to achieve large reductions in time complexity primarily on large graphs (see Section 4). The space complexity of our method is $O(|V||L|)$, which in the worst case is $O(|V|^2)$, as we have to store all the local subgraphs for all the nodes.

Algorithm 1 Localized Naïve Sampling (LNS)

1: **procedure** LOCALIZEDNAIVESAMPLING($\mathcal{G}, T, N_s, \sigma$)
2:     **for** $u = 1 : |\mathcal{V}|$ **do**
3:         $spread[u] = 0$
4:         Assign weights W to all edges of $\mathcal{G}$ equal to the
    means of the corresponding edge distributions
5:         $dijkstraTree[u] = $ DijkstraTree($u, \mathcal{G}, W, T, \sigma$)
6:
    Parallel Block:
7:     **for** $u = 1 : |\mathcal{V}|$ **do**
8:         **for** $n = 1 : N_s$ **do**
9:             Generate a sample of $\{\tau_{vw}\}_{(v,w) \in dijkstraTree[u]}$
10:             **for** each node v in $dijkstraTree[u]$ **do**
11:                 $distance[v] = $ distance of v from u in $dijkstraTree[u]$
12:                 **if** $distance[v] < T$ **then**
13:                     $spread[u] = spread[u] + 1$
14:         $spread[u] = spread[u] / N_s$
15:     **return** $spread[]$

In comparison, the state-of-the-art framework, ConTinEst [3], achieves its superior runtime performance compared to traditional methods by employing a randomized version of Dijkstra's algorithm [2]. This comes, however, at the cost of space complexity. Specifically, the runtime to compute expected influence across all nodes and all samples is $O(N_s(|\mathcal{V}| + |\mathcal{E}|))$, while space complexity is $O(N_s|\mathcal{E}|)$ — a significant memory bottleneck, particularly for large graphs.

Algorithm 2 Dijkstra Trees

```
 1: procedure DIJKSTRATREES(source, 𝒢, W, T, σ)
 2:     distance[source] = 0
 3:     variance[source] = 0
 4:     create vertex set Q
 5:     for u = 1 : |𝒱| do
 6:         if u ≠ source then
 7:             distance[u] = ∞
 8:             variance[u] = ∞
 9:             parent[u] = UNDEFINED
10:         Q.add_with_priority(u, distance[u])
11:     while Q is not empty do
12:         u = Q.extract_min()
13:         if distance[u] − σ * √variance[u] ≥ T then
14:             break
15:         for each out-neighbor v of u in 𝒢 do
16:             alt_distance = distance[u] + W(u, v)
17:             if alt_distance < distance[v] then
18:                 distance[v] = alt_distance
19:                 parent[v] = u
20:                 variance[v] = variance[u] + Var(u, v)
21:                 Q.decrease_priority(v, alt_distance)
22:     dijkstraTree = []
23:     dijkstraTree.add([source, source])
24:     for u = 1 : |𝒱| except source do
25:         if distance[u] − σ * √variance[u] < T then
26:             dijkstraTree.add([u, parent[u]])
27:     return dijkstraTree[]
```

To compute the joint influence spread of a set $\mathcal{S}$ of nodes, for each sample, we mark the nodes that are influenced by each element $s \in \mathcal{S}$. Then, we count all such nodes that are influenced by at least one element of $\mathcal{S}$; this count is exactly the influence spread of $\mathcal{S}$ for that sample. We repeat the process N_s times and take an average to obtain the expected influence spread of $\mathcal{S}$. It can then be shown that, over the local regions where our method operates, $\iota(\mathcal{S}, T)$ satisfies a diminishing returns property referred to as *submodularity* : for $S_1, S_2 \subset \mathcal{V}$ with $S_1 \subseteq S_2$ and $u \in \mathcal{V} \setminus S_2$, it holds that $\iota(S_1 \cup \{u\}, T) - \iota(S_1, T) \geq \iota(S_2 \cup \{u\}, T) - \iota(S_2, T)$. This implies that our method can be used as a subroutine in the greedy algorithm that we describe below (cf. Algorithm 3).

Our goal is to find a set $\mathcal{S}$ of nodes of size $\mathcal{C}$ such that their combined influence spread is maximum. Due to its intractability, the problem calls for an approximation algorithm. For monotonic submodular functions, the greedy algorithm described by Kempe et al. [9] is one such well-known approximation algorithm. The algorithm is iterative, and at the ith iteration, adds to the seed set $\mathcal{S}_{i-1}$ the node $s \in \mathcal{V} \setminus \mathcal{S}_{i-1}$ that maximizes the *marginal gain* $\iota(\mathcal{S}_{i-1} \cup \{s\}, T) - \iota(\mathcal{S}_{i-1}, T)$. We use LNS in each iteration of the greedy algorithm to find such nodes. Because we approximate the influence spread $\iota(\mathcal{S}, T)$ by using random samples drawn from edge distributions, we introduce a sampling error. Fortunately, the greedy algorithm is tolerant to such sampling noise (see [3]).

Algorithm 3 Overall Algorithm

1: **procedure** INFLUENCEMAXIMIZATION($\mathcal{G}, T, \mathcal{C}, N_s, \sigma$)
2:    **for** $u = 1 : |\mathcal{V}|$ **do**
3:       Assign weights W to all edges of $\mathcal{G}$ equal to the means of the corresponding edge distributions
4:       $dijkstraTree[u] = \text{DijkstraTree}(u, \mathcal{G}, W, T, \sigma)$
5:    $\mathcal{S} = \emptyset$
6:    **for** $i = 1 : \mathcal{C}$ **do**
7:       **for** $u = 1 : |\mathcal{V}|$ **do**
8:          Call the Parallel Block in LNS to compute :
9:          $marginal\_spread[u] = \iota(\mathcal{S} \cup \{u\}, T) - \iota(\mathcal{S}, T)$
10:       $s = \arg\max_{u \in \mathcal{V} \setminus \mathcal{S}} marginal\_spread[u]$
11:       $\mathcal{S} = \mathcal{S} \cup \{s\}$
12:    **return** $\mathcal{S}$

As shown in [14], the above greedy approach obtains a seed set which achieves at least a constant fraction $(1 - \frac{1}{e})$ of the optimal spread, provided the influence spread function is a monotonic submodular function. In the proposed method, however, the influence spread function is not sub-modular under the IC model. Thus, the approximation ratio cannot be claimed. However, there are loose bounds that we can claim by showing that $\iota(\mathcal{S}, T)$ is approximately submodular, based on the following definition.

Approximate Submodularity: For given $\epsilon \geq 0$, we say that a function $F : 2^V \to \mathbb{R}$ is ϵ-*approximately submodular* if there exists a submodular function $f : 2^V \to \mathbb{R}$ such that for every $S \subseteq V$:

$$(1 - \epsilon)f(S) \leq F(S) \leq (1 + \epsilon)f(S).$$

Theorem 1 *The influence spread $\iota(\mathcal{S}, T)$ is ϵ-approximately submodular under the continuous IC model.*

Proof. Consider $f : 2^V \to \mathbb{R}$ to be the influence spread function under the continuous IC model when the whole graph is considered for computations. The function f abides to the standard spread definition, which is known to be submodular. In the decomposition we propose, a finite set of graph nodes $\emptyset \subseteq R_S \subsetneq V$ are rendered unreachable by set S (for finite T), when in fact they might be reachable in the IC model. Thus, it follows that $f(S) - \iota(\mathcal{S}, T) \leq |R_S|$, or $f(S) - |R_S| \leq \iota(\mathcal{S}, T)$. It also trivially holds that $\iota(\mathcal{S}, T) \leq f(S)$, which implies $\iota(\mathcal{S}, T) \leq f(S) + |R_S|$. Hence,

$$f(S)\left(1 - \frac{|R_S|}{f(S)}\right) \leq \iota(\mathcal{S}, T) \leq f(S)\left(1 + \frac{|R_S|}{f(S)}\right).$$

Set $\epsilon := max\left\{ \frac{|R_S|}{f(S)} \,\middle|\, S \subseteq V \right\}$. Then, $0 \leq \epsilon < 1$ (as $|R_S| < f(S)$ for every $S \subseteq V$). Also, we obtain

$$(1 - \epsilon)f(S) \leq \iota(\mathcal{S}, T) \leq (1 + \epsilon)f(S).$$

Thus, $\iota(\mathcal{S}, T)$ is ϵ-approximately submodular. $\square$

By Theorem 5 of [8], we obtain the following approximation result.

Corollary 1. *For $\mathcal{C} \geq 2$ and any constant $0 \leq \delta < 1$ with $\epsilon = \frac{\delta}{\mathcal{C}}$, the greedy algorithm obtains a seed set which achieves at least a constant fraction $1 - \frac{1}{e} - 16\delta$ of the optimal value.*

Since $\epsilon = \frac{\delta}{\mathcal{C}}$, we have $\epsilon = O(\delta)$, provided $\mathcal{C}$ is constant. It follows that for lesser values of ϵ (equivalently, for smaller values of $|R_S|$), we achieve better approximations of the influence spread, which are closer to the optimal value. Of course, the values of $|R_S|$ and ϵ are entirely dependent on the depth of the local subtrees we choose. Deeper subtrees offer potentially better approximations, but induce extra computational cost. In Section 4, we show that when we select deep enough trees, without still covering the whole graph, we obtain quality results on par with the state-of-the-art methods while maintaining runtime benefits.

4   Experiments

We base our evaluation on real world networks found in the Stanford Network Analysis Project (SNAP) [11]. Table 1 shows some of the network characteristics. In each network, we associate each directed edge with a transmission function obeying an exponential density, whose scale parameter is drawn uniformly at random from the open-closed interval $(0, 5]$.

Table 1: Network Statistics

Network	# nodes	# edges	density
ego-Facebook	4,039	88,234	21.84
gnutella08	6,301	20,777	3.29
wiki-vote	7,115	103,689	14.57
gnutella04	10,876	39,994	3.68
soc-Epinions1	75,879	508,837	6.71
ego-twitter	81,306	2,468,149	30.35
soc-Slashdot0922	82,168	948,464	11.54

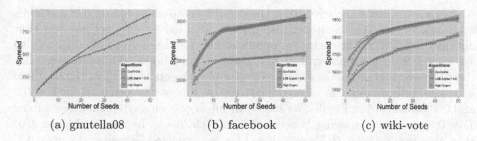

(a) gnutella08   (b) facebook   (c) wiki-vote

Fig. 2: Seed set spread vs. seed set size with $T = 0.2$

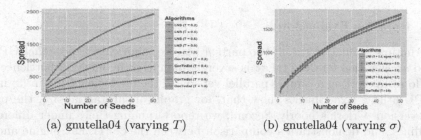

(a) gnutella04 (varying T)   (b) gnutella04 (varying σ)

Fig. 3: Effect of T and σ on spread respectively

4.1 Quality of Seed Sets

To compare the quality of seed sets that are produced by our methodology, we perform an immediate comparison between our method and ConTinEst [3], since the latter has been shown to statistically outperform other approximation methods on real world data. To do so, we need a procedure that can receive these seed sets as input and obtain near ground-truth estimates of their influence spread. We perform the comparison by running Naïve Sampling for 10,000 iterations on each seed set that is obtained. For practical reasons, we restrict this comparison on the four smallest graphs in our dataset, since Naïve Sampling requires at least 8,000 hours to produce results for the larger graphs.

For this set of results, the algorithms are configured as follows: abiding to author guidelines in [3], ConTinEst is set to perform 10,000 sampling rounds for transmission times. For each sampling round, it is also configured to run 5 iterations for the randomization required by the neighborhood size estimation subroutine. LNS is also set to 10,000 iterations, while the σ parameter is tuned to 0.9 to produce large enough local regions.

Results are shown in Figure 2. We also report the spread achieved by greedily selecting the node with highest out-degree each time (High Degree). Finally, the deadline is fixed to $T = 0.2$. One observes that LNS is on par with ConTinEst in terms of seed quality. In fact, the relative error never surpasses 11% across all four graphs. On the average, the relative error of our method is 1.2%. In contrast, a simplistic method such as High Degree, produces solutions that are, on the average, up to 21.2% off in terms of influence spread compared to ConTinEst.

We also evaluate the effect that the deadline constraint has on our method's accuracy. Figure 3(a) compares the spread obtained by our method and that obtained by ConTinEst under various deadlines (between $T = 0.2$ and $T = 1.0$) for $\sigma = 0.9$ on a graph with 10,876 nodes (**gnutella04**). One observes that the proposed method maintains the quality of seed sets even when the deadline increases. Indicatively, the average relative error ranges between 0.2% for $T = 0.2$ and 0.7% for $T = 1.0$. Finally, we study the effect of σ on seed quality. For a fixed $T = 0.8$, and for σ ranging between 0.1 and 0.9, it can be seen in Figure 3(b) that accuracy drops with lower values of σ, as expected. However, for $\sigma \geq 0.4$, our average relative error falls between 0.8% and 3.5%.

4.2   Runtime Evaluation

Our runtime evaluation entails two parts. First, we empirically expose that GPU-based acceleration for ConTinEst is severely hampered by memory bottlenecks. We do so by porting a sample-parallel version of the implementation in [3] into a GPU utility. Experiments show that, for graphs of substantial size, the parallel version performs poorly. Second, we report running times under different deadlines, and demonstrate a comparison between the ConTinEst engine and a node-parallel version of LNS implemented on GPUs. Any implementation mentioned henceforth is carried out on a single 2.6GHz processor, and a GPU card with 4GB RAM on an NVidia K520 Grid platform.

First, we report runtime comparisons between four implementations: two CPU-based implementations of ConTinEst and LNS, a sample-parallel GPU implementation of ConTinEst, and a node-parallel GPU implementation of LNS. Note that ConTinEst, by construction, does not allow node-level parallelism. All implementations are set to retrieve seed sets of size 50.

Figure 4(a) demonstrates results for the three smallest graphs in the dataset. As it can be seen, the CPU implementation of LNS is the slowest one. This is expected as the method is designed for parallel computing and is not a good fit for serial execution. Indicatively, the GPU implementation of LNS is dramatically faster by a factor ranging from x100 to x1000, justifying the above argument. One interesting finding is that for graph sizes like the ones in Figure 4(a), the sample-parallel implementation of ConTinEst is slower by a factor ranging between x2 and x5 compared to the serial implementation. This can be justified by the fact that only a small batch of samples can be ported into the GPU each time, due to the large space complexity of the process. This, in turn, necessitates multiple kernel calls which incur significant communication overhead between the host system and the GPU device. Finally, it can be seen that the GPU-based implementation of LNS and the CPU-based implementation of ConTinEst have similar run-times for the set of smaller networks.

However, as seen in Figure 4(b), the performance of LNS surpasses that of ConTinEst for larger networks ($> 100K$ edges). Specifically, for deadline T fixed to 0.2, LNS is x13.7 and x6.8 times faster than ConTinEst for **soc-Epinions1** and **soc-Slashdot0922**, respectively. Further, the performance of LNS improves significantly as the deadline becomes smaller (stricter constraints), while ConTinEst

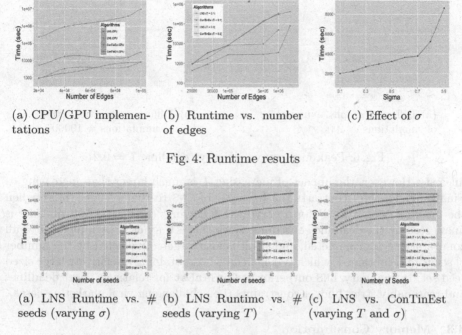

(a) CPU/GPU implementations

(b) Runtime vs. number of edges

(c) Effect of σ

Fig. 4: Runtime results

(a) LNS Runtime vs. # seeds (varying σ)

(b) LNS Runtime vs. # seeds (varying T)

(c) LNS vs. ConTinEst (varying T and σ)

Fig. 5: Effect of T and σ on runtime

largely remains unaffected. In more detail, for T fixed to 0.1, LNS is faster across all social graphs, with improvements ranging between x2.1 and x85.7. In fact, the greatest gains appear for the three largest graphs in our dataset — **soc-Epinions1**, **soc-Slashdot0922** and **ego-twitter** — for which the runtime improvements are x47.9, x85.7 and x5.2, respectively. For **ego-twitter**, ConTinEst consumes $\approx$ 420K seconds, while LNS terminates after $\approx$ 80K seconds. In absolute terms, these savings correspond to days of computations (*i.e.*, savings amounting to 4 days for **ego-twitter**), which is substantial in a realistic campaign planning process. Finally, note that ConTinEst runtime remains similar moving to shorter deadlines, whereas LNS is 66.9% faster, on the average. This stresses the merits of leveraging deadline information prior to influence estimation.

We also report how the choice of σ affects runtime for the GPU-based implementation of LNS. Results are obtained on **soc-Epinions1**. Figure 4(c) shows that runtime grows exponentially with σ. This is expected, since for a larger σ, the borders of the local regions extend far beyond the deadline, possibly to a point where the local regions for multiple vertices cover the entire reachable set of nodes from the corresponding source nodes.

Finally, Figure 5 illustrates the relation between runtime and the number of seeds (logarithmic scale). Results are obtained on **soc-Slashdot0922**. In Figure 5(a), we fix $T = 0.1$ and let $\sigma \in \{0.1, \ldots, 0.7\}$. One observes that LNS runtime is linear in the number of seeds to obtain, irrespective of σ. Further, in Figure 5(b), we fix $\sigma = 0.4$ and let $T \in \{0.1, 0.2, 0.3\}$. Again, T does not affect

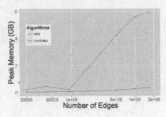
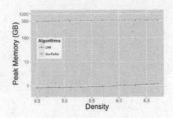

(a) Different graphs, Number
of simulations = 100

(b) Different densities, Number of simulations = 10000

Fig. 6: Peak memory vs. density (deadline $T = 0.2$)

linearity. However, the impact of increasing T is much larger than increasing σ.
Finally, Figure 5(c) contrasts the behavior of LNS to that of ConTinEst when
the deadline varies. One can observe the following: (a) ConTinEst runtime is
unaffected by the deadline constraint, since this is only used as a query after all
computations are completed, and (b) it is also largely unaffected by the number
of seeds to be obtained, which is beneficial when obtaining a relatively large sized
seed set. In summary, LNS outperforms ConTinEst for relatively short deadlines
and reasonably strict budget constraints (i.e., challenging viral marketing cases).

4.3 Memory Consumption

In this section we empirically evaluate to what extent LNS and ConTinEst stress
memory when implemented using their faster variants. We report peak memory
consumption for GPU-based LNS and CPU-based ConTinEst, which are set to
perform 100 sampling rounds. Figure 6(a) confirms the memory intensiveness
of ConTinEst, especially for larger graphs, while it shows that LNS maintains
memory consumption relatively low, even as graph size increases. Specifically,
ConTinEst consumes between 0.21GB and 5.9GB, while peak memory for LNS
ranges between 0.04GB and 0.36GB, which corresponds to an average improvement of 1260%.

Finally, we discuss the effect of graph density on memory consumption for
10,000 iterations on both methods, and a social graph with 70K nodes and 500K
edges. Figure 6(b) demonstrates that ConTinEst requires approximately 500GB
of space to accommodate computations. The bulk of this space is occupied by
the least label list structures, which store information for all 10,000 sampling
iterations to support faster queries. As the density of the graph drops (from 7.0
to 4.5), peak memory remains at similar levels (8.3% drop). In contrast, LNS
starts off at 1.2GB and for the smallest density it only consumes 0.8GB; a 32%
reduction by virtue of producing shallower local subgraphs.

5 Conclusion

We present a novel approximation framework for influence maximization in the
continuous-time domain. Our work addresses two drawbacks of existing meth-

ods: the lack of node-level parallelism, and the memory intensive nature of the dominant methodologies. The proposed approximation algorithm is the first to enable node-parallel influence estimation in the continuous-time setting, while maintaining memory requirements relatively low. By employing commercial-level GPUs we dramatically speed-up computations with minimal impact on accuracy.

6 Appendix

In this section, we explain our criterion for selecting local subgraphs. As mentioned in Section 3, the transmission times $\{f_{uv}(\tau_{uv})\}_{(u,v)\in\mathcal{E}}$ are differently distributed across the edges, but are mutually independent. Consequently, the joint distribution of the transmission times is fully factorized. Also, the variance of a path q is the sum of the variances of the distributions corresponding to the edges on q, i.e., $Var(q) = \sum_{(u,v)\in q} Var(u,v)$, where $Var(u,v)$ is the variance of the distribution f_{uv} associated with the edge $(u,v) \in \mathcal{E}$. We use these properties to define our criterion for selecting the local subgraphs corresponding to each node. By linearity of expectation, if q is a path from node u to node v, the expected time for u to influence v along q equals the sum of the means of the distributions corresponding to the edges on q. It also holds that the expected shortest distance from u to v equals the length of the shortest path from u to v with edge weights equal to the means of the corresponding distributions. Thus, if there is a path q from u to v whose expected length is less than the deadline T, then v is most likely to be within the influence spread of u for an arbitrary given sample, and should be included in the local subgraph of u. As such, running Dijkstra with edge weights being the distribution means allows us to extract these subgraphs.

Unfortunately, the converse is not true, i.e., if the shortest path from u to v has expected length greater than T, it does not mean that v cannot be in the influence spread of u for any sample. Since each sample is a set of random numbers generated from a given set of distributions, it is entirely possible that, for a given sample, the edge weights on a path from u to v are small enough to have v influenced by u for that particular sample. Fortunately, the frequency of such an incident happening decreases as the expected shortest distance between u and v progressively increases beyond the deadline T. Since most of the mass of a probability distribution is concentrated near its mean and is within some multiple of its standard deviation, we use that fact to decide whether v is going to be in the influence spread of u for a significant number of samples. This is the reason behind our selection criterion at Line 25 of Algorithm 2:

$$\textbf{if } distance[u] - \sigma * \sqrt{variance[u]} < T \textbf{ then}$$
$$dijkstraTree.\text{add}([u, parent[u]]).$$

We measure the variance of a node u as the variance of the shortest path between the source node and u. The parameter σ in the above criterion is a free parameter. It suggests how much of the variability in the model we are willing to account for. If we increase σ, we cover wider local neighbourhoods, thereby improving accuracy, but at the cost of runtime. On the other hand, if we decrease σ, we obtain narrower local neighbourhoods, which leads to faster computations, but at the cost of accuracy. Our experiments suggest that, for moderately big graphs, we can choose $\sigma = 0.9$, while for much larger graphs, $\sigma = 0.3$ suffices.

Finally, the subgraph we choose for each node is a portion of the Dijkstra tree, where we only keep the nodes that satisfy the above selection criterion. Replacing a local subgraph by a local subtree leads to significant under-approximations of the influence spread. But, as our empirical evaluations in Section 4 show, it is a good representative region to sample from for deciding whether v lies in the influence spread of u for an arbitrary given sample.

References

1. Chen, W., Wang, C., Wang, Y.: Scalable influence maximization for prevalent viral marketing in large-scale social networks. In: Proceedings of the 16th ACM SIGKDD International Conference on Knowledge Discovery and Data Mining. pp. 1029–1038. KDD'10 (2010)
2. Cohen, E.: Size-estimation framework with applications to transitive closure and reachability. J. Comput. Syst. Sci. 55(3), 441–453 (1997)
3. Du, N., Song, L., Gomez-Rodriguez, M., Zha, H.: Scalable influence estimation in continuous-time diffusion networks. In: Advances in Neural Information Processing Systems. NIPS'13 (2013)
4. Du, N., Song, L., Yuan, M., Smola, A.J.: Learning networks of heterogeneous influence. In: Pereira, F., Burges, C.J.C., Bottou, L., Weinberger, K.Q. (eds.) Advances in Neural Information Processing Systems 25. pp. 2780–2788 (2012)
5. Gomez-Rodriguez, M., Balduzzi, D., Schölkopf, B.: Uncovering the temporal dynamics of diffusion networks. In: Proceedings of the 28th International Conference on Machine Learning. pp. 561–568 (2011), http://www.icml-2011.org/papers/354_icmlpaper.pdf
6. Gomez-Rodriguez, M., Schölkopf, B.: Influence maximization in continuous time diffusion networks. In: Proceedings of the 29th International Conference on Machine Learning. pp. 313–320 (2012)
7. Goyal, A., Lu, W., Lakshmanan, L.V.S.: Simpath: An efficient algorithm for influence maximization under the linear threshold model. In: 2011 IEEE 11th International Conference on Data Mining. pp. 211–220 (2011)
8. Horel, T., Singer, Y.: Maximization of approximately submodular functions. In: Lee, D.D., Sugiyama, M., Luxburg, U.V., Guyon, I., Garnett, R. (eds.) Advances in Neural Information Processing Systems 29, pp. 3045–3053. Curran Associates, Inc. (2016)
9. Kempe, D., Kleinberg, J., Tardos, E.: Maximizing the spread of influence through a social network. In: Proceedings of the 9th ACM SIGKDD International Conference on Knowledge Discovery and Data Mining. pp. 137–146. KDD'03 (2003)
10. Leskovec, J., Krause, A., Guestrin, C., Faloutsos, C., VanBriesen, J., Glance, N.: Cost-effective outbreak detection in networks. In: Proceedings of the 13th ACM SIGKDD International Conference on Knowledge Discovery and Data Mining. pp. 420–429. KDD'07 (2007)
11. Leskovec, J., Krevl, A.: SNAP Datasets: Stanford large network dataset collection. http://snap.stanford.edu/data (Jun 2014)

12. Li, L., Zha, H.: Learning parametric models for social infectivity in multi-dimensional hawkes processes. In: Proceedings of the 28th AAAI Conference on Artificial Intelligence. pp. 101–107. AAAI'14 (2014)
13. Liu, X., Li, M., Li, S., Peng, S., Liao, X., Lu, X.: Imgpu: Gpu-accelerated influence maximization in large-scale social networks. IEEE Transactions on Parallel and Distributed Systems 25(1), 136–145 (2014)
14. Nemhauser, G.L., Wolsey, L.A., Fisher, M.L.: An analysis of approximations for maximizing submodular set functions. Mathematical Programming 14(1), 265–294 (1978)
15. Richardson, M., Domingos, P.: Mining knowledge-sharing sites for viral marketing. In: Proceedings of the 8th ACM SIGKDD International Conference on Knowledge Discovery and Data Mining. pp. 61–70. KDD'02 (2002)
16. Yang, S.H., Zha, H.: Mixture of mutually exciting processes for viral diffusion. In: Proceedings of the 30th International Conference on Machine Learning. vol. 28, pp. 1–9 (2013)

Visual Scenes Mining for Agent Awareness Module

Gang Ma[1,2], Zhentao Tang[2,3], Xi Yang[4], Zhongzhi Shi[1], and Kun Yang[5]

[1] The Key Laboratory of Intelligent Information Processing, Institute of Computing
Technology, Chinese Academy of Sciences. 100190 Beijing, China
{mag,shizz}@ics.ict.ac.cn
[2] University of Chinese Academy of Sciences. 100190 Beijing, China
[3] The State Key Laboratory of Management and Control for Complex Systems, Institute of
Automation, Chinese Academy of Sciences. 100190 Beijing, China
tangzhentao2016@ia.ac.cn
[4] Beijing Advanced Innovation Center For Future Education,
Beijing Normal University. 100875 Beijing, China
xiyang85@bnu.edu.cn
[5] National Institute of Metrology. 100029 Beijing, China
yangkun@nim.ac.cn

Abstract. Most agents obtain knowledges from natural scenes through some single preestablished rules. In practice, those single rules can't achieve the aim to
freely awareness the natural scenes, such as the visual scenes. Inspired by biological visual cortex (V1) and higher brain areas perceiving visual features, in this
paper we propose an improved visual awareness module, called as visual scenes
mining module, for the agent ABGP-CNN in order to directly mine the visual
scenes information. Then ABGP-CNN with the visual scenes mining module is
deployed on a toy car. The visual information mining from the nature scenes is
served as the knowledges of the agent ABGP-CNN to drive the toy car. The toy
car deployed the agent ABGP-CNN can easily understand the special natural visual scenes, and has the ability to plan its behaviors according to the visual information mining from the nature scenes. The application of the agent ABGP-CNN
with visual scenes mining module enhances the capability of communication between the toy car and the natural visual scenes.

Keywords: Agent, Visual Scenes Mining, ABGP-CNN

1 Introduction

Rational agents have an explicit representation for their environment (sometimes called
world model) and objectives they are trying to achieve. Rationality means that the agent
will always perform the most promising actions (based on the knowledge about itself
and sensed from the world) to achieve its objectives. For a rational agent faced with a
complex natural scene, how to get knowledge from scenes to drive their actions? Most
agent designer maybe have a common view that either create a virtual scene or set some
single inflexible rules for agent to recognize surrounding.

There exist numerous agent architectures as mentioned above, such as BDI [1,16],
AOP [19], SOAR [10] and 3APL [6]. In these architectures, the communication of

© Springer International Publishing AG 2017
P. Perner (Ed.): ICDM 2017, LNAI 10357, pp. 166–180, 2017.
DOI: 10.1007/978-3-319-62701-4_13

information between agents and world is based on a single fixed rules as well. With respect to the theoretical foundation and the number of implemented and successfully applied systems, the Belief-Desire-Intention (BDI) architecture designed by Bratman [1] as a philosophical model for describing rational agents should be the most interesting and widespread agent architecture. Of course, there are also a wide range of agents characterized by the BDI architecture. Where one of these types is called ABGP (a 4-tuple $\langle Awareness, Belief, Goal, Plan \rangle$ BDI agent model shown in Figure 1) model proposed by Shi Z. [18].

ABGP model consists of the concepts of awareness, beliefs, goals, and plans. Awareness is an information pathway connecting to the world (including natural scenes and other agents in multi-agent system). Beliefs can be viewed as the agent's knowledge about its setting and itself. Goals make up the agent's wishes and drive the course of its actions. Plans represent agent's means to achieve its goals. However, ABGP model agent still has a disadvantage that just transfers a special fixed-format message from the world. Of course, there are many researchers built some usefull and interesting awareness modules, where one of works may be in [13] proposed ABGP-CNN model by introducing Convolution Neural Networks (CNN) as an environment visual pathway to implement the visual awareness module of the agent model ABGP. The ABGP-CNN model has the ability to directly capture the information from the natural scenes .

The literature [13] had described the functions and application fields of ABGP-CNN through a maze search experiment based on the agents implemented by ABGP-CNN model. However the experiment is carried out only through the software simulation on the computer, which means the experiment environment is very idealized. In practice there are many problems that we can't find out in the software simulation experiment, such that dynamically capturing visual information is affected by the light condition, how to find out a special interesting area from a large scene, etc.. Furthermore, ABGP-CNN as an intelligent model should be applied to the other fields with help of some hardware equipments, which can help human finish some more difficult tasks, such as disaster relief, danger detection, automatic drive, etc..

Motivated by above analysis, in this paper we will firstly improve the awareness module of ABGP-CNN agent, and then deploy the improved ABGP-CNN model on a true equipment. Where the improved ABGP-CNN will be served as a software driver (like human brain) to plan the behaviors of a toy car only through depending on the visual information that the toy car captures by its camera. The experiment can help us further explore the intelligent behaviors of the machine.

2   Agent Model ABGP-CNN

In computer science and artificial intelligence, agent can be viewed as perceiving its environment information through sensors and acting environment using effectors [17]. A agent model for rational agent should especially consider external perception and internal mental state. The external perception as a knowledge is created through interaction between an agent and its world. For the internal mental state, we can consult BDI model conceived by Bratman [1] as a theory of human practical reasoning. Reducing the explanation framework for complex human behavior to the motivational stance is the especially attractive character [15].

An agent model ABGP-CNN (Figure 1) proposed by Ma G. [13] is one of the most typical agent model characterized by BDI architecture, it is represented as a 4-tuple framework as $\langle Awareness, Belief, Goal, Plan \rangle$. Where visual awareness implemented by Convolution Neural Networks (CNN) is an information pathway connecting to the world and a relationship between agents. Belief can be viewed as the agent's knowledge about its environment and itself. Goal represents the concrete motivations that influence an agent's behaviors. Plan is used to achieve agent's goals. Moreover, an important module, policy-driven reasoning in ABGP model, is used to handle a series of events to achieve plans selection.

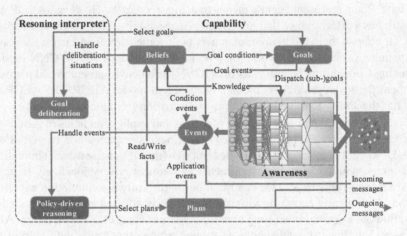

Fig. 1: Agent model ABGP-CNN [13].

For ABGP-CNN, the learning process of recognizing the natural scenes should mainly focus on how to train the CNN as its visual awareness module and how to build appropriate belief base, goals, and plans library. Training CNN includes what the multi-stages architecture is appropriate for the natural object recognition, what learning strategy is better. The aim of building beliefs, goals and plans is to achieve a series of agent's behaviors feedback according to the accepted environment information.

The implementation (Figure 2) of ABGP-CNN model adopts a declarative and a procedural approach to define its core components Awareness, Belief, Goal and Plan as well. The awareness and plan module have to be implemented as the ordinary Java classes that extend a certain framework class, thus providing a generic access to the BDI specific facilities. Belief and Goal module are specified in a so-called Agent Definition File (ADF) using an XML language (Figure 2). Within the XML agent definition files, any developer can use valid expressions to specify any designated properties. Some other information is also stored in ADF such as default arguments for launching the agent or service descriptions for registering the agent at a directory facilitator. Moreover, Awareness and Plan need to be declared in the ADF before they work.

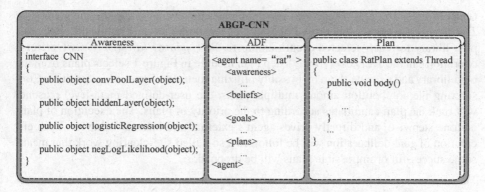

Fig. 2: Implementation of ABGP-CNN model[13].

Awareness is commonly viewed as an information path connecting to the environment. The ADF of ABGP-CNN provides a description of attributes for CNN, such as the number of CNN stages, the number of hidden layers, filter shape, pooling size, etc. Which can be any kinds of ordinary Java objects contained in the awareness set as an XML tuple. Those Java objects are stored as named facts.

Beliefs are some facts known by the agent about its environment and itself, which are usually defined in the ADF and accessed and modified from plans. Generally the facts can be described as an XML tuple with a name and a class through any kind of ordinary Java objects.

Plans in ABGP-CNN model can be considered as a series of concrete actions expecting to achieve some goals or tasks. ABGP-CNN model adopts the plan-library approach to represent the plans of an agent. Each plan contains a plan head defining the circumstances under which the plan may be selected and a plan body specifying the actions to be executed. In general for reusing plans in different agents, we need to decomposes concrete agent functionality into separate plan bodies, which are predefined courses of action implemented as Java classes.

Policy-driven Reasoning mainly make the plans (policies) selection by handling a series of events. A policy will directly or indirectly cause an action a_i to be taken, the result of which is that the system or component will make a deterministic or probabilistic transition to a new state S_i. Kephart et al. [14] outlined a unified framework for autonomously computing policies based upon the notions of states and actions. The agent policy model can be defined as a 4-tuple $P = \{S_t, A, S_g, U\}$, where S_t is the trigger state set at a given moment in time; A is the action set; S_g is the set of goal states; U is the goal state utility function set to assess the merits of the goal state level.

Algorithm 1 shows the functionality of those modules when they execute the interpret reasoning process. Goals deliberation constantly triggers awareness module to purposefully perceive visual information from the world the agent locates (extract visual information features $y : y_j = g_j \cdot tanh(\sum_i k_{i,j} * x_i), y \in D$) and convert those visual information features into the unified internal message events which are placed in the event queue (signal mapping $T : D \mapsto E$). According to goal events the event dispatcher continuously consumes the events from the event queue (corresponds to the

$OptionGenerator(EQ)$ function) and deliberates the events satisfying the goals (like $Deliberate(Options)$ function). Policy-driven reasoning module builds the applicable plan library for each selected event (similar to the $UpdatePlans(SelectedOptions)$ operation). In the $Execute(Plans)$ step Plan module in Figure 1 selects plans from the plan library and execute them by possibly utilizing meta-level reasoning facilities. Considering the competition among multiple plans, the user-defined meta-level reasoner will rank the plan candidates according to the priority of plans. The execution of plans is done stepwise, and directly drives agent's external and internal behavior. Each circulation of goal deliberation will be followed a so-called site-clearing work that means some successful or impossible plans will be dropped.

Algorithm 1: Interpret-Reasoning Process of ABGP-CNN

Initialize agent's states;
while *not achieve goals* **do**
  Deliberate goals;
  if *world information INF or incoming messages IME was perceived* **then**
    create internal event E according to *INF or IME*;
    fill internal event E into event queue EQ;
  end
  $Options \leftarrow OptionGenerator(EQ)$;
  $SelectedOptions \leftarrow Deliberate(Options)$;
  $Plans \leftarrow UpdatePlans(SelectedOptions)$;
  $Execute(Plans)$;
  Plans drive agent's behaviors;
  Drop successful or impossible plans;
end

---

3 Visual Scenes Mining Module

The experiment in [13] showed that CNN employed to construct the visual awareness module of ABGP behaves a high performance. Which means that the recognizing function of ABGP-CNN is mainly related to the visual awareness module. However the experiment is just carried out on the normative MNIST dataset in software simulation way. If CNN is directly used to construct the visual awareness module to recognize the handwrite digital in a complex visual scenes when ABGP-CNN is deployed on the toy car, it will behave a very low recognizing performance and can't achieve to correctly plan the behaviors of toy car. Therefore, we need to construct an improved visual awareness module of ABGP-CNN, called as the visual scenes mining module. In a complex visual scene, there are only some local elements containing interesting points which can guide the behaviors of toy car. We need to firslty locate the area with interesting points, then split it from the entire complex scene. At last the splited interesting area is recognized by CNN. Therefore some preliminaries, Pulse Coupled Neural Network,

Improved 2-Stage Sequential Similarity Detection Algorithm and Convolution Neural Networks, need to be introduced before building a novel visual scenes mining module.

3.1 Pulse Coupled Neural Network (PCNN)

The PCNN neuron in the standard PCNN model consists of three parts: the dendritic tree, the linking modulation, and the pulse generator, as shown in Figure 3.

The role of the dendritic tree is to receive the inputs from two kinds of receptive fields. Depending on the type of the receptive field, it is subdivided into two channels (the linking and the feeding). The linking receives local stimulus from the output of surrounding neurons, while the feeding, besides local stimulus, still receives external stimulus.

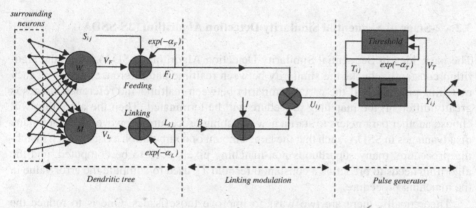

Fig. 3: Architecture of standard PCNN [21].

In the following expressions, the indexes i and j refer to the pixel location in the image, k and l refer to the dislocation in a symmetric neighborhood around one pixel, and n denotes the current iteration (discrete time step). Here n varies from 1 to N (N is the total number of iterations)

$$F_{ij}[n] = e^{-\alpha_F} F_{ij}[n-1] + V_F \sum_{k,l} w_{ijkl} Y_{ij}[n-1] + S_{ij}. \tag{1}$$

$$L_{ij}[n] = e^{-\alpha_L} L_{ij}[n-1] + V_L \sum_{k,l} m_{ijkl} Y_{ij}[n-1]. \tag{2}$$

$$U_{ij}[n] = F_{ij}[n](1 + \beta L_{ij}[n]). \tag{3}$$

$$Y_{ij}[n] = \begin{cases} 1, & U_{ij}[n] > T_{ij}[n]. \\ 0, & otherwise. \end{cases} \tag{4}$$

$$T_{ij}[n] = e^{-\alpha T} T_{ij}[n-1] + V_T Y_{ij}[n]. \tag{5}$$

The dendritic tree is given by Equation 1 and Equation 2. The two main components F and L are called feeding and linking, respectively. w_{ijkl} and m_{ijkl} are the synaptic weight coefficients and S is the external stimulus. V_F and V_L are normalizing constants. α_F and α_L are the time constants; generally, $\alpha_F < \alpha_L$. The linking modulation is given in Equation 3, where $U_{ij}[n]$ is the internal state of the neuron. β is the linking parameter and the pulse generator determines the firing events in the model in Equation 4. $Y_{ij}[n]$ depends on the internal state and threshold. The dynamic threshold of the neuron is Equation 5, where V_T and αT are normalized constant and time constant, respectively. Here is a brief review of the standard PCNN. The detailed description of the implementation of the standard PCNN model on digital computers can be found in literature [8]. More details about PCNN will be found in the literatures [11].

3.2   2-Stage of Sequential Similarity Detection Algorithm (2S-SSDA)

The basic idea of Sequential Similarity Detection Algorithm (SSDA) is that the algorithm constantly checks the similarity between realtime and reference images in each matching procedure. If the partial similarity between realtime and reference images is greatly different, the matching procedure will be terminated. Then the algorithm will choose another parameters to start a new matching procedure. However there are many disadvantages in SSDA, such that the large amount of computation exists in each matching procedure, many superfluous non-matching pixels need to be computed, and the algorithm needs to predefine a constant threshold T_k used to compute the error value in the matching procedure.

Theoretically, there are two ways to improve those issues. One is to reduce the searching space when the template T is used to search the reference image S. The other is to reduce the computational complexity of the similarity between the template T and the sub-image $S^{i,j}$. Next, we will give a 2-stage sequential similarity detection algorithm (2S-SSDA) [5] based above two theoretical inspires.

An image used to the matching operation needs to be preprocessed because of many superfluous information. Where we employ the preprocessing method that the normalized gray value is used as the reference image of probability density. Firstly the median filter method is used to equalize the image histogram. Then the edge features are extracted by Sobel edge detection operator. After extracting the edge features, we need to execute a sampling process on the edge features in order to improve the time efficiency of algorithm. The probability distribution of the processed image through above steps is described as:

$$P\{X = i, Y = j\} = \frac{H(i,j)}{\sum_{k=1}^{N} \sum_{s=1}^{N} H(k,s)} \tag{6}$$

in Equation 6 $H(i,j)$ indicates the gray value of the pixel (i,j) in the gray image. Then we sample a sequence $A(n)$ from the probability distribution of the gray image. n satisfies $n = |\{(i,j)\}H(i,j) > th|$, where th is a threshold. The roulette method can be served to execute the sampling process. In the following the 2-stage of matching processing, rough matching and accurate matching, is executed to capture the best position of the matching sub-image.

In rough matching process, the candidate image set Q must be from the entire matched sub-image set S. Q has to satisfy the following conditions:

- The true matched point (i, j) must be in Q or Q-centered domain.
- In order to reduce the search space in the accurate matching, $|Q|$ should be as small as possible.
- Q should be easy and quick to get.

The distance between the sub-image and the template image is defined as $D(S^{i,j}, T)$:

$$D(S^{i,j}, T) = \sum_{k=1}^{n} d(S^{i,j}(A(k)), T(A(k))). \qquad (7)$$

In the rough matching process, the elements contained in the set Q are determined by matching operation with the search step length h. In general, the candidate sub-image set Q should satisfy $Q = \{S^{i,j} | D(S^{i,j}, T) \leq \theta\}$, where θ is a threshold determined by the actual situation. A large Q will increase the matching time, while a small Q may lead to matching distortion, in most cases $|Q| = 1$.

In the whole process of rough matching, we don't make decision whether a sub-image is the target image when the matching process between the template image and the sub-image is finished, while Th is dynamically generated in the matching process. Where set $Th_1 = D(S^{1,1}, T)$, for the k-th matching, if s satisfies

$$\sum_{t=1}^{s} d(S^{i,j}(A(t)), T(A(t))) > Th_{k-1}; (s < n) \qquad (8)$$

then $Th_k = Th_{k-1}$, and the current matching process is finished. Otherwise $Th_k = D(S^{i,j}, T)$.

In accurate matching process, the surface fitting method will be used to fit the 9 points of neighborhood area (Table 1) centered at the located pixel (i, j) generated from the rough matching process. Where the correlation coefficient function of image gray is defined as:

$$R(x, y) = \sum_{m=1}^{M} \sum_{n=1}^{M} S^{x,y}(m, n) \times T(m, n). \qquad (9)$$

The correlation coefficient of the gray for 9 points can be computed through their actual locations.

From the above analysis we can see when all possible positions $R(x, y)$ achieve the maximum value, $D(S^{x,y})$ will obtain the minimum value, and the coordinate (x_0, y_0) of the point with the minimum value is the best matching location. So the quadratic surface fitting can be used to find the best matching location. Here the quadric surface is fitted by the least square method. Set the analytical formula of the quadratic surface:

$$f(x, y) = a_0 x^2 + a_1 y^2 + a_2 xy + a_3 x + a_4 y + a_5. \qquad (10)$$

Table 1: 9 points of neighborhood area of pixel (i, j).

$(i-1, j-1)$	$(i-1, j)$	$(i-1, j+1)$
$(i, j-1)$	(i, j)	$(i, j+1)$
$(i+1, j-1)$	$(i+1, j)$	$(i+1, j+1)$

The generated mean square error (MSE) between $f(x, y)$ and $R(x, y)$ is:

$$\delta(a_0 \sim a_5) = \sum_{i=1}^{9} [f(x_i, y_i) - R(x_i, y_i)]^2. \qquad (11)$$

The stagnation point (x_0, y_0) of the quadric surface $f(x, y)$ can be obtained through computing the partial derivative of Equation 11. If $D(x_0, y_0) > 0$ and $\frac{\partial^2 f(x_0, y_0)}{\partial x^2} < 0$, (x_0, y_0) is the maximal value point [4]. If and only if there is one maximal value point for $\delta(a_0 \sim a_5)$ and $R(x, y)$ is a single value function, the location (x_0, y_0) is the best matching location when the sub-image $S^{x,y}$ and the template image are fully matched.

3.3 Convolutional Neural Networks

Convolutional Neural Networks (CNN) consisted of multiple layers of small neuron collections has been adopted to recognize natural images [9]. In general, some local or global pooling layers may be included in Convolutional Neural Networks, which combine the outputs of neuron clusters [2]. They also consist of various combinations of convolutional layers and fully connected layers, with point-wise non-linearity applied at the end of or after each layer [3]. Generally, we call the combination with a filter bank layer, a non-linearity transformation layer, and a feature pooling layer, as a stage. Figure 4 shows a typical CNN framework composed of two stages.

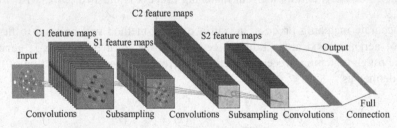

Fig. 4: A typical Convolution Neural Networks framework with two feature stages [13].

In filter bank layer, the input is a 3D array with n_1 2D feature maps of size $n_2 \times n_3$. Each component can be marked as $x_{i,j,k}$, and each feature map is denoted x_i. Where, The output is also a 3D array, and y consists of m_1 feature maps of size $m_2 \times m_3$. A trainable filter (so-called kernel) $k_{i,j}$ in the filter bank has size $l_1 \times l_2$ and connects input feature map x_i to output feature map y_j. The module computes: $y_j = b_j + \sum_i k_{ij} * x_i$, where $*$ is the 2D discrete convolution operator and b_j is a trainable bias parameter [7].

In non-linearity layer, a useful non-linearity function for natural image recognition is the rectified sigmoid $R_{abs} : abs(g_i \cdot tanh())$, where g_i is a trainable gain coefficient. The rectified sigmoid is sometimes followed by a subtractive and divisive local normalization N, which enforces a sort of local competition between adjacent features in a feature map, and between features at the same spatial location. The subtractive normalization operation for a given site $x_{i,j,k}$ computes: $v_{i,j,k} = x_{i,j,k} - \sum_{ipq} w_{pq} \cdot x_{i,j+p,k+q}$, where w_{pq} is a normalized truncated Gaussian weighting window (typically of size 9×9) normalized so that $\sum_{ipq} w_{pq} = 1$. The divisive normalization computes: $y_{i,j,k} = v_{i,j,k}/max(mean(\sigma_{j,k}), \sigma_{j,k})$, where $\sigma_{j,k} = (\sum_{ipq} w_{pq} \cdot v_{i,j+p,k+q}^2)^{1/2}$ [12].

The purpose of feature pooling layer is to build robustness to small distortions, playing the same role as the complex cells in models of visual perception. P_A (*Average Pooling and Subsampling*) is a simplest way to compute the average values over a neighborhood in each feature map. The average operation is sometimes replaced by a P_M (*Max-Pooling*). Traditional CNN use a point-wise $tanh()$ after the pooling layer, but more recent models do not. Some CNNs dispense with the separate pooling layer entirely, but use strides larger than one in the filter bank layer to reduce the resolution [20] . In some recent versions of CNN, the pooling also pools similar feature at the same location, in addition to the same feature at nearby locations.

3.4   Architecture of Visual Scenes Mining Module

In general, the true environment that toy car runs on is a scenes with many complex elements. The interesting points guiding the behaviors of toy car always locate in some local elements of visual scenes. If we want to obtain the interesting information from these local elements. We firstly need to capture the location area of those elements with interesting points with help of location technologies (where PCNN and 2S-SSDA are adopted), then recognize the elements located in the interesting area through some recognition models (CNN is used here). In order to achieve the recognition task in the actual complex scenes, we proposed a novel visual scenes mining module (Figure 5) for ABGP-CNN based on PCNN, 2S-SSDA and CNN.

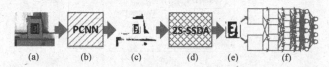

Fig. 5: Visual scenes mining module for the agent ABGP-CNN

Figure 5 illustrates the process that the visual scenes mining module captures and recognizes the interesting area in the actual complex scenes. In order to more explicitly show the function of the visual scenes mining module, the intermediate information products is shown in the architecture of visual scenes mining module. The visual scenes mining module consists of the modules (b), (d) and (f). The original image (a) with an actual complex scenes is processed by PCNN and output as the intermediate image (c).

Then the interesting area (e) is captured by 2S-SSDA from the intermediate image (c). At last CNN (f) will further recognize the interesting area and send the recognized information to the event module of ABGP-CNN, which helps ABGP-CNN make a series of activity plans.

In this paper, the handwrite digital with the value $0 \sim 3$ from MNIST data set is still used as the guidepost to guide the behaviors of toy car, because there are only four types of activities for toy car, namely '0' denotes moving on, '1' moving back, '2' turning left, '3' turning right. Next, we will give some preliminaries about PCNN, 2S-SSDA and CNN in order to have a better description about visual scenes mining module in ABGP-CNN in the following.

4 Application to Toy Car

The application of ABGP-CNN model in the literature [13] was mainly carried out through software simulation about the agent implemented by ABGP-CNN. For further exploring the practical application of ABGP-CNN, here we will deploy the ABGP-CNN model on a self-made toy car as a software control system like human brain. The toy car with ABGP-CNN model can freely plan its behaviors according to its belief, goal, plan and the visual information mined by the visual scenes mining module. The detialed information flow is shown in Figure 6.

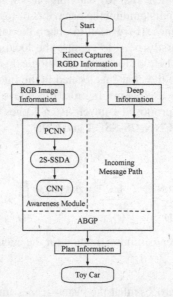

Fig. 6: Information Flow of Toy Car with ABGP Model

The self-made toy car consists of Kinect camera, STM32 control module, wireless receiving module, motor drive module, serial communication module, car body structure and a central controller implemented by a computer running ABGP-CNN model. The architecture of toy car is illustrated in Figure 7.

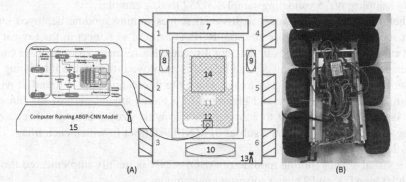

(A)   (B)

Fig. 7: (A) Architecture of toy car. 1~6: Direct current (DC) motors. 7: Visual information collector. 8: Driver module of DC motors 1 and 2. 9: Driver module of DC motors 4 and 5. 10: Driver module of DC motors 3 and 6. 11: Circuit board. 12: Serial communication interface (SCI). 13: Wireless receiving device. 14: Main control module. 15: Computer running ABGP-CNN Model. (B) Actual toy car.

All modules in the architecture of toy car work collaboratively. Visual information collecting depends on Kinect with the RGB and deep camera. The RGB-D (RGB and deep) information from Kinect will be sent to the computer running ABGP-CNN model through serial communication interface (The visual information is transported in wire way because of large amounts of data). The visual scenes mining module in ABGP-CNN further recognizes the visual information from SCI, and a series of behaviors of the toy car will be planned immediately based on beliefs, goals, and the recognized visual information from the visual scenes mining module. The behaviors as a command controlling the toy car will be sent from the *Outgoing Message* interface in ABGP-CNN in radio wave way, then the wireless receiving device in the architecture of toy car will receive and transport the command to the main control module.

In the architecture of toy car, the serial communication interface (SCI) is implemented by RS-232, which provide the real-time wired communication between the main control module and the computer running ABGP-CNN model. STM32F103ZET6 is employed to construct the main control module, which receives and processes the control signals from SCI RS-232. The processed control signals will be served to control the speed and steering of the DC motors in Pulse-Width Modulation (PWM) square wave way, which achieves to control the speed and direction of toy car.

5   Experiment Analysis

In the experiment that ABGP-CNN is deployed on a toy car, the toy car is designed to have 4 basic behaviors moving on, moving back, turning left and turning right. Therefore, we construct a sub-MNIST dataset, called *mnist0_3*, extracting 4 types of hand-written digits with flag '0' denoting moving on, '1' moving back, '2' turning left and '3' turning right from the original MNIST dataset. The dataset *mnist0_3* consists of 20679 training samples, 4075 validating samples, 4257 testing samples.

When the ABGP-CNN agent with visual scenes mining module deployed on the toy car recognizes the nature scenes, the RGB-D camera of Kinect in the toy car will collect in real time the images in the actual scenes. Each image from RGB-D camera will be converted to a gray image, and PCNN is employed to search a interesting area (handwrite digit area) on the gray image. Once the interesting area is found, it will be split out from the gray image and binarized as a binary image. At last the binarized image is linearly compressed to a smaller image with size of 28×28, because the training image of CNN is from a subset of MNIST dataset in which each image is the size of 28×28 as well.

The visual scenes mining module of ABGP-CNN is mainly implemented through Open NI, Open CV and Theano software environment.

- Open NI provides the function interface for extracting the Kinect RGB-D image, namely the depth information of each pixel. Which mainly tells us the distance between the toy car and the target object.
- Open CV mainly provides some function interfaces for generating gray images and template matching methods.
- Theano is served to implement CNN.

In order to validate that the visual scenes mining module can work correctly when ABGP-CNN is deployed on toy car, the experiment is carried out on three different light environments (Table 2), incandescent lamp light, fluorescent lamp light and natural light. In those three different light conditions, we design an actual road sand table (the architecture is same as [13]) with 32 different four types of guideposts (handwritten digits '0' denoting moving on, '1' moving back, '2' turning left and '3' turning right) from *mnist0_3* dataset. The toy car with ABGP-CNN model running on the road sand table can correctly plan its behaviors through depending on the visual scenes mining module recognizes the guideposts in the path.

Table 2: Recognition rate of ABGP-CNN for 1000 tests. ABGP-CNN with the visual scenes mining module deployed on toy car on condition of incandescent lamp light (ILL), fluorescent lamp light (FLL) and natural light (NL). Original ABGP-CNN deployed on toy car (OATC) and software simulation platform(OASP).

Experiment Type	ILL	FLL	NL	OATC	OASP
Recognition Rate	73.25%	83.47%	85.60%	52.30%	99.73%

From the experimental results in Table 2 can be seen, the recognition rate has a great fluctuation under the different light conditions, because the optical camera is sensitive to light conditions. The closer the light is to natural light, the lower the optical camera is interfered, and the higher the recognition rate. The experiment of the original ABGP-CNN deployed on the software simulation platform shows that the recognition rate of

ABGP-CNN has reached the excellent performance of 99.73%. While the recognition rate of the original ABGP-CNN directly deployed on toy car drops to the low performance of 52.30%. However the recognition rate of the ABGP-CNN with visual scenes mining module keep a high performance of 85.60% when it is deployed on an actual toy car on the condition of natural light. Though the ABGP-CNN with visual scenes mining module can achieve a high performance, we still have the following conclusions why it can't reach a higher performance.

- External environment. The sharpness of guideposts and the optical characteristics of camera result to the deviation of collecting images.
- Image processing. The binarized threshold setting is affected by the changes of the external environment. The image compressing in the two-dimensional plane may produces the linear compression error. The above two processes may decrease the recognition rate.
- CNN training process. The training samples of CNN is from the $mins0\_3$, while the samples used to recognition test is collected in real time from the external environment. Because we lack of the training samples collecting in real time from the actual external environment.

If we want to get the better recognition rate for ABGP-CNN in the actual application, the ABGP-CNN with the visual scenes mining module can be optimized further based on the above three reasons.

6   Conclusion and Future Work

The agent built by ABGP-CNN can plan its behaviors through the environment visual information. However all applications for ABGP-CNN model are carried out only through the software simulation. In this paper we deploy ABGP-CNN on an actual toy car and improve the visual scenes mining module, such that ABGP-CNN can directly mine the visual information from the true natural scenes. ABGP CNN with the novel visual scenes mining module can directly conduct the behaviors of toy car, which enhances the capability of communication between the toy car and true environment, and improves the intelligence of toy car. The current visual scenes mining module does not still satisfy high accuracy behaviors plan of toy car. For future work, we will focus on improving awareness accuracy of visual scenes mining module in true environment.

Acknowledgments

This work is supported by the National Basic Research Program of China (973) (No. 2013CB329502), the National Natural Science Foundation of China (No. 61035003, 61202212, 61072085), the National Key Technology Research and Development Program (No. 2014BAK02B07).

References

1. Bratman, M.E.: Intention, Plans, and Practical Reason. Cambridge University Press (1987)
2. Ciresan, D.C., Meier, U., Masci, J., Maria Gambardella, L., Schmidhuber, J.: Flexible, high performance convolutional neural networks for image classification. In: Proceedings of the 21th International Joint Conference on Artificial Intelligence. vol. 22, pp. 1237–1242 (2011)
3. Ciresan, D., Meier, U., Schmidhuber, J.: Multi-column deep neural networks for image classification. In: Proceedings of the 25th Conference on Computer Vision an Pattern Recognition. pp. 3642–3649. IEEE (2012)
4. Department of Applied Mathematics in Tongji University: Advanced Mathematics. Higher Education Press, Beijing (2014)
5. Duan, X.: Improvement of Matching Algorithm Based on Gray Image. Master's thesis, Central South University (2012)
6. Hindriks, K.V., de Boer, F.S., van der Hoek, W., Meyer, J.J.C.: Agent programming in 3apl. Journal of Autonomous Agents and Multi-Agent Systems 2(4), 357–401 (1999)
7. Jarrett, K., Kavukcuoglu, K., Lecun, Y.: What is the best multi-stage architecture for object recognition? In: IEEE 12th International Conference on Computer Vision. pp. 2146–2153 (2009)
8. Johnson, J.L., Padgett, M.L.: Pcnn models and applications. IEEE Transactions on Neural Networks 10(3), 480–498 (May 1999)
9. Korekado, K., Morie, T., Nomura, O., Ando, H., Matsugu, M., Iwata, A.: A convolutional neural network vlsi for image recognition using merged/mixed analog-digital architecture. Knowledge-Based Intelligent Information and Engineering Systems pp. 169–176 (2003)
10. Lehman, J.F., Laird, J., Rosenbloom, P.: A gentle introduction to soar, an architecture for human cognition. Invitation to Cognitive Science 4, 212–249 (1996)
11. Lindblad, T., Kinser, J.: Image Processing Using Pulse-Coupled Neural Networks. Springer-Verlag Berlin Heidelberg (2005)
12. Lyu, S., Simoncelli, E.P., Hughes, H.: Nonlinear image representation using divisive normalization. In: Proceedings of the 21th Conference on Computer Vision an Pattern Recognition. pp. 23–28 (2008)
13. Ma, G., Yang, X., Lu, C., Zhang, B., Shi, Z.: A visual awareness pathway in cognitive model abgp. High Technology Letters 22(41), 395–403 (2016)
14. O., K.J., E., W.W.: An artificial intelligence perspective on autonomic computing policies. In: Proceedings of the Fifth IEEE International Workshop on Policies for Distributed Systems and Networks. pp. 3–12 (2004)
15. Pokahr, A., Braubach, L.: The active components approach for distributed systems development. International Journal of Parallel, Emergent and Distributed Systems 28(4), 321–369 (2013)
16. Rao, A.S., Georgeff, M.P.: Bdi agents: From theory to practice. In: Proceedings of the First International Conference on Multi-Agent Systems. pp. 312–319 (1995)
17. Shi, Z., Wang, X., Yue, J.: Cognitive cycle in mind model cam. International Journal of In-telligence Science 1(2), 25–34 (2011)
18. Shi, Z., Zhang, J., Yue, J., Yang, X.: A cognitive model for multi-agent collaboration. International Journal of Intelligence Science 4(1), 1–6 (2013)
19. Shoham, Y.: Agent-oriented programming. Artificial Intelligence 60(1), 51–92 (1993)
20. Simard, P.Y., Steinkraus, D., Platt, J.C.: Best practices for convolutional neural networks applied to visual document analysis. In: the 12th International Conference on Document Analysis and Recognition. pp. 958–963 (2003)
21. Wang, Z., Ma, Y., Gu, J.: Multi-focus image fusion using pcnn. Pattern Recognition 43(6), 2003 – 2016 (2010)

Predicting Hospital Re-admissions from Nursing Care Data of Hospitalized Patients

Muhammad K Lodhi[1], Rashid Ansari[1], Yingwei Yao[2], Gail M Keenan[2], Diana Wilkie[2], Ashfaq A Khokhar[3]

[1] University of Illinois at Chicago, Chicago, IL, USA
{mlodhi3, ransari}@uic.edu
[2] University of Florida, Gainesville, FL, USA
{yyao, gkeenan,diwilkie}@ufl.edu
[3] Iowa State University, Ames, IA, USA
ashfaq@iastate.edu

Abstract. Readmission rates in the hospitals are increasingly being used as a benchmark to determine the quality of healthcare delivery to hospitalized patients. Around three-fourths of all hospital re-admissions can be avoided, saving billions of dollars. Many hospitals have now deployed electronic health record (EHR) systems that can be used to study issues that trigger readmission. However, most of the EHRs are high dimensional and sparsely populated, and analyzing such data sets is a Big Data challenge. The effect of some of the well-known dimension reduction techniques is minimized due to presence of non-linear variables. We use association mining as a dimension reduction method and the results are used to develop models, using data from an existing nursing EHR system, for predicting risk of re-admission to the hospitals. These models can help in determining effective treatments for patients to minimize the possibility of re-admission, bringing down the cost and increasing the quality of care provided to the patients. Results from the models show significantly accurate predictions of patient re-admission.

Keywords: electronic health records (EHR) · predictive modeling · Re-admission

1   Introduction

Sparse and high dimensional datasets pose a serious challenge to existing data mining and machine learning methods, mainly because of their size and exponential complexity w.r.t dimensions. Due to these characteristics, the gap between our ability to process and analyze data and the rate at which it is accumulating is rapidly widening [1]. These data sets stem from diverse application areas, such as electronic health records (EHRs), biology, astronomy, web data, and medical imaging. Due to the presence of non-linear variables and their varying degree of importance in different domains, the problem is complex and extremely challenging. Different data mining techniques have been used to extract knowledge available in some of these data sets, albeit with

© Springer International Publishing AG 2017
P. Perner (Ed.): ICDM 2017, LNAI 10357, pp. 181–193, 2017.
DOI: 10.1007/978-3-319-62701-4_14

limited success until now [2]. Various algorithms [3, 4] have been introduced, that use row-wise enumeration method instead of traditional column-wise enumeration method to address the dimensionality problem, however, they have their own limitations as they work best for dense high dimensional datasets, with significantly lower number of rows compared to number of columns. Different dimension reduction algorithms, such as Principal Component Analysis [5], Multi-Dimensional scaling [6] and Independent component analysis [7], are too restrictive due to their reliance on global linearity assumption. The context of various variables is also abated using these conventional dimensional reduction methods.

In this paper, we target the analysis of a high dimensional and sparse dataset that stems from nursing care EHRs. In such a dataset, thousands of variables consisting of vitals, drugs, tests, treatments, etc. exist (high dimensionality), yet only a limited number of them are tracked for any individual patient (sparse). While nursing care data are an important part of the EHR, it usually goes unnoticed when planning for the improvement of healthcare delivery systems [8]. Effective use of different data mining methods can be extremely helpful in provisioning of better care to the patient, and developing more effective care, and consequently help in decreasing the healthcare cost.

Historically, the data stored in EHRs has been used to monitor the progress of the patients, though recently, a lot of research is being performed to build predictive models using this data. We believe that EHR systems are also a perfect candidate to study big data issues due to size and heterogeneity of the data. Other industries have been using the big data methods to save costs. On the other hand, more than a trillion dollars are wasted annually in healthcare industry partly due to latest technology not being used to its fullest in healthcare [9]. Big data techniques can have a huge impact in reducing the healthcare costs that are expected to continue to markedly increase in the coming years [10].

One of the reasons of the increasing costs of healthcare are the patient re-admissions to the hospitals. Reducing repeat hospitalizations can greatly reduce these costs. As with many other issues in the healthcare system, repeat hospitalizations often occur due to poor treatment provided to the patients [11, 12], more specifically, they are often caused by premature discharges [13] or communication breakdown between the patients and healthcare team while the patient is being discharged. These readmissions result in higher costs to taxpayers [14], costing as much as $45 billion annually [15, 16]. Medicare, along with other healthcare payers, are concerned with the cost of unnecessary readmissions as Medicare alone spends roughly $15 billion annually on repeat hospitalizations [17] and almost 20% of the patients are readmitted within 30 days after being discharged from the hospitals [18]. According to a report, almost 76% of repeat admissions can be avoided by improving care before and after the patient is discharged [19]. By decreasing these preventable repeat hospitalizations, overall productivity of the hospitals and staff can improve considerably [20, 21].

Avoidable re-admissions are a huge burden on hospital resources, including the workforce. In this study, our aim is to determine different nursing and patient factors that contribute towards patient re-admission to the hospital. Our objective is to construct predictive models that can predict whether a patient is at risk of being re-admitted in near future. In particular, we focus on readmitted patients suffering from pain problems. Pain is a common problem that a patient has to endure even though patient comfort is of utmost importance. A plethora of research has been conducted to lessen pain problems for the patients, though no significant improvements have been made in this regard [22, 23].

To tackle the issue of re-admissions, a lot of research is being done. Several techniques and predictive models, which consider several patient factors, including socio-economic status, marital status, sex, and age, among others, have been developed to predict readmission [24, 25]. Some recent research studies have used administrative data to predict patient readmission within a year [26-28] or even within a month after being discharged from the hospital [24, 29-32]. A few studies have only concentrated on a select group of patients [25, 27, 29, 31] or only on a single hospital [24, 25, 33]. Despite all the work, most of the predictive models have poor predictive capability and are too complex to be utilized in daily practice [26]. Furthermore, none of the aforementioned research studies have considered the effect of nursing care on patient re-admission.

2   Data Description

The data for this experiment has been obtained from the HANDS database [34], deployed in 9 units at four different hospitals. The HANDS is an EHR system, designed specifically to record nursing care provided to the patients. Nursing diagnoses were based on NANDA-I [35], outcomes on the Nursing Outcome Classification (NOC) [36] and nursing interventions on the Nursing Interventions Classifications (NIC) [37] terminologies. The data were collected for a period of three years from 2005 till 2008.

In the three year period, there were a total of 42,403 episodes (from 34,927 unique patients).For our analysis purposes, a continuous stay of the patient in a hospital spanning over single or multiple units, is considered as an episode. The episode ends if a patient is discharged or if the patient dies. An episode consists of single or multiple nurse shifts. In our study, we have considered only those episodes with at least two nurse shifts. In every shift, a nurse documents a plan of care (POC). The POC consists of multiple nursing diagnoses (NANDAs), different identified outcomes (NOCs) associated with NANDAs, their initial and expected score (assigned to each unique NOC with value between 1 and 5, 1 being the worst), and interventions (NICs) to achieve the expected outcome. The POC also consists of patients and nurses demographics.

In our dataset, a total of 5298 patients (~15% of the patients in the dataset) were re-admitted, after being discharged, at least once. 2618 of these 5298 patients had either Acute or Chronic Pain (or both) diagnosis in their plan of care (POC) in both the original and re-admission episode. On the other hand, there were 15,956 patients, diagnosed with Pain, that were admitted only once. All patient deaths in hospitals have been excluded in this work. Both the sets were further reduced by considering only patients with NOC: Pain Control. 980 of 2618 patients that were admitted again after being discharged had a NOC of Pain Control, whereas, 5095 of 15956 patients, admitted only one time, had Pain Control as an outcome. Note that the number of single admissions might not be completely accurate, as the patient could have been re-hospitalized to another hospital unit not using HANDS database, or the patient might have been re-admitted after the study period. A few characteristics of the dataset are given in Table 1.

Table 1. Dataset Characteristics

Full Dataset Characteristics	
Total number of admissions	42,203
Number of unique individuals	34,927
Number of Patients re-admitted	5298
All patients diagnosed with Pain	18574
Number of Pain Patients re-admitted	2618
Number of re-admitted Patients with NOC: Pain Control	980
Number of single admission patients with NOC: Pain Control	5095
Variables used in prediction	
Age (years) mean (SD)	59.0 (18.4)
Length of Stay (hours) mean (SD)	91.5 (98.5)
Average Nurse Experience (years) mean (SD)	1.7 (2.4)

3   Dimension Reduction and Feature Selection

In this study, we have conducted predictive modeling at the episode-level. The target variable is whether a patient was re-admitted after being discharged. Hence, the target variable is a binary variable.

There are over 700 features in the HANDS database system but only a very few patient records (if any) have a value for each of these features. These features include, but are not limited to, nurse's and patient's information, the various nursing diagnoses, nursing outcomes, different nursing interventions and other organizational data. To ascertain the attributes that are clinically relevant to our aim of predicting re-admission, and to reduce the number of dimensions of our data, we have used association mining as a dimension-reduction method.

A list of a few rules with significant confidence level and support determined via association mining technique are listed below:

- If Final Pain Control NOC Rating >= 4 → Not Re-admitted (confidence: 0.64, support: 0.23)
- If patient is between ages of 18 and 50 → Not Re-admitted (confidence: 0.70, support: 0.18)
- If First rating <= 3 → Readmitted (confidence: 0.53, support: 0.21)

- If there is no intervention from the Patient Education class is present → Readmitted (confidence: 0.56, support: 0.18)
- If there is at least one intervention from the Information Management class in the patient's care plan → Not Re-admitted (confidence: 0.65, support : 0.21)

Using these rules, we have determined the list of features that we use for our predictive modeling experiment. The variables such as the age of the patient, their stay in the hospital and the experience of the nurse that took care of the patient have been discretized, along with clustering nursing diagnoses and interventions into their respective domains and classes.

Age has been discretized into four groups. These groups, young (18-49), middle-aged (50-64), old (65-84), and very old (85+), are established on theoretical rationale [38] and the data frequency distribution.

We derive LOS for each episode from our database. It was calculated by adding the number of hours for all the nurse shifts in the episode and was grouped into three categories: short (up to 48 hours), medium (48-119 hours), and long stay (120+ hours). Currently, in most places, visits less than 48 hours are called observation visits, considered as short stay. Average LOS of patient in the hospital is typically about 120 hours, and in our study, a stay of between 48 and 120 hours is considered as a medium stay. Anything over 120 hours is, consequently, considered as a long stay [39].

Average nurse experience has also been derived from the database. For nurse experience, a nurse with at least two years of experience in her current position was considered to be an experienced nurse, and nurses with less than two years' experience were considered inexperienced. The episode was categorized as care provided by an experienced nurse team if more than 50% of the nurses providing care in that episode had at least 2 years of experience. These categories were based on professional criteria [40].

NOC being met or not met was calculated using two variables, Expected and Final NOC rating. A NOC is "met" when the final NOC rating is the same or better than the Expected rating, which is set by the nurse that first enters a NOC into a patient's care plan, often when the patient is admitted. Otherwise, NOC is "not met".

The time of Discharge is essentially the completion time of the last POC in the episode and the time is classified into three categories as follows: morning (7am - 3pm), afternoon (3pm – 11pm) and evening (11pm - 7am). These categories were based on timings of nurse shifts in the HANDS database and represent the nursing day, evening and night shifts that are typical in hospitals with 8 hour shifts.

Along with these patient and nurse staff variables, the NANDA-I diagnoses and NIC interventions that appeared in the POCs were also considered as predictive variables. The NANDAs and NICs were clustered together by domains and classes, based upon the nursing literature [35, 37]. We included 10 of 12 NANDA-I terms, that had frequencies of more than 5% in our sample episodes. (Activity/Rest, Comfort, Coping/Stress Tolerance, Elimination, Health Promotion, Life Principles, Nutrition, Perception, Role Relationships, and Safety/Protection). Our data sample included terms from all 7 NIC domains (Behavioral, Community, Family, Health System, Safety, Physiological: Basic, and Physiological: Complex). 8 of 19 NANDA classes (Activity/Exercise, Cardiovascular/Pulmonary Responses, Cognition, Hydration, Infection, Physical Comfort, Physical Injury, and Pulmonary System) were included. Finally, of the 30 different NIC classes, 16 (Activity & Exercise Management, Cognitive Therapy, Communication Enhancement, Drug Management, Electrolyte and Acid/Base Management, Immobility Management, Information Management, Nutrition Support, Patient Education, Physical Comfort Promotion, Psychological Comfort Promotion, Respiratory Management, Risk Management, Self-Care Facilitation, Skin/Wound Management, and Tissue Perfusion Management) had frequencies higher than our threshold of 5% in our data sample. A particular NANDA-I or NIC domain and class was assumed to be either present or absent in an episode. In this sparse dataset, the NANDA-I and NIC classes and domains having extremely low frequencies (less than 5%) were excluded to reduce the impacts of spurious correlations.

4   Data Modeling

Our key objective for this study was to construct predictive models that can predict whether a patient having pain problems will be re-admitted, after being discharged from a hospital unit. The secondary objective was to assess the feasibility of constructing predictive models using data from a nursing database system.

After extracting and refining the data, multiple models were built on the dataset using different prediction tools and their performances were compared. The models were based on C4.5 Decision Trees (DT) [41], k-nearest neighbors (k-NN) [42], support-vector machines (SVM) [43], and Naïve-Bayes [44].

5   Experimental Results

The analysis was performed to build models for predicting re-hospitalization of patients suffering from Pain problems based on a number of patient and nurse features, nursing diagnoses, and nursing interventions. The performance of the models was evaluated using 10-fold cross-validation [45]. For the experiment, we chose a sample of 2300 patients including both single and multiple admissions. The results are presented in Table 2.

Table 2. Model Comparison for experiment predicting re-admission

Model	Accuracy	Recall	Precision	F Measure	AUC
Decision Tree	73.7	77.4	72.5	0.75	0.78
Naïve-Bayes	69.3	72.7	67.4	0.70	0.71
K-NN (K = 2)	64.5	84.1	60.8	0.71	0.62
K-NN (K = 5)	66.1	85.1	60.5	0.71	0.69
K-NN(K= 10)	64.9	82.3	59.1	0.69	0.67
SVM	65.1	80.7	59.1	0.68	0.65

The preliminary results indicate the Decision Tree and Naïve-Bayes algorithms have a relatively high prediction accuracy compared to k-NN and SVM models. The decision tree has the best accuracy at 73.7%, followed by Naïve-Bayes with an accuracy of 69.3%. K-NN models have accuracy ranging between 64.5-66.1% and SVM has an accuracy of 65.1%. F-measure is 0.75 for the decision tree model and around 0.7 for rest of the models. Area under the curve (AUC) was 0.78 for the decision tree.

Figures 1 - 4 give the decision trees after partitioning the tree based on topmost node or the best predictor; age. These figures show the different important variables for these four different groups of patients. Figure 1 shows the decision tree for the young patients. Around 70% of the younger patients had a single admission. The next best predictor is the length of stay. When the LOS is short, around 76.5% of the patients are not re-admitted; for medium LOS, 67.9% have only a single admission; on the other hand, only 57.1% of the young patients with a long episode were not re-admitted.

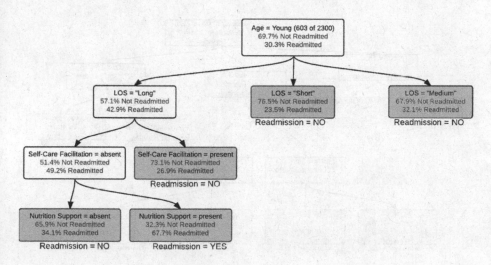

Fig. 1. Readmission Decision Tree for Young Patients

The decision tree for middle-aged patients is depicted in Figure 2. The next best predictor for middle-aged patients is also patient's LOS. When the LOS for middle-aged patient is "short", 69.7% of them are not re-admitted. These set of patients have Final NOC Rating as the next predictor. If the Final Rating is 3 or less, only 43.2% of the patients are not re-admitted. On the other hand, if the Final Rating is 4 or 5, 72.4% of the patients had a single visit only. Whenever the LOS is "medium", 55.3% of the patients are not re-admitted. The next predictor is also the rating in the Final shift for this set. For patients with "long" LOS, 55.7% of the patients were not re-admitted. For these patients, the next predictor was Nutrition NANDA domain. Whenever the Nutrition domain was present, 63.6% of the patients had a single hospital visit, whereas when Nutrition was absent, only 36.9% of the patients were not hospitalized again.

The old patients' decision tree is given in Figure 3. The number of old patients that had a single admission (50.9%) only and those who were re-admitted (49.1%) was almost equal. When the predictor Behavioral NIC domain was absent, it is observed that 59.3% of the patients were re-admitted as compared to 46.8% when the Behavioral NIC domain was present. For the patients that did not have Behavioral NIC domain, whenever the Expected rating for NOC: Pain Control was 2 or below, all of the patients were re-admitted, though there were only 23 such cases. For patients with behavioral domain present, whenever the NOC: Pain Control was met, around 37.5% of the patients were re-admitted, whereas 54.5% of the patients were re-admitted when the NOC: Pain Control was not met for patients having interventions from the Behavioral domain.

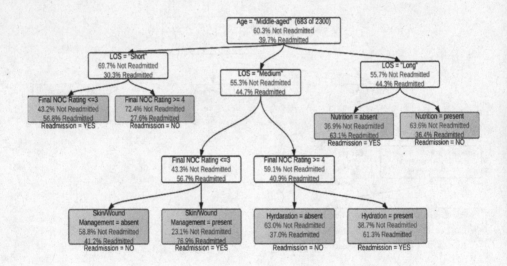

Fig. 2. Readmission Decision Tree – Middle-aged Patients

Around 37.9% of the very old patients were not re-admitted (Figure 4). Whenever a very old patient was discharged during afternoon or evening hours, there was a 66% and 60% chance respectively of the patient being re-admitted again. On the other hand, if the patient was discharged during morning hours, the chance of re-admission was only 35.3%.

The best accuracy results were obtained using the decision tree model. Also, the model generated by the decision tree is easily understandable. Naïve-Bayes model also had a high accuracy, however, the Naïve-Bayes models have independent features assumption [46], and it is often not clear if the features are truly independent. Therefore, we propose to use decision tree predictive modeling on our data.

6   Discussion and Conclusion

Lowering the re-hospitalization rate is one of the main actions that can help achieve a reduction in healthcare costs. The re-admission problem needs to be handled as many hospitals are facing financial issues [47-49]. Different strategies can be implemented using results from predictive modeling. The capability to recognize patients at high risk of re-admission is the key first step to improve quality of care for the patients [50], potentially leading to interventions tailored to individual patients to lower the risk of re-admission. Unfortunately, most of the current work cannot be utilized properly due to different complexities.

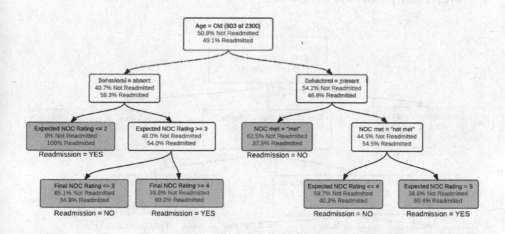

Fig. 3. Readmission Decision Tree for Old Patients

In this work, we constructed models to predict hospital re-admission of patients suffering from pain problems using nursing data. Unlike some previous studies, our data were not gathered through questionnaires and interviews, as patients have been known to under-report hospital re-admissions [51]. Decision tree model had the best accuracy of all the models tested and therefore will be used for further analysis. Our preliminary findings suggest that patient demographics, different nursing diagnoses and interventions, among other variables can be used to predict whether a patient will be coming back for treatment. The model had a reasonable accuracy of 73.7%.

The model has some limitations due to data issues. It was developed using data from a nursing EHR system which was not deployed in all units or hospitals. There is a strong probability that some patients that have been counted as a single admission patient, might have been re-admitted to a different hospital unit in which the EHR system was not deployed or was not a part of our original study. Furthermore, a patient might have been re-admitted to a hospital after the study period. Nevertheless, we believe that these differences would not have a considerable effect on the accuracy of the decision tree model.

Notwithstanding a few shortcomings, the predictive modeling techniques have vital implications for developing effective strategies for preventing repeat hospitalization of the patients, since the results of the models can be used to identify at-risk patients of future re-admission the hospital. Further, our use of nursing care data in this analysis has revealed the potential importance of utilizing nursing care variables to identify risk factors of re-admission. This makes sense to us since nurses are the main front line providers of care in the hospital setting. A lot of money can be saved by reducing the re-admission rate at the hospitals, thus careful identification of the risk factors is important. Apart from this benefit, the findings from the current study can be used to improve the care provided to the patients.

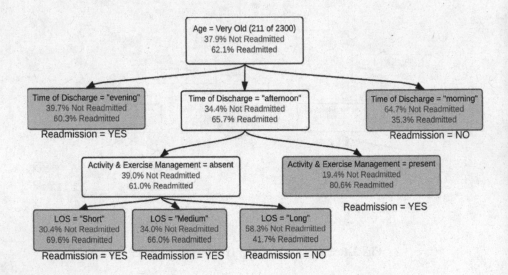

Fig. 4. Re-admission Decision Tree for Very Old Patients

7   References

1.   Hilbert, M. and P. López, *The world's technological capacity to store, communicate, and compute information.* science, 2011. **332**(6025): p. 60-65.

2.   Chen, Y., D. Hu, and G. Zhang, *Data Mining and Critical Success Factors in Data Mining Projects*, in *Knowledge Enterprise: Intelligent Strategies in Product Design, Manufacturing, and Management.* 2006, Springer. p. 281-287.

3.   Pan, F., et al., *CARPENTER: Finding closed patterns in long biological datasets*, in *International Conference on Knowledge Discovery and Data Mining.* 2003.

4.   Liu, H., et al., *Mining Frequent Patterns from Very High Dimensional Data: A Top-Down Row Enumeration Approach*, in *2006 SIAM International Conference on Data Mining (SDM'06).* 2006: Bethesda, MD. p. 280-291.

5.   Jolliffe, I., *Principal component analysis.* 2002: Wiley Online Library.

6.   Cox, T.F. and M.A. Cox, *Multidimensional scaling.* 2000: CRC Press.
7.   Hyvärinen, A., J. Karhunen, and E. Oja, *Independent component analysis.* Vol. 46. 2004: John Wiley & Sons.
8.   Harper, E. and J. Sensmeier. *Why is Big Data Important to Nurses?* Himss 2015 [cited 2015 September 10]; Available from: http://www.himss.org/News/NewsDetail.aspx?ItemNumber=43374.
9.   Kavilanz, P.B. *Health care's big money wasters.* 2009 August 10, 2009 [cited 2014 April 29]; Available from: http://www.money.cnn.com/2009/08/10/news/economy/healthcare_money_wasters/.
10.   Cuckler, G., *National Health Expenditures Projections 2012-2022,* C.f.M.a.M.S. 2014, Editor. 2014.
11.   Smith, P.C., *Performance measurement for health system improvement: experiences, challenges and prospects.* 2009: Cambridge University Press.
12.   Billings, J., et al., *Impact of socioeconomic status on hospital use in New York City.* Health affairs, 1993. **12**(1): p. 162-173.
13.   Goodman, D.C., et al., *After hospitalization: a Dartmouth atlas report on post-acute care for Medicare beneficiaries.* The Dartmouth Institute, September, 2011. **28**.
14.   Yam, C., et al., *Measuring and preventing potentially avoidable hospital readmissions: a review of the literature.* Hong Kong medical journal= Xianggang yi xue za zhi/Hong Kong Academy of Medicine, 2010. **16**(5): p. 383-389.
15.   Herzog, R. *5 Ways Healthcare Providers Can Reduce Costly Hospital Readmissions.* 2013 March 31, 2013 [cited 2015 August 29]; Available from: http://hitconsultant.net/2013/03/31/5-ways-healthcare-providers-can-reduce-costly-hospital-readmissions/.
16.   Vest, R.J., et al., *Determinants of preventable readmissions in the United States: A systematic review.* Implementation Science, 2010. **5**(88).
17.   *Reducing Hospital Readmission with Enhanced Patient Education,* K.P. Education, Editor. 2010.
18.   Jencks, S.F., M.V. Williams, and E.A. Coleman, *Rehospitalizations among patients in the Medicare fee-for-service program.* The New England Journal of Medicine, 2009. **360**: p. 1418-28.
19.   Commission, M.P.A., *Report to the Congress: promoting greater efficiency in Medicare.* 2007: Medicare Payment Advisory Commission (MedPAC).
20.   Minott, J. *Reducing Hospital Readmissions.* 2008 [cited 2015 August 28]; Available from: http://www.academyhealth.org/files/publications/ReducingHospitalReadmissions.pdf .
21.   Foster, D. and G. Harkness. *Healthcare reform: Pending Changes to Reimbursement for 30-day Readmissions.* August 2010 [cited 2015 August 31]; Available from: http://www.communitysolutions.com/assets/2012_Institute_Presentations/acareimbue sementchanges051812.pdf.
22.   *A controlled trial to improve care for seriously ill hospitalized patients. The study to understand prognoses and preferences for outcomes and risks of treatments (SUPPORT). The SUPPORT Principal Investigators.* Jama, 1995. **274**(20): p. 1591-8.
23.   Yao, Y., et al., *Current state of pain care for hospitalized patients at end of life.* Am J Hosp Palliat Care, 2013. **30**(2): p. 128-36.
24.   Hasan, O., et al., *Hospital readmission in general medicine patients: a prediction model.* Journal of general internal medicine, 2010. **25**(3): p. 211-219.
25.   Mudge, A.M., et al., *Recurrent readmissions in medical patients: a prospective study.* Journal of Hospital Medicine, 2011. **6**(2): p. 61-67.
26.   Billings, J., et al., *Case finding for patients at risk of readmission to hospital: development of algorithm to identify high risk patients.* Bmj, 2006. **333**(7563): p. 327.

27.   Cui, Y., et al., *Development and validation of a predictive model for all-cause hospital readmissions in Winnipeg, Canada.* Journal of Health Services Research and Policy, 2015. **20**(2): p. 83-91.

28.   Howell, S., et al., *Using routine inpatient data to identify patients at risk of hospital readmission.* BMC Health Services Research, 2009. **9**: p. 96.

29.   Holloway, J., S. Medendorp, and J. Bromberg, *Risk factors for early readmission among veterans.* Health services research, 1990. **25**(1 Pt 2): p. 213.

30.   Meldon, S.W., et al., *A Brief Risk-stratification Tool to Predict Repeat Emergency Department Visits and Hospitalizationsin Older Patients Discharged from the Emergency Department.* Academic Emergency Medicine, 2003. **10**(3): p. 224-232.

31.   Rowland, K., et al., *The discharge of elderly patients from an accident and emergency department: functional changes and risk of readmission.* Age and ageing, 1990. **19**(6): p. 415-418.

32.   van Walraven, C., et al., *Derivation and validation of an index to predict early death or unplanned readmission after discharge from hospital to the community.* Canadian Medical Association Journal, 2010. **182**(6): p. 551-557.

33.   Phillips, R.S., et al., *Predicting emergency readmissions for patients discharged from the medical service of a teaching hospital.* Journal of general internal medicine, 1987. **2**(6): p. 400-405.

34.   Keenan, G., et al., *Maintaining a consistent big picture: Meaningful use of a Web-based POC EHR system.* International Journal of Nursing Knowledge, 2012. **23**(3): p. 119-133.

35.   Association, N.A.N.D., *NANDA Nursing Diagnoses.* 2007: North American Nursing Diagnosis Association.

36.   Moorhead, S., M. Johnson, and M. Maas. *Iowa Outcomes Project, Nursing outcomes classification (NOC).* 2004. St. Louis, MO: Mosby.

37.   Bulechek, G.M., H.K. Butcher, and J.M. Dochterman, *Nursing interventions classification (NIC).* 2008: Mosby.

38.   Gronbach, K.W., *The Age Curve: How to Profit from the Coming Demographic Storm.* 2008.

39.   *Hospital utilization (in non-federal short-stay hospitals).* 2014: Centers for Disease Control and Prevention.

40.   Benner, P., *From novice to expert.* American Journal of Nursing, 1982. **82**(3): p. 402-407.

41.   Quinlan, J., *C4.5: Programs for machine learning,* M. Kaufmann, Editor. 2003: San Francisco, CA.

42.   Aha, D., D. Kibler, and M. Albert, *Instance-based learning algorithms.* Machine Learning, 1991. **6**(1): p. 37-66.

43.   Cortes, C. and V. Vapnik, *Support-vector networks.* Machine Learning, 1995. **20**(3): p. 273.

44.   Pearl, J. *Bayesian networks.* in *The handbook of brain theory and neural networks.* 1998. MIT Press.

45.   Witten, I.H. and E. Frank, *Data Mining: Practical Machine Learning Tools and Techniques.* 2 ed, ed. J. Gray. 2005: Elsevier.

46.   Lewis, D., *Naive (Bayes) at Forty: The Independence Assumption in Information Retrieval,* in *Proceedings of 10th European Conference on Machine Learning.* 1998. p. 4-15.

47.   Whitmarsh, C. *Hospitals Facing Economic Challenges.* [cited 2015 September 6]; Available from: http://www.businesslife.com/articles.php?id=1104.

48.   Gugliotta, G., *Rural hospitals, beset by financial problems, struggle to survive,* in *The Washington Post.* 2015.

49. Campbell, D., *NHS cuts: One in three hospitals face financial crisis as result of cash squeeze*, in *The Guardian*. 2013.
50. Desikan, P., et al. *Predictive Modeling in Healthcare: Challenges and Opportunities*. [cited 2014 September 27]; Available from: http://lifesciences.ieee.org/publications/newsletter/november-2013/439-predictive-modeling-in-healthcare-challenges-and-opportunities.
51. Norrish, A., et al., *Validity of self-reported hospital admission in a prospective study*. American journal of epidemiology, 1994. **140**(10): p. 938-942.

Activity Prediction in Process Management using the WoMan Framework

Stefano Ferilli[1,2](✉), Domenico Redavid[3], and Sergio Angelastro[1]

[1] Dipartimento di Informatica – Università di Bari, Bari, Italy
{stefano.ferilli, sergio.angelastro}@uniba.it
[2] Centro Interdipartimentale di Logica ed Applicazioni – Università di Bari, Bari, Italy
[3] Artificial Brain S.r.l., Bari, Italy
redavid@abrain.it

Abstract. In addition to the classical exploitation of process models for checking process enactment conformance, a very relevant but almost neglected task concerns the prediction of which activities will be carried out next at a given moment during process execution. The outcomes of this task may allow to save time and money by taking suitable actions that facilitate the execution of those activities, may support more fundamental and critical tasks involved in automatic process management, and may provide indirect indications on the correctness and reliability of a process model. This paper proposes an enhanced declarative process model formalism and a strategy for activity prediction using the WoMan framework for workflow management. Experimental results on different domains show very interesting prediction performance.

Keywords: Process Mining, Activity Prediction, Process Model

1 Introduction & Background

A *process* consists of actions performed by agents [1, 2]. A *workflow* is a formal specification of a process. It may involve sequential, parallel, conditional, or iterative execution [12]. A process execution, compliant to a given workflow, is called a *case*. It can be described as a list of *events* (i.e., identifiable, instantaneous actions, including decisions upon the next activity to be performed), associated to *steps* (time points) and collected in *traces* [13]. Relevant events are the start and end of process executions, or of activities [2]. A *task* is a generic piece of work, defined to be executed for many cases of the same type. An *activity* is the actual execution of a task by a *resource* (an agent that can carry it out).

Process Management techniques are useful in domains where a production process must be monitored (e.g. in the industry) in order to check whether the actual behavior is compliant with a desired one. When a process model is available, new process enactments can be automatically supervised. The complexity of some application domains requires to learn automatically the process models, because building them manually would be very complex, costly and error-prone.

P. Perner (Ed.): ICDM 2017, LNAI 10357, pp. 194–208, 2017.
DOI: 10.1007/978-3-319-62701-4_15

The area of research aimed at inferring workflow models from specific examples of process executions is known as *process discovery* [2] or *process mining* [14, 9]. *Declarative* process mining approaches [11] learn models expressed in terms of a set of constraints, instead of monolithic models (usually expressed as some kind of graph, e.g. Petri Nets). More precisely, given a set of tasks T and a set of cases $C \subseteq T^*$, the aim of process mining is discovering a workflow model that fulfills the following requirements [2, 1, 8, 14]:

Completeness it can generate ('explain', 'cover') all event sequences in C;
Irredundancy it generates as few event sequences of $T^* \setminus C$ as possible;
Minimality it is as simple and compact as possible.

Having to deal with real-world environments, an additional requirement may be the ability to deal with noise [14] (i.e., the presence of wrong process executions in the training examples).

The WoMan framework [5, 4] lies at the intersection between *Declarative* Process Mining and Inductive Logic Programming (ILP) [10]. Indeed, it pervasively uses First-Order Logic as a representation formalism, that provides a great expressiveness potential and allows one to describe contextual information using relationships [6]. Differently from all previous approaches in the literature, it is *fully incremental*: not only can it refine an existing model according to new cases whenever they become available, it can even start learning from an empty model and a single case, while others need a (large) number of cases to draw significant statistics before learning starts. This allows to carry out continuous adaptation of the learned model to the actual practice efficiently, effectively and transparently to the users [4]. Experiments proved that WoMan can handle efficiently and effectively very complex processes, thanks to its powerful representation formalism and process handling operators [5, 4].

While the most classical task in Process Management is supervision of a process enactment in order to check its compliance with given model, in some context an extremely important task may be activity prediction. It may be stated as follows: given a process model and the current (partial) status of a new process execution, guess which will be the next activity that will take place in the execution. Its importance follows from the applications that it may serve. Let us give a few examples.

- Given an intermediate status of a process execution, knowing how the execution will proceed might allow the environment, or the (human or automatic) supervisor, to take suitable actions that facilitate the next activities. In industrial environments, this may bring significant savings in terms of time and money. In smart environments, considering the daily routines of people at home or at work as a process may allow the environment to provide more comfortable support to the users, improving their quality of life.
- Also, having a reliable list of expected activities to be carried out next can support the activity recognition task, which is one of the most critical requirements for automatic process management. In fact, being able to determine which high-level process-related activities are being carried out in terms

of the low-level data obtained from the sensors placed in the environment is a very complex and not yet fully solved issue.

– Another, very relevant and interesting, application of process-related predictions is in the assessment of the quality of a model. Indeed, since models are learned automatically exactly because the correct model is not available, only an empirical validation can be run. In literature, this is typically done by applying the learned model to new process enactments. Getting correct predictions when using a model may be interpreted as an indirect indication that the model is correct.

Despite its relevance, the task of activity prediction has received very little attention so far in the literature. This is possibly due to the fact that, when using traditional graph-based formalisms for expressing process models, determining the next activities may be quite simple. Indeed, reporting the current status of the process execution on the graph (e.g., as a marking of tokens in Petri Nets) allows to determine quite straightforwardly which tasks are currently enabled. Using declarative approaches the issue becomess less straightforward. Also, in traditional domains the rules that determine how the process must be carried out may be quite strict, so that predicting the process evolutions becomes a trivial consequence of conformance checking. Other, less traditional application domains (e.g., the cited routines of people), involve much more variability, and obtaining reliable predictions becomes both more difficult and more useful.

In this paper, we show how the activity prediction task can be carried out effectively using WoMan. Interesting preliminary results were obtained in [7] on various application domains using the formalism proposed in [6]. Here we extend the formalism in [6] and the approach in [7] by considering additional information in the models in order to improve the prediction performance. Full details about the extended formalism and the prediction approach are given in this paper for the first time.

This paper is organized as follows. The next section presents the WoMan (extended) formalism, while Section 3 describes in details its approach to activity prediction. Then, Section 4 reports and discusses the experimental outcomes. Finally, in Section 5, we draw some conclusions and outline future work issues.

2   The WoMan Formalism

WoMan representations [6] are based on the Logic Programming formalism, and works in Datalog, where only constants or variables are allowed as terms. Following foundational literature [1, 8], trace elements in WoMan are 7-tuples, represented in WoMan as facts

$$\texttt{entry}(T, E, W, P, A, O, R).$$

that report information about relevant events for the case they refer to:

1. T is the event timestamp (all events in a case must have different timestamps),

2. E is the type of the event (one of **begin_process**, **end_process**, **begin_activity**, **end_activity**, and **context_description**),
3. W is the name of the workflow the process refers to,
4. P is a unique identifier for each process execution,
5. A is the name of the activity,
6. O is the progressive number of occurrence of that activity in that process,
7. R (optional) specifies the agent that carries out activity A.

Activity begin and end events allow to properly handle time span and to identify concurrency in task execution, avoiding the need for inferring it by means of statistical (possibly wrong) considerations [13]. When $E =$ **context_description**, A is used to describe contextual information at time T, in the form of a conjunction of FOL atoms built on domain-specific predicates.

Given a set of training cases $\mathcal{C}$, WoMan learns a model consisting of a set of atoms built on several predicates, each expressing a different kind of constraint[4]. The core of the model, established in its very first version, is expressed by predicates **task/2** and **transition/4**.

- **task**(t, C_t) : task t occurred in training cases C_t.
- **transition**(I, O, t, C_t) : transition[5] t, occurred in training cases C_t, is enabled if all input tasks in $I = [i_1, \ldots, i_n]$ are active; if fired, after stopping the execution of all tasks in I (in any order), the execution of all output tasks in $O = [o_1, \ldots, o_m]$ is started (again, in any order). If several instances of a task can be active at the same time, I and O are multisets, and application of a transition consists in closing as many instances of active tasks as specified in I and in opening as many activations of new tasks as specified in O.

task/2 atoms express the tasks that are allowed in the process. **transition/4** atoms express the allowed connections between activities in a very modular way. Transitions can be seen as 'consumers' of their input tasks, and 'producers' of their output tasks. In this perspective, the completion of an activity during a case can be seen as the production of a resource, that is to be consumed by some transition.

Compared to classical representations, in which the overall topology of the graph is fixed, this representation breaks the process models in several small pieces, that might in principle be recombined together in many ways (when different transitions have input multisets whose intersection is non-empty). To enforce irredundancy, WoMan exploits a number of additional information items. A fundamental one is the C_t parameter. First, and most important, it allows

---

[4] In the following, the extended part of the formalism with respect to [6] is marked by an asterisk '*'

[5] Note that this is a different meaning than in Petri Nets. A convenient notation for expressing transitions is

$$t : I \Rightarrow O \; [C_t]$$

where the C_t parameter can be omitted if irrelevant.

WoMan to check that all transitions involved in a new execution were all involved in the same (at least one) training case [4]. Second, it allows WoMan to compute the probability of a task or transition t, as the relative frequency $|C_t|/n$ where $n = |\mathcal{C}|$ is the number of training cases. This can be used for process simulation, for activity prediction and for noise handling (ignoring all tasks/transition in the model whose probability does not pass a specified noise threshold). Third, it allows WoMan to bound the number of repetitions of loops. Indeed, C_t is a multiset, because if a task or transition t was executed k times in case c, then C_t includes k occurrences of c. So, WoMan knows the maximum number of times that a task or transition can be executed in the same case.

Another limitation to the possible combinations of transitions is expressed using the following predicate:

* `transition_provider`$([\tau_1, \ldots, \tau_n], t, q)$: transition t, involving input tasks $I = [i_1, \ldots, i_n]$, is enabled provided that each task $i_k \in I, k = 1, \ldots, n$ was 'produced' as an output of transition τ_k, where the τ_k's are placeholders (variables) to be interpreted according to the Object Identity assumption ("terms (even variables) denoted with different symbols must be distinct (i.e., they must refer to different objects)"); several combinations can be allowed, numbered by progressive q, each encountered in cases C_{tq}.

that partitions the input multiset of a transition according to the producers of the activities to be consumed[6].

Additional constraints concern the agents that may run the activities:

– `task_agent`(t, A) : an agent, matching the roles A, can carry out task t.
– `transition_agent`$([A'_1, \ldots, A'_n], [A''_1, \ldots, A''_m], t, C_{tq}, q)$: transition t, involving input tasks $I = [i_1, \ldots, i_n]$ and output tasks $O = [o_1, \ldots, o_m]$, may occur provided that each task $i_k \in I, k = 1, \ldots, n$ is carried out by an agent

---

[6] Let us see this through an example. Consider a model that includes, among others, the following transitions:

$$t_1 : \{x, y, z\} \Rightarrow \{a, b\} \quad ; \quad t_2 : \{x, y\} \Rightarrow \{a\} \quad ; \quad t_3 : \{x\} \Rightarrow \{a, d\}$$

and suppose that the current set of activities to be 'consumed' is $\{x, y, z\}$. If an activity a is started, any of the above transitions might be the 'consumer'. Suppose that WoMan also knows the producers of these activities: $\{x/p22, y/p21, z/p22\}$, and that the model includes the following atoms related to transitions t_1, t_2 and t_3:

`transition_provider`$([\tau_1, \tau_1, \tau_2], t_1, 1)$.
`transition_provider`$([\tau_1, \tau_2, \tau_2], t_1, 2)$.
`transition_provider`$([\tau_1, \tau_2, \tau_1], t_1, 3)$.
`transition_provider`$([\tau_1, \tau_1], t_2, 1)$.
`transition_provider`$([\tau_1], t_3, 1)$.

In this case, transition t_2 is not a valid consumer, since it would require that both x and y were produced by the same transition τ_1, while they were actually produced by two different transitions ($p22$ and $p21$, respectively). Conversely, pattern #3 of transition t_1 is compliant with the available producers, which makes it an eligible candidate. Also transition t_3 is enabled.

matching roles A'_k, and that each task $o_j \in O, j = 1, \ldots, m$ is carried out by an agent matching roles A''_j; several combinations can be allowed, numbered by progressive q, each encountered in cases C_{tq}.

WoMan can handle taxonomies of agent roles. Each A'_k or A''_j is an expression in disjunctive normal form:

$$(r_{11} \wedge \ldots \wedge r_{1n_1}) \vee \ldots \vee (r_{m1} \wedge \ldots \wedge r_{mn_m})$$

where each r_{ij} is an individual or a role in the taxonomy, meaning that the agent must match all roles in at least one disjunct. The conjuncts are introduced to handle multiple inheritance. The generalization/specialization relationship is handled, in that a role is considered as matched by an agent if the agent matches any of its subclasses in the taxonomy. During the mining phase, generalizing means replacing one or more roles/instances with one of their superclasses.

The new version of the WoMan formalism added the following predicates to deal with time constraints:

* `task_time`$(t, [b', b''], [e', e''], d)$: task t must begin at a time $i_b \in [b', b'']$ and end at a time $i_e \in [e', e'']$, and has average duration d;
* `transition_time`$(t, [b', b''], [e', e''], g, d)$: transition t must begin at a time $i_b \in [b', b'']$ and end at a time $i_e \in [e', e'']$; it has average duration d (from the beginning of the first activity in I to the end of the last activity in O), and requires an average time gap g between the end of the last input task in I and the activation of the first output task in O;
* `task_in_transition_time`$(t, p, [b', b''], [e', e''], d)$: task t, when run in transition p, must begin at a time $i_b \in [b', b'']$ and end at a time $i_e \in [e', e'']$, and has average duration d;

where i_b, b', b'', i_e, e', and e'' are relative to the start of the process execution, i.e. they are computed as the timestamp difference between the **begin_process** event and the event they refer to.

In addition to the exact timestamp of events, WoMan internally associates each activity in a case to a unique integer identifier, called *step*, assigned by progressive start timestamp. So, the above constraints may be expressed also in terms of steps, as follows:

* `task_step`$(t, [b', b''], [e', e''], d)$: task t must start at a step $s_b \in [b', b'']$ and end at a step $s_e \in [e', e'']$, along an average number of steps d;
* `transition_step`$(t, [b', b''], [e', e''], g, d)$: transition t must start at a step $s_b \in [b', b'']$ and end at a step $s_e \in [e', e'']$; it takes place along an average number of steps d (from the step of the first activity in I to the step of the last activity in O), and requires an average gap of g steps between the end of the last input task in I and the beginning of the first output task in O;
* `task_in_transition_step`$(t, p, [b', b''], [e', e''], d)$: task t, when run in transition p, must start at a step $s_b \in [b', b'']$ and end at a step $s_e \in [e', e'']$, along an average number of steps d;

Finally, WoMan can expresses pre- and post-conditions for tasks (in general), transitions, and tasks in the context of a specific transition. Specifically, conditions on transitions define when a transition may take place; task conditions define what must be true for a given task in general, task in transition conditions define further constraints for allowing a task to be run in the context of a specific transition (provided that its general conditions are met). They are defined as FOL rules of the following form:

- $act\_T(A,S,R)$:- ... meaning that "activity A, of type T, can be run by agent R at step S of a case execution provided that ...";
- $trans\_T(S)$:- ... meaning that "transition T can be run at step S of a case execution provided that ...";
- $act\_T\_in\_trans\_P(A,S,R)$:- ... meaning that "activity A, of type T, can be run by agent R in the context of transition P at step S of a case execution provided that ...";

where the premises '...' are conjunctions of atoms based on contextual and control flow information. Conditions are not limited to the current status of execution. They may involve the status at several steps using two predicates:

- $activity(s,t)$: at step s (unique identifier) t is executed;
- $after(s',s'',[n',n''],[m',m''])$: step s'' follows step s' after a number of steps ranging between n' and n'' and after a time ranging between m' and m''.

Due to concurrency, predicate $after/3$ induces a partial ordering on the set of steps. The difference between pre- and post- conditions is that premises in the former refer only to steps up to S, while in the latter they may refer to any step, both before and after S.

3 Workflow Exploitation: Supervision and Prediction

The previous section has already pointed out that different transitions may be composed in different ways with each other, and that, as a consequence, many transitions may be eligible for application at any moment, and when a new activity takes place there may be some ambiguity about which one is actually being fired. This is clear from the example in footnote 6, where each of the two proposed options would change in a different way the status of the process, as follows: firing t_1 would consume x, y and z, leaving no activity to be consumed and causing the system to wait for a later activation of b, 'produced' by t_1; firing t_2 is inhibited, because the transition providers do not match the required pattern of variables (if it were enabled, firing it would consume x and y, leaving $\{z/p22\}$ to be consumed and causing the completion of transition t_2); firing t_3 would consume x, leaving $\{y/p21, z/p22\}$ to be consumed and causing WoMan to wait for a later activation of d, 'produced' by t_3. Another example, taken from [7], is the following (for the sake of simplicity, here we will assume that the constraints on the producers are all fulfilled). Consider a model that includes, among others, the following transitions:

$$t_1 : \{x\} \Rightarrow \{a, b\} \quad ; \quad t_2 : \{x, y\} \Rightarrow \{a\} \quad ; \quad t_3 : \{w\} \Rightarrow \{d, a\}$$

Suppose that the current set of activities to be 'consumed' is $\{x, y, z\}$, and that activity a is started. It might indicate that either transition t_1 or transition t_2 have been fired. Also, if an activity d is currently being executed due to transition t_3, the current execution of a might correspond to the other output activity of transition t_3, which we are still waiting to happen to complete that transition. Each of the these options would change in a different way the process evolution, as follows: firing t_1 would consume x, leaving $\{y, z\}$ to be consumed and causing the system to wait for a later activation of b; firing t_2 would consume x and y, causing the completion of transition t_2 and leaving $\{z\}$; firing t_3 would not consume any element in the marking, but would cause the completion of transition t_3.

We call each of these alternatives a *status*. This ambiguity about different statuses that are compliant with a model at a given time of process enactment must be properly handled when supervising the process enactment. Since it can be resolved only at a later time, all the corresponding alternate evolutions of the status must be carried on by the system, and each new event must be handled with respect to each alternate status. Then, in some cases, the same ambiguity issues will arise, and more alternate evolutions will be generated; in other cases, the new event will point out that some current alternate statuses were wrong, and will cause them to be cancelled. So, as long as the process enactment proceeds, the set of alternate statuses that are compliant with the activities carried out so far can be both expanded with new branches, and pruned of all alternatives that become incompatible with the activities carried out so far.

This procedure is carried out by WoMan's supervision module, **WEST** (Work-flow Enactment Supervisor and Trainer). As explained in [7], to handle this ambiguity, WEST maintains the set S of alternate statuses. Given a status $S \in \mathcal{S}$ and a new event e, the compliance check of the latter to the former checks may yield 3 possible outcomes:

ok : e is compliant with S;

error : there is a syntactic inconsistency in e (e.g., the termination of an activity that was never started, or the completion of a case while activities are still running); or

warning : indicating a deviation from the model that does not violate syntactic constraints; more specifically, the following types of warnings are available:

1. the pre-/post-conditions of a task, a transition or a task in the context of a transition are not fulfilled;
2. unexpected agent running a certain activity, in general or in the context of a specific transition;
3. a known task or transition, not expected at the current point of process execution, was run;
4. a new task or transition was run;

5. a task or transition was run more times than expected;
6. a task, transition or task in the context of a specific transition started or ended out of the expected time or step bounds.

Each warning carries a different degree of severity, expressed as a numeric weight. E.g., an unexpected task or transition also implies that the agent that runs it was not expected, and so has a greater severity degree than the unexpected agent alone. The degrees of severity currently embedded in WoMan for each type of warning were heuristically determined (a discussion and experimentation on how to determine these weights in order to improve performance is outside the scope of this paper).

Each status in $\mathcal{S}$ is represented as a 5-tuple $\langle M, R, C, T, W \rangle$ recording the following information:

M the 'Marking', i.e., the multiset of produced (i.e., terminated) activities associated with their provider identifier, not yet consumed (i.e. used to fire a transition);

R (for 'Ready') the multiset of activities that have not been started yet, but that are expected because they are in the output of some transition that was fired but not yet completed, each associated to the identifier of transition that produces it;

C the set of training cases that are compliant with that status;

T the sequence of (hypothesized) transitions that have been fired to reach that status;

W the multiset of warnings raised by the various events that led to this status (of course, each status may have a different multiset of warnings).

Algorithm 1 shows how the set of statuses is maintained as a consequence of the compliance check of a new event.

Given a status $S \in \mathcal{S}$, the set of candidate activities to be expected next in the case is made up of the activities in the 'Ready' component of that status and the activities that are reported in the output component of any transition that is enabled in that status. So, in principle, each status may expect a different set of activities to be carried out next. This ambiguity is more than in classical, graph-based, process model formalisms. Indeed, in those formalism, one always knows at which point of the graph the status of the current execution is located (e.g., based on the marking in Petri Nets), and thus which are the enabled tasks. On one hand, this increased ambiguity makes the activity prediction task harder, but, on the other hand, the availability of several alternate statuses, and of the associated information, allows WoMan to compute more refined statistics and to perform more elaborate reasoning to support activity prediction.

The module of WoMan, that is in charge of activity prediction, is **SNAP** (Suggester of Next Activity in Process). It exploits $\mathcal{S}$ (maintained by WEST) to determine which are the expected next activities and to rank them by some sort of likelihood, according to Algorithm 2. Specifically, components M and R of the statuses in $\mathcal{S}$ are used to determine the candidate activities, and components C and W are used to rank them by likelihood. Indeed, each status $S \in \mathcal{S}$ may have

---

Algorithm 1 Maintenance of the structure recording valid statuses in WEST

Require: $\mathcal{M}$: process model
Require: $\mathcal{S}$: set of currently compliant statuses compatible with the case
Require: $Running$: set of currently running activities
Require: $Transitions$: list of transitions actually carried out so far
Require: $\langle T, E, W, P, A, O, R \rangle$: log entry
  if $E = $ begin_activity **then**
    $Running \leftarrow Running \cup \{A\}$
    for all $S = \langle M, R, C, T, P \rangle \in \mathcal{S}$ **do**
      $\mathcal{S} \leftarrow \mathcal{S} \setminus \{S\}$
      if $A \in R$ **then**
        $\mathcal{S} \leftarrow \mathcal{S} \cup \{\langle M, R \setminus \{A\}, C, T, P \rangle\}$
      for all $p : I \Rightarrow O\ [C_p] \in \mathcal{M} \ni' A \in O$ **do**
        if $\exists$**transition_provider**$(Q, p, q) \in \mathcal{M} : $matches$(Q, M)$ **then**
          $P' \leftarrow P \cup P_{A,p,S}$ /* $P_{A,p,S}$ warnings raised by running A in p given S */
          $R' = \{t'/p \mid t' \in O \wedge t' \neq A\} \cup R$
          $\mathcal{S} \leftarrow \mathcal{S} \cup \{\langle M \setminus I, R', C \cap C_p, T\&\langle p \rangle, P' \rangle\}$
  if $E = $ end_activity **then**
    if $A \notin Running$ **then**
      Error
    else
      $Running \leftarrow Running \setminus \{A\}$
      for all $S = \langle M, R, C, T, P \rangle \in \mathcal{S}$ **do**
        select transition $t : I \Rightarrow O \in T$ that produced A
        $S \leftarrow \langle M \cup \{A/t\}, R, C, T, P \rangle$
      if a transition t has been fully carried out **then**
        $Transitions \leftarrow Transitions \ \& \langle t \rangle$
        for all $S = \langle M, R, C, T, P \rangle \in \mathcal{S}$ **do**
          if $T \neq Transitions$ **then**
            $\mathcal{S} \leftarrow \mathcal{S} \setminus \{S\}$

---

where matches(Q, M) checks that provider constraint Q is fulfilled by marking M.

---

a different set of warnings and of compliant training cases, and this variability can be exploited to rank the next activities that are expected in the process execution by likelihood. The algorithm is organized in phases. In the first phase, all evolutions $\mathcal{S}'$ of the current set of statuses $\mathcal{S}$ that are compliant with the new event are computed. Then, the multiset N of all candidate activities to be performed next are collected from the 'Ready' component of the evolved statuses. To make the prediction more selective, only those statuses whose overall weight of warnings does not pass a given threshold can be considered. In the third step, each candidate activity is associated with a score that represents an estimation of its likelihood, computed based on the following parameters:

- number of occurrences of that activity in the computed evolutions (activities that appear more often are more likely to be carried out next);
- number of cases supporting that activity in the computed evolutions (activities expected in the statuses supported by more training cases are more likely to be carried out next); and

---

Algorithm 2 Activity Prediction in WoMan using SNAP

Require: $\mathcal{M}$: process model
Require: $\mathcal{S}$: set of currently compliant statuses compatible with the case
Require: E : current event of trace
Require: ϵ: threshold to filter only more compliant statuses
   if $E = \text{end\_activity} \vee E = \text{begin\_process}$ **then**
      $\mathcal{S}' \leftarrow \emptyset$
      for all $S = \langle M, R, C, T, P \rangle \in \mathcal{S}$ **do**
         for all $p : I \Rightarrow O\ [C_p] \in \mathcal{M}$ **do**
            if $\exists \texttt{transition\_provider}(Q,p,q) \in \mathcal{M} : \text{matches}(Q,M)$ **then**
               $P' \leftarrow P \cup P_{p,S}$ /* $P_{p,S}$ warnings raised by firing p given S */
               $\mathcal{S}' \leftarrow \mathcal{S}' \cup \{\langle M \setminus I, R \cup O, C \cap C_p, T \& \langle t \rangle, P' \rangle\}$
   if $E = \text{begin\_activity}$ **then**
      $\mathcal{S}' \leftarrow \mathcal{S}$
   $N = \{a \mid \exists S = \langle M, R, C, T, P \rangle \in \mathcal{S}' : \delta(S) < \epsilon \wedge a \in R\}$ /* multiset of candidate
   next activities */
   $Ranking \leftarrow \{\}$
   for all $a \in N$ **do**
      $\mathcal{S}_a = \{\langle M, R, C, T, P \rangle \in \mathcal{S}' \mid a \in R\}$
      $\delta_a = \sum_{S \in \mathcal{S}_a} \delta(S)$ /* overall discrepancy of all statuses involving a */
      $C_a = \bigcup_{\langle M, R, C, T, P \rangle \in \mathcal{S}_a} C$ /* overall set of cases supporting the execution of a in
   all statuses */
      $score \leftarrow (|C_a| \cdot |\mathcal{S}_a| \cdot |a|_N) / \delta_a$
      $Ranking \leftarrow Ranking \cup \{\langle score, a \rangle\}$
Ensure: $Ranking$

---

where:

- matches(Q, M) checks that provider constraint Q is fulfilled by marking M.
- $|\cdot|$ denotes the cardinality of a set or multiset;
- $|\cdot|_M$ denotes the number of occurrences of an element in a multiset M;
- $\delta(\cdot)$ is the discrepancy of a status, computed as the sum of the weights of the warnings raised by the status.

- overall discrepancy of the statuses that expect that activity (activities expected in statuses that raised less warnings are more likely to be carried out next).

The final prediction is obtained by ranking the candidate activities by decreasing score (higher positions indicating more likelihood).

4 Evaluation

The performance of the proposed activity prediction approach was evaluated on several datasets, concerning different kinds of processes associated with different kinds and levels of complexity. The datasets related to Ambient Intelligence concern typical user behavior. Thus, they involve much more variability and

Table 1. Dataset statistics

	cases	avg events	avg activities	tasks	transitions
Aruba	220	62.67	30.34	10	92
GPItaly	253	734.56	366.28	8	79
White	158	232.71	115.35	681	4083
Black	87	243.01	120.51	663	3006
Draw	155	209.17	103.59	658	3434

subjectivity than in industrial process, and there is no 'correct' underlying model, just some kind of 'typicality' can be expected:

Aruba from the CASAS benchmark repository[7]. It includes continuous recordings of home activities of an elderly person, visited from time to time by her children, in a time span of 220 days. Each day is mapped to a case of the process representing the daily routine of the elderly person. Transitions correspond to terminating some activities and starting new activities. The resources (persons) that perform activities are unknown.

GPItaly from one of the Italian use cases of the GiraffPlus project[8] [3]. It concerns the movements of an elderly person (and occasionally other people) in the various rooms of her home along 253 days. Each day is a case of the process representing the typical movements of people in the home. Tasks correspond to rooms; transitions correspond to leaving a room and entering another.

The other concerns chess playing, where again the 'correct' model is not available:

Chess from the Italian Chess Federation website[9]. 400 reports of actual top-level matches were downloaded. Each match is a case, belonging to one of 3 processes associated to the possible match outcomes: *white* wins, *black* wins, or *draw*. A task is the occupation of a square by a specific kind of piece (e.g., "black rook in a8"). Transitions correspond to moves: each move of a player terminates some activities (since it moves pieces away from the squares they currently occupy) and starts new activities (that is, the occupation by pieces of their destination squares). The involved resources are the two players: 'white' and 'black'.

Table 1 reports statistics on the experimental datasets: number of cases, average number of events and activities per case, number of tasks and transitions in a model learned using the whole dataset as a training set. There are more cases for the Ambient Intelligence datasets than for the chess ones. However, the chess datasets involve many more different tasks and transitions, many of which are rare or even unique. The datasets are different also from a qualitative viewpoint. Aruba cases feature many short loops and some concurrency (involving up to 2

---

[7] http://ailab.wsu.edu/casas/datasets.html
[8] http://www.giraffplus.eu
[9] http://scacchi.qnet.it

Table 2. Activity prediction statistics

	Pred	Recall	Rank	Tasks	1st	Quality
Aruba	0.88	0.97	0.86	6.3	0.84	0.78
GPItaly	1.0	0.99	0.98	8.2	0.88	0.97
black	0.53	0.98	1.0	11.09	0.91	0.51
white	0.55	0.98	1.0	10.9	0.91	0.5
draw	0.65	0.98	1.0	10.6	0.91	0.64
chess	0.58	0.98	1.0	10.9	0.91	0.55

activities), optional and duplicated activities. The same holds for GPItaly, except for concurrency. The chess datasets are characterized by very high concurrency: each game starts with 32 concurrent activities (a number which is beyond the reach of many current process mining systems [5]). This number progressively decreases (but remains still high) as long as the game proceeds. Short and nested loops, optional and duplicated tasks are present as well. The number of agent and temporal constraints is not shown, since the former is at least equal, and the latter is exactly equal, to the number of tasks and transitions.

The experimental procedure was as follows. First, each dataset was translated from its original representation to the input format of WoMan. Then, a 10-fold cross-validation procedure was run for each dataset, using the learning functionality of WoMan (see [4]) to learn models for all training sets. Finally, each model was used as a reference to call WEST and SNAP on each event in the test sets: the former checked compliance of the new event and suitably updated the set of statuses associated to the current case, while the latter used the resulting set of statuses to make a prediction about the next activity that is expected in that case (as described in the previous section).

Table 2 reports average performance, on the measures reported in the columns, for the processes on the rows ('chess' refers to the average of the chess sub-datasets)[10]. *Pred* is the ratio of cases in which SNAP returned a prediction. Indeed, when unknown tasks or transitions are executed in the current enactment, it assumes a new kind of process is enacted, and avoids making predictions. *Recall* is the ratio of cases in which the correct activity (i.e., the activity that is actually carried out next) is present in the ranking, among those in which a prediction was made. *Rank* reports its position in the ranking, normalized into $[0, 1]$ (1 meaning it is the first, and 0 meaning it is the last), *Tasks* is the average length of the ranking (the shorter, the better), and *1st* is the lower bound of the Rank interval associated to the first activity (*Rank > 1st* means that the activity is the first on average). $Quality = Pred \cdot Recall \cdot Rank \in [0, 1]$ is a compound index that provides an immediate indication of the overall activity prediction performance. When it is 0, it means that predictions are completely unreliable; when it is 1, it means that WoMan always makes a prediction, and that such a prediction is correct (i.e., the correct activity is at the top of the ranking).

---

[10] This can be considered as a baseline: fine-tuning the weights for the different kinds of warnings might result in even better performance.

As expected, the number of predictions is proportional to the number of tasks and transitions in the model. Indeed, the more variability in behaviors, the more likely it is that the test sets contain behaviors that were not present in the training sets. WoMan is almost always able to make a prediction in the Ambient Intelligence domain, which is extremely important in order to provide continuous support to the users. While much lower, the percentage of predictions in the chess domain still covers more than half of the match, the worst performance being on 'black' (the one with less training cases). In all cases, when WoMan makes a prediction, it is extremely reliable: the correct next activity is almost always present in the ranking (97-99% of the times). For chess processes, this means that WoMan is able to distinguish cases in which it can make an extremely reliable prediction from cases in which it prefers not to make a prediction at all. Also, the correct activity is always in the range associated with the top place. For the chess processes, in particular, it is always at the very top.

In conclusion, the experimental outcomes confirm that WoMan is effective on processes having very different degrees and kinds of complexity. In the Ambient Intelligence domain, this means that it may be worth spending some effort to prepare the environment in order to facilitate that activity, or to provide the user with suitable support for that activity. In the chess domain, our figures are better than those of other attempts purposely devised to apply Machine Learning to make the machine able to play autonomously.

5   Conclusions

In addition to the classical exploitation of process models for checking process enactment conformance, a very relevant but almost neglected task concerns the prediction of which activities will be carried out next at a given moment during process execution. The outcomes of this task may allow to save time and money by taking suitable actions that facilitate the execution of those activities, may support more fundamental and critical tasks involved in automatic process management, and may provide indirect indications on the correctness and reliability of a process model. This paper proposed an extended formalism and approach to make these kinds of predictions using the WoMan framework for workflow management. Experimental results on different domains show very good prediction performance, also on quite complex processes. This makes us confident that it can be successfully applied to industrial domains, as well.

Given the positive results, we plan to carry out further work on this topic. First of all, we plan to check the prediction performance on other domains, e.g. Industry 4.0 ones. Also, we will investigate how to further improve the prediction accuracy by means of more refined strategies. Finally, we would like to embed the prediction module in other applications, in order to guide their behavior.

Acknowledgments

This work was partially funded by the Italian PON 2007-2013 project PON02_00563_3489339 'Puglia@Service'.

References

1. R. Agrawal, D. Gunopulos, and F. Leymann. Mining process models from workflow logs. In *Proceedings of the 6th International Conference on Extending Database Technology (EDBT)*, 1998.
2. J.E. Cook and A.L. Wolf. Discovering models of software processes from event-based data. Technical Report CU-CS-819-96, Department of Computer Science, University of Colorado, 1996.
3. S. Coradeschi, A. Cesta, G. Cortellessa, L. Coraci, J. Gonzalez, L. Karlsson, F. Furfari, A. Loutfi, A. Orlandini, F. Palumbo, F. Pecora, S. von Rump, Štimec, J. Ullberg, and B. tslund. Giraffplus: Combining social interaction and long term monitoring for promoting independent living. In *Proc. of the 6th International Conference on Human System Interaction (HSI)*, pages 578–585. IEEE, 2013.
4. S. Ferilli. WoMan: Logic-based workflow learning and management. *IEEE Transaction on Systems, Man and Cybernetics: Systems*, 44:744–756, 2014.
5. S. Ferilli and F. Esposito. A logic framework for incremental learning of process models. *Fundamenta Informaticae*, 128:413–443, 2013.
6. Stefano Ferilli. The woman formalism for expressing process models. In *Advances in Data Mining*, volume 9728 of *Lecture Notes in Artificial Intelligence*, pages 363–378, 2016.
7. Stefano Ferilli, Floriana Esposito, Domenico Redavid, and Sergio Angelastro. Predicting process behavior in woman. In *AI*IA 2016: Advances in Artificial Intelligence*, volume 10037 of *Lecture Notes in Artificial Intelligence*, pages 308–320, 2016.
8. J. Herbst and D. Karagiannis. An inductive approach to the acquisition and adaptation of workflow models. In *Proceedings of the IJCAI'99 Workshop on Intelligent Workflow and Process Management: The New Frontier for AI in Business*, pages 52–57, 1999.
9. IEEE Task Force on Process Mining. Process mining manifesto. In *Business Process Management Workshops*, volume 99 of *Lecture Notes in Business Information Processing*, pages 169–194. 2012.
10. S. Muggleton. Inductive logic programming. *New Generation Computing*, 8(4):295–318, 1991.
11. M. Pesic and W. M. P. van der Aalst. A declarative approach for flexible business processes management. In *Proceedings of the 2006 international conference on Business Process Management Workshops*, BPM'06, pages 169–180. Springer-Verlag, 2006.
12. W.M.P. van der Aalst. The application of petri nets to workflow management. *The Journal of Circuits, Systems and Computers*, 8:21–66, 1998.
13. W.M.P. van der Aalst, T. Weijters, and L. Maruster. Workflow mining: Discovering process models from event logs. *IEEE Trans. Knowl. Data Eng.*, 16:1128–1142, 2004.
14. A.J.M.M. Weijters and W.M.P. van der Aalst. Rediscovering workflow models from event-based data. In *Proc. 11th Dutch-Belgian Conference of Machine Learning (Benelearn 2001)*, pages 93–100, 2001.

Hierarchical Text Classification of Autopsy Reports to Determine MoD and CoD through Term-Based and Concepts-Based Features

Ghulam Mujtaba[1,3], Liyana Shuib[1], Ram Gopal Raj[1], Mohammed Ali Al-Garadi[1], Retnagowri Rajandram[2], Khairunisa Shaikh[2]

[1]Faculty of Computer Science and Information Technology, University of Malaya, Kuala Lumpur, Malaysia
[2]Faculty of Medicine, University of Malaya, Kuala Lumpur, Malaysia
[3]Department of Computer Science, Sukkur Institute of Business Administration, Sukkur, Pakistan
{mujtaba, mohammedali,khairunisashaikh}@siswa.um.edu.my
{liyanashuib,ramdr,rretnagowri}@um.edu.my

Abstract. Nowadays, text classification has been extensively employed in medical domain to classify free text clinical reports. In this study, text classification techniques have been used to determine cause of death from free text forensic autopsy reports using proposed term-based and SNOMED CT concept-based features. In this study, detailed term-based features and concept-based features were extracted from a set of 1500 forensic autopsy reports belonging to four manners of death and 16 different causes of death. These features were used to train text classifier. The classifier was deployed in cascade architecture: the first level will predict the manner of death and the second level will predict the CoD using proposed term-based and SNOMED CT concept-based features. Moreover, to show the significance of our proposed approach, we compared the results of our proposed approach with four state-of-the-art feature extraction approaches. Finally, we also presented the comparison of one-level classification versus two-level classification. The experimental results showed that our proposed approach showed 8% improvement in accuracy as compared to other four baselines. Moreover, two-level classification showed improved accuracy in determining CoD compared to one-level classification.

Keywords: Text Classification, Feature Extraction, Term-Based Features, Concepts-Based Features, Hierarchical Classification, Forensic Autopsy Reports, Postmortem Reports

© Springer International Publishing AG 2017
P. Perner (Ed.): ICDM 2017, LNAI 10357, pp. 209–222, 2017.
DOI: 10.1007/978-3-319-62701-4_16

1   Introduction

Text classification is the process of automatically predicting one or more predefined classes that are relevant to particular text document [1, 2]. The classification of text document is typically carried out with the help of supervised machine learning algorithms. These algorithms typically require the text input to be represented as a fixed-length vector [1, 2]. Hence, to convert textual document into such fixed-length vectors, various feature extraction and feature representation techniques are applied. The feature extraction technique typically captures very informative category patterns from set of labeled corpus that can be used to classify the new or unlabeled text documents [3]. Examples of popular feature extraction techniques include Bag of Words (BoW) and n-gram. In fixed-length vector, each row represents each text document in the collected corpus, whereas, each column represents one feature that was extracted during feature extraction process. The actual values in the vector are actually the feature weights. These feature weights are assigned using feature representation techniques such as binary representation, term frequency with inverse document frequency, etc. Once, this fixed-length vector is prepared then this will serve as input to text classification algorithm such as support vector machine (SVM), decision tree (J48), random forest (RF), etc. These algorithms learn the classification patterns from the fixed-length vector, and construct text classification model. These trained models can be used in future to predict the predefined category of the new and unlabeled text document.

Text classification is widely used in many application areas such as email classification, classification of web documents [4], sentiment analysis [5], and the classification of medical documents [6, 7]. Here, we have applied text classification to determine the manner of death (MoD) and cause of death (CoD) from forensic autopsy reports. Forensic autopsy also known as postmortem examination is the morphological and anatomical examination of the deceased to ascertain the MoD and CoD. There are four MoDs such as accident, suicide, homicide, and natural death. Each of these MoDs has many related CoDs. For instance, the accident related CoDs include, death due to Craniocerebral injury (ICD-10 code is S06), death due to abdominal injury (ICD-10 code is S38), etc. In forensic autopsy, pathologists collect various findings from dead body, death scene, deceased history, police, witnesses and relatives to determine accurate MoD and CoD. These autopsy examination findings are recorded into a report, called forensic autopsy report. The initial autopsy report is prepared in 2 to 3 days. However, it may take more than a month to prepare final report [8]. Hence, the process of determining the MoD and CoD from postmortem findings is resource intensive and time consuming. Therefore, automatic classification of MoD and CoD using text classification techniques can be helpful to reduce the time and resources.

Many recent studies have used text classification to classify medical reports [9-13]. For instance, Koopman et.al [10] used text classification to predict CoD from death certificates. However, there are very few studies in which text classification is used to predict MoD and CoD from autopsy reports. For instance, Mujtaba et.al [6] employed conventional text classification techniques to determine the CoD from collected corpus of free text forensic autopsy reports. Here, authors obtained the highest accuracy of 78% using unigram feature extraction technique and SVM text classification technique.

Hence, this accuracy may not be fair enough to deploy such models in real-time systems. Hence, more accurate models are required. In another study, Mujtaba et.al [7] proposed a semi-automated expert-driven feature engineering approach to automatically classify CoD from autopsy reports. Here, authors obtained the highest accuracy of 90%. However, the limitation of this study is that the features were engineered with the help of experts and hence, the technique is highly dependent on the efforts of experts and consequently it requires time and efforts to engineer such learning vectors. These limitations motivated us to contribute in the same area.

In this paper, we used hierarchical classification to determine the CoD from autopsy reports. In the first level classification MoD is predicted, whereas, in the second level classification CoD related to predicted MoD is determined. We also showed experimental results to show the effectiveness of two-level classification as compared to one-level classification. Moreover, this study also proposed the use of term-based and concept-based features to determine the CoD from plaintext autopsy reports. The term-based features were extracted from the text of autopsy reports, whereas, the concept-based features were derived from the term-based features using SNOMED CT API. Moreover, three different text classifiers; support vector machine (SVM), decision tree (J48), and random forest (RF), were used to obtain the best classification results. Finally, the results obtained by proposed features are compared with four state-of-the-art feature extraction techniques to show its significance. These four techniques are; bag of words (BoW) [14], entropy optimized feature-based bag of words (EO-BoW) [15], paragraph vector [16], and hybrid of BoW and skip-grams [17]. Compared to these four state-of-the-art feature extraction techniques, our proposed approach obtained a relative improvement of more than 8% in terms of accuracy.

This paper is organized in the following manner. In Section 2, the related work is presented. In Section 3, the methodology of this work is described. This section also includes the dataset description, preprocessing of data, feature extraction, classification task, and the experimental setup. Section 4 presents the results and discussion. Finally, Section 5 concludes this work.

2   Related Work

In recent years, text classification has widely been employed in medical field [6, 7, 9, 10, 12, 18-22] to categorize clinical reports into predefined categories. In such studies, authors typically apply text classification techniques, or propose novel features, or novel feature extraction techniques, or novel text classification techniques to classify free text clinical reports. For instance, Jouhet et.al [18] applied SVM and naïve bayes classifiers to develop a machine learning model to classify plain text pathology reports. The authors reported that the use of text classification techniques is appropriate for classifying pathological reports. Danso et.al [19] proposed the use of linguistics and statistical features to predict the CoD from verbal autopsy reports. Authors concluded that the use of combined statistical and linguistic features extracted from verbal autopsy reports can significantly improve the accuracy of text classifier in determining the CoD. Danso et.al [20] compared various text classifier algorithms, and feature representation techniques for determining CoD using verbal autopsy reports. Authors reported that SVM obtained the most promising result in text classification algorithms and term frequency with inverse document frequency (TFiDF) and normalized TFiDF (NTFiDF) were proven better feature representation schemes. From above studies, it can be concluded that the text classification is viable solution to automatically classify the clinical reports.

Perhaps, the most related works to our proposed work is that of [6, 7, 9, 10]. Yeow et.al [9] employed case based reasoning (CBR) together with naïve bayes classifier to predict CoD from forensic autopsy reports and obtained 80% accuracy. Experimental results showed the feasibility of proposed system to predict CoD. However, there are two major limitations highlighted in the study [9]. First, the authors did not use the complete and detailed reports for CoD prediction. They just used a summary of whole autopsy reports for feature extraction and prediction. Second, various pathologists may use different set of vocabulary while reporting the autopsy findings. Hence, author did not consider the issue of synonymy and polysemy to derive similar concepts from the base features. Another, most relevant work to our proposed work is that of [6]. Mujtaba et.al [6] used the dataset of complete and detailed autopsy reports to determine CoD and obtained 78% accuracy. In their study, authors compared and applied different feature extraction techniques, term weighing techniques and classification techniques on autopsy reports to predict CoD. According to experimental results SVM outperformed other text classifiers. Moreover, unigram features outperformed bigram features and trigram features. Furthermore, term frequency and TFiDF term weighing techniques outperformed binary representation and NTFiDF representation. However, the weakness of [6] is that the authors only compared the old and conventional text classification, and feature extraction techniques. The new state-of-the-art feature extraction technique may produce better results.

In their recent study Mujtaba et.al [7] proposed semi-automated expert-driven feature engineering technique to predict CoD from postmortem reports. Here, authors used the features derived by experts and compared such features with features derived by state-of-the-art automated feature engineering techniques. According to experimental results, expert-driven feature engineering approach obtained the highest accuracy of 90% and outperformed other automated feature engineering techniques. The limitation of [7] is that in expert-driven approach, features were extracted and ranked by experts themselves. Therefore, such technique is highly dependent upon the experts and expert knowledge. Moreover, the authors reported that the expert-driven performed better because in many autopsy reports the different vocabulary words were used by different pathologist and they suggested considering the issue of synonymy and polysemy during feature engineering process. Therefore, there is a strong need to overcome the dependency of experts and to develop a technique that is fully automated, robust and produce the results like or better than the expert-driven approach. Such need motivated us to contribute in this domain. Hence, motivated by work proposed in [10], we proposed to extract term-based and concept–based features from autopsy reports and use these features to predict the MoD and CoD.

3   Materials and Methods

The complete flow of this research study is shown in Fig 1. Here, we have shown all the steps, which were taken to predict the MoD and CoD from autopsy reports. In subsequent sections, all steps have thoroughly been discussed.

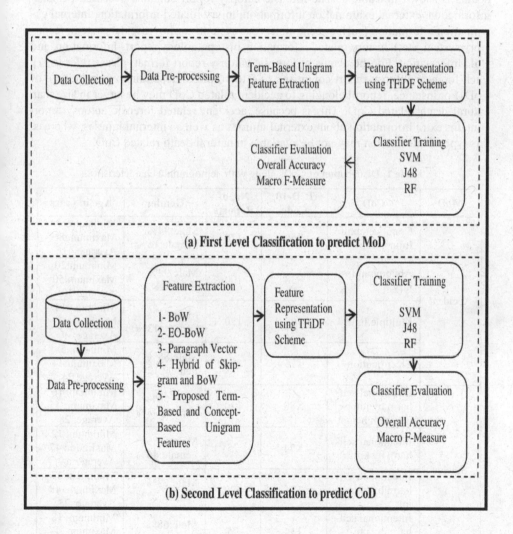

(a) First Level Classification to predict MoD

(b) Second Level Classification to predict CoD

Fig 1. Flowchart of research methodology

3.1   Data Collection

The experiments involved 1500 autopsy reports. These autopsy reports were collected from famous and largest hospital situated in Kuala Lumpur, Malaysia. These reports belong to four manners of death (MOD) i.e. accident, suicide, homicide, and natural. Of these 1500 reports, 450 reports belong to accident, 390 belong to suicide, 370 belong to homicide, and 300 belong to natural MoD. Each MoD contains the reports of four distinct cause of death (CoD), hence, the collected dataset comprised of 16 CoDs belonging to four MoDs. The detail distribution of CoDs along with some demographic details is shown in Table 1. The forensic autopsy report contains deceased personal information, external examination information, injury related information, internal examination information (such as central nervous system, cardio-vascular system, respiratory system, etc.), history related information, histopathology reports information, and CoD information. For the details of forensic autopsy report format, please refer [6, 7]. Each forensic autopsy report comprised of 5 to 10 pages depending upon the type of CoD. For instance, report belonging to accident related CoD may be larger in size than natural death related CoD. This is because, accident related forensic autopsy report contains extra information about external injuries as well as internal injuries, whereas, this type of information may not be available in natural death related CoD.

Table 1. Distribution of CoDs along with demographic characteristics.

MoD	CoD	ICD-10 Code	No. of Reports	Gender	Age in years
Accident	Craniocerebral Injury	S06	120	Male:84% Female:16%	Minimum:6 Maximum:86 Average:41
	Abdominal Injury	S38	100	Male:92% Female:8%	Minimum:20 Maximum:50 Average:30
	Multiple Injury	T07	120	Male:87% Female:13%	Minimum:14 Maximum:87 Average:39
	Electrocution	T75	100	Male:88% Female:12%	Minimum:5 Maximum:44 Average:24
Suicide	Intentional self-harm by jumping from height	X80	120	Male:66% Female:34%	Minimum: 16 Maximum: 45 Average: 28
	Intentional self-harm by knife	X74	75	Male:54% Female:46%	Minimum: 12 Maximum: 47 Average: 19
	Intentional self-harm by hanging	T71	120	Male:78% Female:22%	Minimum: 17 Maximum: 48 Average: 23
	Intentional self-harm by poisoning	T14	75	Male:68% Female:32%	Minimum: 18 Maximum: 43 Average: 24

MoD	CoD	ICD-10 Code	No. of Reports	Gender	Age in years
Homicide	Assault by handgun discharge	X93	75	Male:89% Female:11%	Minimum: 19 Maximum: 53 Average: 32
	Assault by sharp object	X99	110	Male:71% Female:29%	Minimum:17 Maximum: 59 Average: 33
	Assault by blunt object	Y00	110	Male:79% Female:21%	Minimum: 18 Maximum: 57 Average: 26
	Assault by un-specified means	Y09	75	Male:73% Female:27%	Minimum:13 Maximum:58 Average: 23
Natural	Acute Myocardial Infarction	I23	75	Male:63% Female:37%	Minimum:23 Maximum:64 Average:36
	Ischemic Heart Diseases	I24	75	Male:82% Female:18%	Minimum: 25 Maximum: 57 Average: 39
	Chronic Heart Disease	I25	75	Male:76% Female:24%	Minimum:27 Maximum:65 Average:37
	Pulmonary Tuberculosis	Z11	75	Male:83% Female:17%	Minimum: 26 Maximum: 69 Average: 34

3.2   Data Preprocessing

In pre-processing, four basic steps were performed. First, spell checker was used to correct all the misspelled words using PyEnchant and the NLTK library [34]. Second, the whole report was tokenized into sentences after converting it into lower case. Third, all the stopwords were removed from each sentence using the stop word list [35]. Finally, those sentences, which were common in all autopsy reports were removed because of their low discriminative power. For instance, the sentences "*gallbladder was intact, contained bile and wall was not thickened*" and "*pituitary, thyroid and right adrenal glands were unremarkable*" were repeating in almost all kinds of forensic autopsy reports. Hence, such sentences may not prove useful in the task of classification and hence, such sentences were removed.

3.3   Feature Extraction methods

After preprocessing phase, two types of features were extracted from collected autopsy reports. These are: term-based features and concept-based features. Term-based features were taken directly from the text of forensic autopsy reports. In term-based features, unigram features were extracted from forensic autopsy reports using term frequency with inverse document frequency (TFiDF) term weighing scheme. The unigram features with TFiDF were used because in their comparative study, Mujtaba et al. [6] found the suitability of unigram features represented by TFiDF or term frequency to determine CoD from autopsy reports.

The concept-based features were derived from original term-based feature, where concepts are related to standard medical lexicons (e.g., the SNOMEDCT ontology). For instance, the term "bleeding" may belong to various similar concepts such as "hemorrhage", "bleeding hemorrhoids", and "hemarthrosis". Each of these concepts has unique concept ids. For instance, the term "bleeding" has a concept id of 131148009, and other relevant concepts ids are 50960005, 51551000, and 81808003 respectively. The concept-based features were extracted because different autopsy reports were prepared by different pathologists. Hence, they may use different vocabulary while preparing the forensic autopsy reports. For instance, few pathologists have used the term bleeding and hemorrhage interchangeably. Hence, by adding the concept-based features, the feature set will be rich and it will resolve the issue of synonymy and polysemy.

Once all features were extracted, all forensic autopsy reports were transformed into master feature vector, where each row represents the autopsy report and each column represents a unique feature. The actual values in numeric vector are the TFiDF values. This master feature numeric vector is fed as an input to the text classifier.

3.4   Classification Methods

For the prediction of CoD from forensic autopsy reports, hierarchical classification method was used. In hierarchical classification task, the first level classifier was used to predict the MoD, and the second level classifier was used to predict the CoD. We translated this into the strategy of machine learning (1) a single multi-class classifier to predict the MoD; and (2) four multi-class classifiers for the prediction of CoDs. Of these four, one will be trained to predict the accident related CoDs, second will be trained to predict suicide related CoDs, third will be trained to predict homicide related CoDs, and fourth will be trained to predict natural death related CoD. Hence, the classifiers will be organized in two level, cascade architecture: an autopsy report is first processed by first level classifier which will predict MoD. For instance, if the first level classifier predicts the MoD 'accident' then 'accident' related second level classifier will process the forensic autopsy report to predict accident related CoD.

For the first level classification, all 1500 reports were labeled into four class labels i.e. accident, suicide, homicide, and natural. Afterwards, the features were extracted from these reports after applying preprocessing steps. In the first level classification, only the term-based unigram features were extracted and these features were represented by TFiDF feature representation scheme to create a master feature vector. This vector was fed as an input to the text classifier. Three text classifiers i.e. SVM, J48, and RF were used to predict MoD on the first classification level. For the second level classification task, reports were labeled into respective CoDs and proposed features were extracted as described in section 3.3 to construct master feature vector. For the prediction of CoD, SVM, J48, and RF classifiers were also used on second level classification tasks. The SVM, J48 and RF were used because Mujtaba et al. [7] compared five text classifiers for the prediction of CoD and found that RF, J48, and SVM showed the highest accuracy. Moreover, we also compared the efficiency of proposed term-based and concept-based features with four state-of-the-art feature extraction approaches such as BoW, EO-BoW, paragraph vector, and the integration of word2vec and bag of words to better understand the behavior of our proposed features.

3.5   Experimental Setup

In this study, we have carried out 3 analyses (1 master feature vector x 3 text classifiers) in first level classification to predict the MoD from forensic autopsy reports. Moreover, in second level classification, 60 analyses (4 datasets x 5 master feature vectors created from five different feature extraction techniques x 3 text classifiers) were performed to predict 16 different CoDs from four types of forensic autopsy reports (i.e. accident, suicide, homicide, and natural). Hence, in total 63 analyses were performed to predict MoD and CoD from autopsy reports. For feature extraction we used Python with NLTK library and Java with Weka and SNOMED CT library. All the classification experiments were performed in Weka Toolkit. All experiments were carried out using default parameters and 10-fold cross validation [23]. To evaluate the performance of all 63 analyses macro F-measure and the overall accuracy metrics were used. Macro F-measure were used due to imbalance class distribution in the dataset [24].

4   Results and Discussion

The classification results for the first level classification for the prediction of MoD are shown in Table 2. Table 2 shows the text classifier, its overall accuracy and F-measure$_M$. Here, it can be seen that the highest accuracy of 95.41% and 93.57% was obtained by SVM and RF classifier respectively. The J48 classifier obtained the lowest accuracy of 88.99%.

Table 2. Results of first level classification

Classifier	Overall Accuracy	F-Measure$_M$
SVM	95.41%	0.95
RF	93.57%	0.933
J48	88.99%	0.87

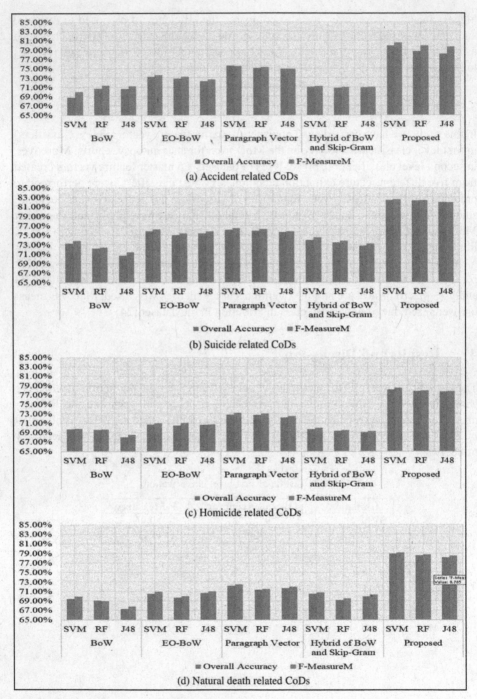

(a) Accident related CoDs

(b) Suicide related CoDs

(c) Homicide related CoDs

(d) Natural death related CoDs

Fig 2. Second level hierarchical results showing overall accuracy and F-measure$_M$ of feature extraction techniques and classification techniques

The second level classification results are shown in Fig 2. Here, it can be seen that the performance was superior in suicide related CoDs as compared to accident, homicide and natural death related CoDs. This is because, all four suicide related CoDs are quite different from one another, and hence are easier for classifier to predict. However, in accident, homicide, and natural death categories, the CoDs are very similar in nature and hence, the classification performance is lower than suicide related CoDs.

In second level classification results, it can be seen that the proposed term-based features and concept-based features obtained the highest classification results, followed by paragraph vector and EO-BoW. Moreover, the BoW and hybrid of BoW and Skip-gram obtained the lowest performance results. The possible reason behind the success of term-based and concept-based features may be the one that is various autopsy reports are prepared by different pathologists. These pathologists may use different vocabulary to prepare the autopsy reports. Hence, in proposed features, this issue is resolved by concept-based features. These concept-based features were derived from the basic term-based features to avoid the issue of synonymy and polysemy. There may be two possible reasons behind the lowest performance of BoW model. First, it does not consider the ordering of the words and second, it ignores word semantics [1-3]. The hybrid of skip-gram and BoW model also produced the lowest results. The possible reason may be because this model uses the voting approach between BoW and skip-gram model. Hence, the BoW results may affect the classification result of skip-gram. The paragraph vector and EO-BoW performed better than BoW and hybrid of BoW and skip-gram. This is because the paragraph vector learns vector representation for variable length paragraphs of text. The vector representation is learnt to predict the surrounding words in contexts sample from the paragraph. Hence, this method captures more semantics of paragraph than skip-gram, Bow, or EO-BoW.

In text classifiers SVM outperformed RF and J48. The possible reason for the out-performance of SVM may be that SVM can produce better results with all available features in the master feature vector because SVMs are not prone to over-fitting [25]. There was a very slight difference between the results of SVM and RF. The possible reason may be that the prediction results of RF can be affected by having a huge number of non-discriminative features but having very few discriminative features. Such combination of features may result in the generation of trees containing less powerful and redundant features in forest. Hence, such tree in forest may generate the incorrect classification results[26]. In most of the cases, J48 classifier performed lower than SVM and RF. However, in few cases, its performance was better than RF. The possible reason behind the lowest performance of J48 may be that the all attributes in master feature vector represent the continues data that makes it hard to find the optimal thresholds needed to construct the J48 decision tree [27]. Hence, to optimize the overall accuracy, and F-measure$_M$, SVM decision model built with proposed term-based and concept-based features is recommended.

4.1   Effect of MoD Classifier

This two level classification method (first to predict MoD, and then to predict CoD) was designed purposefully to enhance the CoD prediction effectiveness. To prove its effectiveness, one-level classification and two-level classification was compared. In both, one-level and two-level classifications, SVM classification model with proposed term-based and concept-based features were used. Fig 3 shows the results obtained from both one-level and two-level classification model. As can be seen here, in many cases two level classification model was very useful in reducing the number of false positives. However, in few cases such as Z11, S06, T71, S38, T14 and T71 the difference is very minor. This may be because these CoDs are very different from other CoDs. For instance, T71 (self-harm through hanging) and T14 (self-harm through poising) are quite different from all other CoDs such as X80, X74, etc. Hence, it is easier for classifier to predict such CoDs accurately.

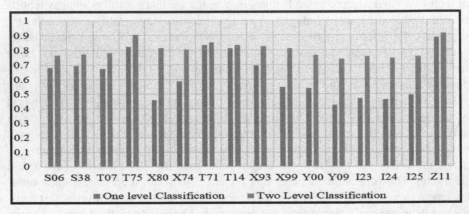

Fig 3. One-level versus two-level classification

5   Conclusion

In this study, an effective approach of text classification was proposed to determine the MoD and CoD from forensic autopsy reports by using term-based and concept-based features. In the proposed system a hierarchical classification was used to determine MoD and CoD. The first level classification was used to determine the MoD and the second level classification was used to determine the CoD. On first level classification, 95% accuracy was obtained using SVM text classifier, whereas, on second level classification averagely 80.15% accuracy was obtained, which is approximately 8% more than the four compared baselines. The performance of one-level classification and two-level hierarchical classification was also compared and it was found that two-level hierarchical classification showed improvement in accuracy as compared to one level classification model. In our future work, we will explore the deep learning algorithms and ontology-based approaches to obtain more accurate prediction results.

References

1. K. Nigam, A. K. McCallum, S. Thrun, and T. Mitchell, "Text classification from labeled and unlabeled documents using EM," *Machine learning,* vol. 39, pp. 103-134, 2000.
2. F. Sebastiani, "Machine learning in automated text categorization," *ACM computing surveys (CSUR),* vol. 34, pp. 1-47, 2002.
3. D. D. Lewis, "Feature selection and feature extraction for text categorization," in *Proceedings of the workshop on Speech and Natural Language,* 1992, pp. 212-217.
4. A. Markov, M. Last, and A. Kandel, "The hybrid representation model for web document classification," *International Journal of Intelligent Systems,* vol. 23, pp. 654-679, 2008.
5. M. A. Al-garadi, K. D. Varathan, and S. D. Ravana, "Cybercrime detection in online communications: The experimental case of cyberbullying detection in the Twitter network," *Computers in Human Behavior,* vol. 63, pp. 433-443, 2016.
6. G. Mujtaba, L. Shuib, R. G. Raj, R. Rajandram, and K. Shaikh, "Automatic Text Classification of ICD-10 Related CoD from Complex and Free Text Forensic Autopsy Reports," in *Machine Learning and Applications (ICMLA), 2016 15th IEEE International Conference on,* 2016, pp. 1055-1058.
7. G. Mujtaba, L. Shuib, R. G. Raj, R. Rajandram, K. Shaikh, and M. A. Al-Garadi, "Automatic ICD-10 multi-class classification of cause of death from plaintext autopsy reports through expert-driven feature selection," *PloS one,* vol. 12, p. e0170242, 2017.
8. S. H. James, J. J. Nordby, and S. Bell, *Forensic science: an introduction to scientific and investigative techniques*: CRC press, 2002.
9. W. L. Yeow, R. Mahmud, and R. G. Raj, "An application of case-based reasoning with machine learning for forensic autopsy," *Expert Systems with Applications,* vol. 41, pp. 3497-3505, 2014.
10. B. Koopman, G. Zuccon, A. Nguyen, A. Bergheim, and N. Grayson, "Automatic ICD-10 classification of cancers from free-text death certificates," *International journal of medical informatics,* vol. 84, pp. 956-965, 2015.
11. R. Dias, R. Salvini, A. Nierenberg, and B. Lafer, "Machine learning approach with baseline clinical data forecasting depression relapse in bipolar disorder," *Bipolar Disorders,* vol. 18, pp. 103-103, Jul 2016.
12. K. Farooq and A. Hussain, "A novel ontology and machine learning driven hybrid cardiovascular clinical prognosis as a complex adaptive clinical system," *Complex Adaptive Systems Modeling,* vol. 4, p. 21, Jul 2016.
13. M. Galli, I. Zoppis, A. Smith, F. Magni, and G. Mauri, "Machine learning approaches in MALDI-MSI: clinical applications," *Expert Review of Proteomics,* vol. 13, pp. 685-696, Jul 2016.
14. Z. S. Harris, "Distributional structure," *Word,* vol. 10, pp. 146-162, 1954.
15. N. Passalis and A. Tefas, "Entropy optimized feature-based bag-of-words representation for information retrieval," *IEEE Transactions on Knowledge and Data Engineering,* vol. 28, pp. 1664-1677, 2016.
16. Q. V. Le and T. Mikolov, "Distributed Representations of Sentences and Documents," in *ICML,* 2014, pp. 1188-1196.
17. F. Enríquez, J. A. Troyano, and T. López-Solaz, "An approach to the use of word embeddings in an opinion classification task," *Expert Systems with Applications,* vol. 66, pp. 1-6, 2016.
18. V. Jouhet, G. Defossez, A. Burgun, P. Le Beux, P. Levillain, P. Ingrand, *et al.,* "Automated classification of free-text pathology reports for registration of incident cases of cancer," *Methods of information in medicine,* vol. 51, p. 242, 2012.

19. S. Danso, E. Atwell, and O. Johnson, "Linguistic and statistically derived features for cause of death prediction from verbal autopsy text," in *Language processing and knowledge in the web*, ed: Springer, 2013, pp. 47-60.
20. S. Danso, E. Atwell, and O. Johnson, "A comparative study of machine learning methods for verbal autopsy text classification," *arXiv preprint arXiv:1402.4380,* 2014.
21. M. F. Siddiqui, A. W. Reza, and J. Kanesan, "An automated and intelligent medical decision support system for brain MRI scans classification," *PloS one,* vol. 10, p. e0135875, 2015.
22. M. A. Al-garadi, M. S. Khan, K. D. Varathan, G. Mujtaba, and A. M. Al-Kabsi, "Using online social networks to track a pandemic: A systematic review," *Journal of biomedical informatics,* vol. 62, pp. 1-11, 2016.
23. R. Kohavi, "A study of cross-validation and bootstrap for accuracy estimation and model selection," in *Ijcai*, 1995, pp. 1137-1145.
24. M. Sokolova and G. Lapalme, "A systematic analysis of performance measures for classification tasks," *Information Processing & Management,* vol. 45, pp. 427-437, 2009.
25. T. Joachims, "Text categorization with support vector machines: Learning with many relevant features," in *European conference on machine learning*, 1998, pp. 137-142.
26. B. Xu, X. Guo, Y. Ye, and J. Cheng, "An Improved Random Forest Classifier for Text Categorization," *JCP,* vol. 7, pp. 2913-2920, 2012.
27. S. Dreiseitl, L. Ohno-Machado, H. Kittler, S. Vinterbo, H. Billhardt, and M. Binder, "A comparison of machine learning methods for the diagnosis of pigmented skin lesions," *Journal of biomedical informatics,* vol. 34, pp. 28-36, 2001.

Collaborative Filtering Fusing Label Features Based on SDAE

Huan Huo[1], Xiufeng Liu[2], Deyuan Zheng[1], Zonghan Wu[1], Shengwei Yu[1], and Liang Liu[1]

[1] School of Optical-Electrical and Computer Engineering,
University of Shanghai for Science and Technology, Shanghai, China
[2] Technical University of Denmark, 2800 Kgs. Lyngby, Denmark

Abstract. Collaborative filtering (CF) is successfully applied to recommendation system by digging the latent features of users and items. However, conventional CF-based models usually suffer from the sparsity of rating matrices which would degrade model's recommendation performance. To address this sparsity problem, auxiliary information such as labels are utilized. Another approach of recommendation system is content-based model which can't be directly integrated with CF-based model due to its inherent characteristics. Considering that deep learning algorithms are capable of extracting deep latent features, this paper applies Stack Denoising Auto Encoder (SDAE) to content-based model and proposes DLCF(Deep Learning for Collaborative Filtering) algorithm by combing CF-based model which fuses label features. Experiments on real-world data sets show that DLCF can largely overcome the sparsity problem and significantly improves the state of art approaches.

Keywords: recommendation system; collaborative filtering; deep learning; label feature

1 Introduction

Recommendation system, as a hot topic in recent years' research, helps people acquire useful information through massive overloading Internet information. In this field, collaborative filtering, due to its capability of digging the latent features of users and items by using matrix decomposition technique, is widely applied to recommendation system and has made a lot of progress. However, one natural drawback of recommendation system is that it cannot address the sparsity problem of rating matrix. In order to overcome the shortcoming of conventional collaborative filtering, adding auxiliary information into sparse matrix is an effective choice. At present, many innovative collaborative recommendation approaches have made their own efforts in this field, such as [1] considering the label features of items, [2] introducing hybrid recommendation, and so on. These approaches can alleviate the sparsity problem of rating matrix to some extent. However, label matrix can be very sparse in the most cases, so merely depending on introducing original label information is not enough to overcome the shortcoming of collaborative filtering. Another approach of recommendation system

© Springer International Publishing AG 2017
P. Perner (Ed.): ICDM 2017, LNAI 10357, pp. 223–236, 2017.
DOI: 10.1007/978-3-319-62701-4_17

is content-based recommendation. [3] maintains a feature vector or an attribute set to establish recommendation system by depicting the portraits of users or items. The drawback of content-based recommendation is that it cannot extract deep features automatically and dip the deep latent features of the items and the potential interests of users. Combined with the industry consideration towards Internet privacy, this problem can be even worse. Therefore, content-based recommendation needs to combine with collaborative filtering to establish hybrid recommendation in order to achieve better performance of recommendation.

How to increase the information of recommended data and dig the deep latent features of content information is the key of improving the performance of recommendation algorithm. Content-based recommendation algorithm usually uses the model such as LDA(Latent Dirichlet Allocation) when extracting features. This kind of model performs well in conventional content-based recommendation. On the other hand, deep learning is capable of learning deep features automatically. In this field, CNN(Convolutional Neural Network)[5] and RNN(Recurrent Neural Networks)[6] are widely used in the field of image recognition and natural language processing, which have achieved great performance. This ability of deep learning algorithms is very suitable for combining with collaborative filtering in the application of content-based model.

This paper proposes DLCF(Deep Learning for Collaborative Filtering) algorithm. Its framework generates a new matrix of label-items by using SDAE(Stacked Denoising Autoencoders)[7] training the feature vectors of items and lables, transforming sparse label matrix which contains less amount of information into label matrix which contains information of deep latent features, increasing the original data information substantially; then conducting collaborative filtering processing combining with the original rating matrix.

The contribution of this paper mainly consists of three aspects: 1)Applying deep learning algorithm to content-based model, increasing the original data information substantially by extracting deep features; 2)Combining the feature vectors of items and labels with the original rating matrix by building auxiliary matrix, playing the role of data label to the most extent, thus integrating data label, content and rating matrix into a whole framework; 3)Conducting experiments on real-world data sets for the performance of algorithm, pointing out the effect of model parameters on the performance of algorithm.

The rest of this paper is organized as follows. Section 2 reviews previous related work. Section 3 provides the details of DLCF algorithm. Section 4 conducts experiments on real-world data sets.

2   Related Work

Collaborative filtering is the most widely-used recommendation technique at present. [8] firstly proposes LFM(Latent Factor Model) based on SVD(Singular Value Decomposition); [9] adds probability distribution into LFM and introduces PMF(Probabilistic Matrix Factorization); [10] goes further to extend PMF into Bayesian PMF, and train the data by Markov Chain Monte Carlo; [11] and [12] try to combine collaborative filtering model with content-based model. Considering the auxiliary information, [13] proposes that labels of items can be applied to collaborative filtering; [14] partially solves the sparsity of matrix and cold start problem by the category information of items; [15] proposes social label recommendation by analyzing the relationship between the objects in the label system; [16] proposes a kind of tag-based Recommendation system method called Tag-CF. However, these approaches can only utilize the auxiliary information provided by the data sets, but cannot mine out the deep latent information.

In recent years, deep learning is applied to image recognition and natural language processing. But the attempt to apply deep learning to recommendation system has also emerged. [17] uses RBM(Restricted Boltzmann Machine) to take the place of Matrix Decomposition Model in order to implement collaborative filtering; [18] develops further based on [17] by combining the correlation between users and items; [19] uses DBN(Deep Belief Networks) to dig deep content information; [20] proposes a kind of relational SDAE model;[21] proposes a kind of content-based music recommendation system called DeepMusic. But all these algorithms haven't been applied to collaborative filtering.

The core of DLCF algorithm proposed in this paper is to apply the deep learning model to content-based recommendation model and combine with collaborative filtering algorithm. The deep learning model SDAE [22] is a model formed by stacking multiple DAEs (Denoising Auto Encoder) , which can extract the deep features of content information, and have strong interpretability and lower model complexity. SDAE can extract the new feature dimension from the original content information and train the label content matrix. It greatly increases the amount of information available in the original data. We then integrate the new content label matrix into PMF, thus perfectly combing the content-based recommendation model and collaborative filtering model. Meanwhile, this paper adopts SGD (Stochastic Gradient Descent) to train the model parameters by minimizing the loss function, which overcomes the problem of slow iteration of deep learning algorithm, and solves the problem of sparsity in the conventional collaborative filtering algorithm, as well as the problem of lack of useful information in the content-based recommendation model.

2.1 Probabilistic Matrix Factorization

PMF model introduces prior probability distribution into conventional matrix factorization model. Assuming that the conditional probability of observed rating data is:

$$p(R|U,V,\sigma_R^2) = \prod_{u}^{M} \prod_{i}^{N} (\varkappa(R_{u,i}|U_u^T V_i, \sigma_R^2))^{I_{u,i}^R} \tag{1}$$

Where $\varkappa(x|\mu,\sigma^2)$ represents the probability density function of normal distribution which has the mean of μ and the variance of σ^2. $I_{u,i}^R$ means that if the user $u's$ rating towards the item i exists, the value of I equals to 1, otherwise it equals to 0. We assume the mean equals to 0 and the variants of u and i are σ_U^2 and σ_V^2, respectively:

$$p(U|\sigma_U^2) = \prod_u^M \varkappa(U_u|0, \sigma_U^2 I)$$

$$p(V|\sigma_V^2) = \prod_i^N \varkappa(V_i|0, \sigma_V^2 I) \tag{2}$$

The model is shown in Fig. 1.

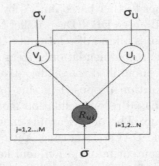

Fig. 1. PMF Schematic Diagram

2.2 Stacked Auto Encoder

DAE, namely Denoising Auto Encoder, consists of encoder and decoder. Each encoder has its corresponding decoder, and processes the noise of data through the course of encoding and decoding. Fig.2 shows that SDAE is a kind of Feed Forward Neural Network which stacks multiple DAEs similar to multilayer perceptions. Each layer's output is the input of its next layer. SDAE uses greedy layer-wise training strategy to train each layer in the network successively, and then pre-train the whole deep learning network. The idea of SDAE is to stack multiple DAEs in order to form a deep learning framework. This model trains the middle layers by the loss input and recovered output.

Formula (3) is the training model of SDAE. In this model, Z_c is the matrix consisting of several label vectors, which is the last output of SDAE. Z_0 represents the initial loss input matrix. Z_L is the middle layer of the model. The target matrix we need to achieve through the training is $Z_{L/2}$, which represents the matrix containing deep content information through the training from Z_0 to Z_C. w_l and b_l represent the weight matrix of the lth layer and the bias vector in SDAE model, respectively. λ represents regular parameter. $|| \cdot ||_F$ is Frobenius norm.

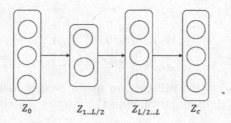

Fig. 2. SDAE Consisting of Multiple Layer DAE

$$min_{\{w_l\},\{b_l\}} = ||Z_c - Z_L||_F^2 + \lambda \sum_l (||w_l||_F^2 + ||b_l||_2^2) \qquad (3)$$

3 DLCF Algorithm

3.1 Preparing Work

According to the clustering algorithm [23], we first generated the user and the item clusters, as shown in algorithm 1, where θ is the threshold that we can decide:

```
Algorithm 1: preprocessing algorithm
1. Compute rating frequency of user f_u and item f_i
2. Sort users and items based on f_u and f_i in reverse order
3. Compute similarity of users and items
4. t ← 1, U ← φ, I ← φ
5. FOR j = 1, 2, ···
6.    IF u_j ∉ U
7.       U_t ← u_j ∪ u_k|s_{u_j,u_k} > θ_j, u_k ∉ U
8.       U← U_t ∪ U, t ← t+1
9.    ENDIF
10.ENDFOR
11.The same as cluster I
```

The clusters we have built contains (user cluster)-(item) preference information and (item cluster)-(user) preference information. By utilizing the local preferences information, we can speed up the greedy layer-wise training strategy in SDAE, and then pre-train the whole deep learning network.

3.2 Training label matrix

Firstly, based on SDAE model and its theory, output matrix Z_c and loss input matrix Z_0 are observed variables. Then Z_c is defined as follows: For each layer l

in SDAE:

$$W_{l,*n} \sim \varkappa(0, \lambda_w^{-1} I_{K_l})$$
$$b_l \sim \varkappa(0, \lambda_w^{-1} I_{K_l}) \tag{4}$$

Where $W_{l,*n}$ represents the nth column in the weight matrix of the lth layer. I_K represents the Kth diagonal value in the unitary matrix. For Z_c and Z_L:

$$Z_{l,j*} \sim \varkappa(\sigma(Z_{l-1,j*} W_l + b_l), \lambda_s^{-1} I_{K_l})$$
$$Z_{c,j*} \sim \varkappa(Z_{L,j*}, \lambda_n^{-1} I_B) \tag{5}$$

Where $\sigma(\cdot)$ represents sigmoid function. λ_w, λ_n and λ_s are all model parameters. Based on the above definitions, maximizing the maximum posterior probability of W_l, b_l, Z_l, Z_c, is the same as minimizing the joint log-likelihood function of the above variables. Thus the loss function of the model is defined as:

$$\epsilon = -\frac{\lambda_w}{2} \sum_l (||W_l||_F^2 + ||b_l||_2^2)$$
$$-\frac{\lambda_n}{2} \sum_j ||Z_{L,j*} - Z_{c,j*}||_2^2$$
$$-\frac{\lambda_s}{2} \sum_l \sum_j ||\sigma(Z_{l-1,j*} W_j + b_j) - Z_{l,j*}||_2^2 \tag{6}$$

Assume $T_{i,j}$ is defined as a boolean. If item j includes label i, the value of $T_{i,j}$ equals to 1, otherwise it equals to 0. Based on $Z_{L/2,j*}$ and original label matrix $T_{i,j}$, we can find latent factor vectors of labels and items t_i and v_j:

$$\gamma = -\frac{\lambda_t}{2} \sum_i ||t_i||_2^2 - \frac{\lambda_v}{2} \sum_j ||v_j - Z_{L/2,j*}^T||_2^2$$
$$-\sum_{i,j} \frac{c_{i,j}}{2} (T_{i,j} - t_i^T v_j)^2 \tag{7}$$

Where λ_t and λ_v are regular parameters, respectively. $c_{i,j}$ is set to 1 when item j includes label i, otherwise it is set to a small value, such as 0.001 or 0.

3.3   Establishing User-label Matrix

We define matrix G as pre-processed label-item matrix, which is used to establish DLCF. We get it by feature vectors of labels and items, namely t_i and v_j. $R_{u,i}$ represents the rating of item i by user u. $G_{i,t}$ represents the grade of item i for label t, which equals to 1 if i includes t, otherwise it equals to 0. By jointing matrix R and G, we can get target matrix H:

$$H_{u,t} = \frac{1}{N} \sum_i^n R_{u,i} G_{i,t} \tag{8}$$

3.4 Establishing DLCF Model

After establishing user-label matrix H, we use rating matrix R and H to improve conventional CF model, as shown in Fig.3. U and V represent the latent feature vectors included by users and items, respectively. Q represents the relationship between labels and latent features, which bridges the information circulation of R and H:

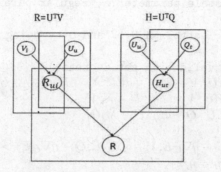

Fig. 3. Establishing Joint Matrix

After fusing H into PMF, we establish new loss function:

$$
\begin{aligned}
E = {} & \frac{1}{2} \sum_u \sum_i I_{u,i}^R (R_{ui} - U_u^T V_i)^2 \\
& + \frac{\omega_U}{2} \sum_u \|U_u\|_F^2 + \frac{\omega_V}{2} \sum_i \|V_i\|_F^2 \\
& + \frac{\alpha}{2} \sum_u \sum_t (H_{ut} - U_u^T Q_t)^2 \\
& + \frac{\varphi_U}{2} \sum_u \|U_u\|_F^2 + \frac{\varphi_Q}{2} \sum_t \|Q_t\|_F^2
\end{aligned}
\tag{9}
$$

where $\omega_U = \frac{\sigma_R}{\sigma_U}$ and $\omega_V = \frac{\sigma_R}{\sigma_V}$

$$
\begin{aligned}
E = {} & \frac{1}{2} \sum_u \sum_i I_{u,i}^R (R_{ui} - U_u^T V_i)^2 \\
& + \frac{\alpha}{2} \sum_u \sum_t (H_{ut} - U_u^T Q_t)^2 \\
& + \frac{\lambda_U}{2} \sum_u \|U_u\|_F^2 + \frac{\lambda_V}{2} \sum_i \|V_i\|_F^2 \\
& + \frac{\lambda_Q}{2} \sum_t \|Q_t\|_F^2
\end{aligned}
\tag{10}
$$

where $\lambda_U = \omega_U + \varphi_U, \lambda_V = \omega_V, \lambda_Q = \omega_Q$.

3.5   SGD Training Algorithm

We use SGD(Stochastic Gradient Descent) to train E.

```
Algorithm 2: DLCF-SGD training algorithm
Input: rating matrix R, the dimension of the feature vectors K,
learning rate η, scale parameter α, regular parameters λ_U,λ_V,λ_Q
Output: U, V
1. Initialization: establish U, V and Q by using a small value
stochastically.
2. While(error on validation set decrease):
```
3. $\qquad \nabla_U E = I(U^T V - R)V + \alpha(U^T Q - H)Q + \lambda_U U$
$\qquad \nabla_V E = [I(U^T V - R)]^T U + \lambda_V V$
$\qquad \nabla_Q E = \alpha(U^T Q - H)^T U + \lambda_Q Q$
4. $\qquad$ Set $\eta = 0.1$
5. $\qquad$ While $E(U - \eta\nabla_U E, V - \eta\nabla_V E, Q - \eta\nabla_Q E) > E(U, V, Q)$
6. $\qquad\qquad$ Set $\eta = \eta/2$
7. $\qquad$ End while
8. $\qquad U = U - \eta\nabla_U E$
$\qquad V = V - \eta\nabla_V E$
$\qquad Q = Q - \eta\nabla_Q E$
```
9. End while
10.Return U,V
```

4   Experiment Results and Analysis

4.1   Data sets

We use the data from Douban Reading (https://book.douban.com/), a social network well-known for users to rate different published books in China. Each book in this website has been rated from 1 to 5 by users. Moreover, each book has the features labeled by users, which can be applied to the feature vectors of content-based model. The data from this website is perfect for the application of DLCF algorithm. The data set includes 384341 users, 89798 books, and 13456139 rating data. The data record is formatted as (UserID, BookID, Rating, Labels). In this experiment, we choose the data sets with different sparsity. For these data sets, we choose 80% as training data and 20% as testing data. During the experiment, we divide the data into 5 parts stochastically for cross-check.

4.2   Algorithm Evaluation Criteria

In general, there are two ways to evaluate the recommendation system. One is evaluating the distance between predicted rating and users' actual rating, which is very common. The other is to evaluate the accuracy of the prediction.

In this paper, we use RMSE (Root Mean Square Error) as the criteria measuring the distance between the predicted rating and the users' rating, as shown in equation 11.

$$RMSE = \sqrt{\frac{1}{||\tau||} \sum_{(u,i)\in\tau} (r_{u,i} - \hat{r})^2} \tag{11}$$

Where τ represents the set including the existing rating of item i by user u.

To evaluate the accuracy of the algorithm prediction, we use recall@R as the metric. By choosing the testing users, recall@R sorts the recommended results and chooses the top R most favored by the users. Then we calculate the ratio of the number of the items in top R results to total number of the items the users like. The greater the ratio is, the more effective the performance of the algorithm is.

$$recall@R = \frac{number\_of\_items\_that\_users\_like\_among\_top\_R}{total\_number\_of\_items\_that\_users\_like} \tag{12}$$

4.3 Comparison Model and Experimental Settings

To evaluate the performance of DLCF algorithm, we choose three algorithms to compare. These algorithms are PMF [9] (collaborative filtering with matrix decomposition), Tag-CF [16](algorithm combining label information), and DBN [19](deep learning algorithm without tag). Firstly, we compare the above three algorithms with DLCF horizontally. Then we observe vertical comparison performance of DLCF under different experimental settings. In such case, we conduct comprehensive evaluation of the performance of each algorithm.

Before making further experimental comparison, we study the parameter α in (10), which represents the effect factor of user-label matrix in the whole model. If we set α to 0, it represents that user-label matrix is not put into consideration. In this case, algorithm is degraded to conventional collaborative filtering without considering label. In the experiment depicted in this paper, when the other parameters are fixed, RMSE can be minimized on condition that α equals to 0.9.

Table 1: The Effect of Parameter α on RMSE

α	RMSE
0	0.98
0.1	0.93
0.5	0.87
0.9	0.82
1.2	0.85
2	0.99
10	1.33

Other parameters $\lambda_U = \omega_U + \varphi_U, \lambda_V = \omega_V, \lambda_Q = \omega_Q, \omega_U = \sigma_R/\sigma_U, \omega_V = \sigma_R/\sigma_V$ which represent the regular parameters in the model, are compounded by other latent parameters. Theoretically, we need to take the value of each latent parameter, and then calculate the value of $\lambda_U, \lambda_V, \lambda_Q$. However, in practice, we can set $\lambda_U, \lambda_V, \lambda_Q$ to small values respectively. For example, $\lambda_U = \lambda_V = \lambda_Q = 0.001$. Then we adjust them by cross validation in the experiment. The result proved that such an approach does not produce a significant impact on the performance of the algorithm.

In the next comparison experiment, we divide experimental data into two parts according to their different sparsity. Then we compare the performance of each algorithm based on these two kinds of evaluation methods.

Table 2: Sparsity of Dataset-a and Dataset-b

	Dataset-a	Dataset-b
Sparsity	97.82%	90.61%

4.4   Horizontal Comparison

The purpose of horizontal comparison is comparing the performance of different algorithms under the same scenario. For the evaluation metric recall@R, we compare the performance of each algorithm with different R. For metric RMSE, we compare the performance by choosing different values of the feature vector decomposition dimension K. The comparisons above will be experimented on different data sets with different sparsity.

Firstly, aiming at the performance of these four kinds of algorithms on recall@R, we can obviously find that collaborative filtering without auxiliary information performs the worst. The performance difference between DBN and Tag-CF is not significant because both of them apply part of auxiliary information. DLCF performs significantly better than the other three algorithms because of fully utilizing of the latent information. Meanwhile, we can also find out that these algorithms perform better in the dense data sets than in the sparse ones.

Next, evaluating the RMSE for the four algorithms, we find that the situation is quite similar to that of recall@R. DLCF still outperforms the rest three. Similarly, these algorithms perform worse in the sparse data sets than in the dense ones. We can also find that, on the dense data sets, the performance difference between DBN and Tag-CF is not so significant compared with the one on the sparse data sets. Together with above horizontal comparison, we can find that the key of the performance of the algorithm is whether the algorithm can use existing data of the latent information as much as possible.

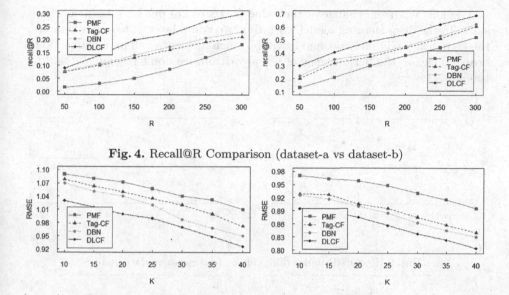

Fig. 4. Recall@R Comparison (dataset-a vs dataset-b)

Fig. 5. RMSE Comparison (dataset-a vs dataset-b)

4.5 Vertical Comparison

Vertical comparison mainly study the effect of key parameters on DLCF. Through several experiments in vertical comparison, the performance difference based on recall@R is more significant than the one based on RMSE. So this part focuses on showing the experiment results based on recall@R. Considering the characteristics of deep learning and collaborative filtering, after the repeating experiments, this paper chooses middle layer $L/2$ and model parameter λ_n in SDAE as the key parameters.

Firstly, we conduct the experiment aiming at the parameter λ_n. λ_n is used in SDAE to train the middle layer, which is the key parameter to generate new label matrix in DLCF. By observing the result, we can find that the value of λ_n is neither the bigger the better nor the smaller the better, and there is a range where the algorithm performs well. When the value of λ_n is very small (usually less than 1), the algorithm performs badly. In this case, the increase of the value λ_n may improve the performance of the algorithm. However, when the value of λ_n reaches after three digits, the improvement is significant by continuing increasing the value of λ_n. Similarly, the model performs significantly better in the dense data sets than in the sparse ones.

At last, we look at the result from the experiment where we set the middle layer $L/2$ as observed variable in SDAE. It is obvious that the algorithms perform significantly better on the dense datasets than on the sparse ones. For the middle layer, when $L/2$ equals to 1, the performance is worse than the results where $L/2$ equals to 2 or 3 because of the fewer layers. But for the results where $L/2$ equals to 2 or 3, we can find that the performance difference between them is very small. With the change of the value R, there are ups and downs on both sides. In the deep learning model, each time we add a layer in the middle layers, the complexity will increase, and it will be much harder to call parameters. So in order to ensure the performance of the algorithm, we don't recommend adding too many layers.

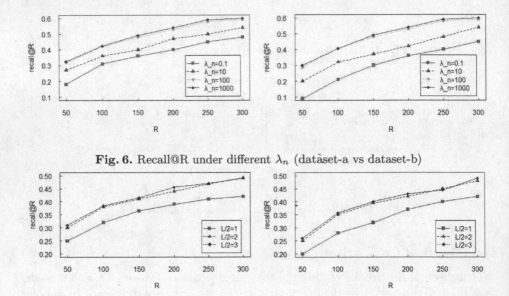

Fig. 6. Recall@R under different λ_n (dataset-a vs dataset-b)

Fig. 7. Recall@R under different $L/2$ (dataset-a vs dataset-b)

5 Conclusions

Conventional collaborative filtering can't overcome the problem caused by sparse matrix. Even if introducing label information, label matrix also has the sparsity problem, which hasn't ideal performance. But conventional content-based recommendation algorithm is not suitable for fusing with collaborative filtering because of its own characteristics. The ability of extracting deep latent information in deep learning model is sufficiently verified in the field of image recognition and natural language processing.

This paper takes advantage of the characteristics of deep learning model and improves the information of original data substantially by processing original label information of items. By combining content-based recommendation model and collaborative filtering, we propose DLCF algorithm and fully apply deep learning model to the recommendation system. Meanwhile, the experiments on real-world data sets show that it can achieve better performance than conventional recommendation model. On the other hand, introducing deep learning model undoubtedly puts forward higher requirements for the calling of the parameters, and the complexity is much higher than conventional model. Future work will focus on addressing these problems in order to improve the interpretability and engineering significance of the algorithm.

6 Acknowledgment

This work is supported by National Natural Science Foundation of China (61003031, 61202376), Shanghai Engineering Research Center Project (GCZX14014), Shanghai Key Science and Technology Project in IT(14511107902), Shanghai Leading Academic Discipline Project(XTKX2012) and Hujiang Research Center Special Foundation(C14001).

References

1. Kim, B.S., Kim, H., Lee, J., et al.: Improving a Recommendation System by Collective Matrix Factorization with Tag Information. In: Soft Computing and Intelligent Systems (SCIS), 2014 Joint 7th International Conference on and Advanced Intelligent Systems (ISIS), 15th International Symposium on. IEEE, pp.980–984(2014)
2. Grivolla, J., Badia, T., Campo, D., et al.: A Hybrid Recommendation Combining User, Item and Interaction Data. In: Computational Science and Computational Intelligence (CSCI), 2014 International Conference on. IEEE, pp.297–301(2014)
3. Shen , Y., Fan , J.: Leveraging Loosely-tagged Images and Inter-object Correlations for Tag Recommendation. In: Proceedings of the 18th ACM, International Conference on. Multimedia, pp.5–14(2010)
4. Blei, D.M., Ng, A.Y., Jordan M.I.: Latent Dirichlet Allocation. The Journal of Machine Learning Research. 3, 993–1022(2003)
5. Krizhevsky, A., Sutskever, I., Hinton, G.E.:Imagenet Classification with Deep Convolutional Neural Networks. In: Advances in Neural Information Processing Systems, pp.1097–1105(2012)
6. Graves, A., Fernndez, S., Gomez, F., et al.: Connectionist Temporal Classification: Labelling Unsegmented Sequence Data with Recurrent Neural Networks. In: Proceedings of the 23rd International Conference on Machine Learning, pp.369–376(2006)
7. Vincent, P., Larochelle, H., Lajoie, I., et al.: Stacked Denoising autoencoders: Learning Useful Representations in a Deep Network with a Local Denoising Criterion. The Journal of Machine Learning Research. 11,3371–3408(2010)
8. Funk, S.: Netflix Update: Try This at Home. http://sifter.org/~simon/journal/20061211.html
9. Salakhutdinov, R., Mnih, A.: Probabilistic Matrix Factorization. In: NIPS(2011)
10. Salakhutdinov, R., Mnih, A.: Bayesian Probabilistic Matrix Factorization. In: Proceedings of the 25th International Conference on Machine Learning, pp.880–887(2008)
11. Hu, L., Cao, J., Xu, G., et al.: Personalized Recommendation via Cross-domaintriadic Factorization. In: Proceedings of the 22nd international conference on World Wide Web, pp.595–606(2013)

12. Li, W.J., Yeung, D.Y., Zhang, Z.: Generalized Latent Factor Models for Social Network Analysis. In: Proceedings of the 22nd International Joint Conference on Artificial Intelligence (IJCAI), pp.1705(2011)

13. Vig, J., Sen, S., Riedl, J.: The Tag Genome: Encoding Community Knowledge to Support Novel Interaction. ACM Transactions on Interactive Intelligent Systems (TiiS). 2(13),13(2012)

14. Pirasteh, P., Jung, J.J., Hwang, D.: Item-based Collaborative Filtering with Attribute Correlation:A Case Study on Movie Recommendation. Intelligent Information and Database Systems, pp.245–252. Springer International Publishing(2014)

15. Zhang, B., Zhang, Y., Gao, K.N., Guo, P.W., Sun, D.M.: Combining Relation and Content Analysis for Social Tagging Recommendation. Journal of Software. 23(3),476–488(2012)

16. Kim, B.S., Kim, H., Lee, J., et al.: Improving a Recommendation System by Collective Matrix Factorization with Tag Information. In: Soft Computing and Intelligent Systems (SCIS), 2014 Joint 7th International Conference on and Advanced Intelligent Systems (ISIS), 15th International Symposium on IEEE, pp.980-984(2014)

17. Salakhutdinov R, Mnih A, Hinton G.: Restricted Boltzmann Machines for Collaborative Filtering. In: Proceedings of the 24th International Conference on Machine Learning, pp.791–798(2007)

18. Georgiev, K., Nakov, P.: A Non-iid Framework for Collaborative Filtering with Restricted Boltzmann Machines. In: Proceedings of the 30th nternational Conference on Machine Learning(ICML-13), pp.1148-1156(2013)

19. Wang, X., Wang, Y.: Improving Content-based and Hybrid Music Recommendation Using Deep Learning. In: Proceedings of the ACM International Conference on Multimedia, pp.627–63(2014)

20. Wang, H., Shi, X., Yeung, D.Y.: Relational Stacked Denoising Autoencoder for Tag Recommendation. In: AAAI, pp.3052–3058(2015).

21. Van den Oord, A., Dieleman, S., Schrauwen, B.: Deep Content-based Music Recommendation. In: Advances in Neural Information Processing Systems, pp.2643–2651(2013)

22. Vincent, P., Larochelle, H., Bengio, Y., et al. Extracting and Composing Robust Features with Denoising Autoencoders. In: Proceedings of the 25th International Conference on Machine Learning, pp.1096-1103(2008)

23. Xin, W., Congfu X.. SBMF: Similarity-Based Matrix Factorization for Collaborative Recommendation. In: Proceedings of the 26th International Conference on Tools with Artificial Intelligence, pp.379-383(2014)

Interestingness Classification of Association Rules for Master Data

Wei HAN[1], Julio BORGES[1], Peter NEUMAYER[2], Yong DING[1], Till
RIEDEL[1] and Michael BEIGL[1]

[1] Karlsruhe Institute of Technology (KIT), TECO
Karlsruhe, Germany
Email: {firstname.surname}@kit.edu

[2] SAP SE
Walldorf, Germany
Email: peter.neumayer@sap.com

Abstract. High quality of master data is crucial for almost every company and it has become increasingly difficult for domain experts to validate the quality and extract useful information out of master data sets.
However, experts are rare and expensive for companies and cannot be
aware of all dependencies in the master data sets. In this paper, we introduce a complete process which applies association rule mining in the
area of master data to extract such association dependencies for quality
assessment. It includes the application of the association rule mining algorithm to master data and the classification of interesting rules (from
the perspective of domain experts) in order to reduce the result association rules set to be analyzed by domain experts. The model can learn
the knowledge of the domain expert and reuse it to classify the rules. As
a result, only a few interesting rules are identified from the perspective
of domain experts which are then used for database quality assessment
and anomaly detection.

1 Introduction

High quality of master data is crucial for almost every company. An error in
the master data set results in an error throughout the organization, which can
lead to considerable fines or damaged credibility with customers and partners.
Traditionally, rule based approaches are used to check the correctness of master
data records and improve the quality of master data. Due to the complex domains
of master data, domain experts are needed to formulate and maintain such rules
[1]. However, domain experts are rare, expensive, and in practice not easily
convinced to thoroughly formalize their entire knowledge. Furthermore, even
domain experts cannot be aware of all dependencies in the master data set [1].
In addition, it is time-consuming to keep the rule set updated with regard to the
master data [1].

© Springer International Publishing AG 2017
P. Perner (Ed.): ICDM 2017, LNAI 10357, pp. 237–245, 2017.
DOI: 10.1007/978-3-319-62701-4_18

The Data Mining technology "Association Rule Mining", can help us solve the above problems. Association rule mining is a method for discovering interesting relations between variables in large databases automatically. It is intended to identify strong rules discovered in databases using measures of interestingness such as support and confidence. However, the algorithms of association rule mining usually produce a huge number of rules, so that domain experts have considerable difficulties manually analyzing so many rules to identify the truly interesting ones. Furthermore, objective measures which involve analyzing the rules structure, predictive performance, and statistical significance are insufficient to identify the user-defined interesting rules [2].

Therefore, the post-processing of association rule mining is introduced in this paper in order to determine the rule interestingness from a subjective perspective (domain experts). In the post-processing, a classification model is constructed to learn the knowledge of domain experts and classify the association rules into "*interesting*" and "*uninteresting*". As a result, the truly interesting rules are selected, and the number of rules is greatly reduced. Moreover, a complete process which applies association rule mining in the area of master data is introduced and implemented in our project.

The rest of the paper is organized as follows. In section 2, we propose the system design and approach. In section 3, we show the results of the proposed approach. Section 4 contains an evaluation of the presented results. Finally, we draw a conclusion of this paper in section 5.

2 Approach

In this paper, we introduce a complete process for the application of association rule mining in the area of master data and for discovering the interesting rules from the perspective of domain experts. In section 2.1, an overview of our system is presented. It includes three main parts: the pre-processing of the master data set, the application of association rule mining and the post-processing of the discovered rules. In section 2.2, we describe our post-processing approach in details.

2.1 System Design

Figure 1 depicts the flow diagram of our complete system design. As input data, the MARA table cf. [3] is used, which contains 451452 materials and 227 dimensions. First, a pre-processing for the MARA table is suggested and performed in our case. Three main steps are carried out in the pre-processing. One step is that we utilize the information theory (entropy) [4] to delete the columns with less information, which are considered as redundant columns. Then, the discretization of numeric attributes is carried out, since the algorithms of association rule mining cannot be used for numeric attributes [5]. Finally, we transform the processed MARA into an appropriate format, so that association rule mining algorithms can be used directly in the transformed table.

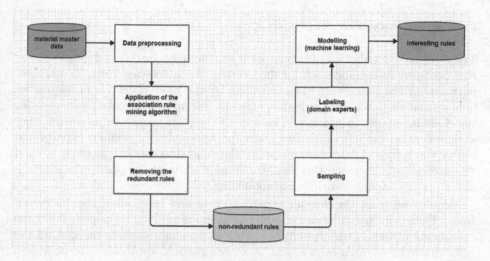

Fig. 1. System Design Workflow

After the pre-processing, the algorithms of association rule mining are applied with interestingness measures (minimal support and minimal confidence) to generate the fulfilled rules. In this paper we use the Apriori-algorithm [6] of the PAL library [7] and run it in SAP HANA databank. However, the algorithm causes a serious problem: a huge number of rules is produced, most of which are trivial, redundant and uninteresting. Therefore, the post-processing approach is needed to help us reduce the number of rules and filter out the truly interesting rules. Before the post-processing, statistical methods are used to analyze the rules structure and to delete the redundant rules with the objective measures, which involve analyzing the rules structure, predictive performance, and statistical significance. As mentioned in [8], a rule R1 is more general than a rule R2, provided that R2 can be generated by adding additional items to the antecedent or consequent of R1. The non-redundant rules are those that are most general (with equal confidence). Based on the non-redundant rules, the post-processing is performed.

In the post-processing, the aim is to identify the interesting rules from a subjective perspective (domain experts). Such interesting rules are concerned in the two ways described below [2]: **1) Unexpectedness**: Rules are interesting if they are unknown to the domain experts or contradict the domain experts existing knowledge (or expectations). **2) Actionability**: Rules are interesting if users can do something with them to their advantage, for example, domain experts can discover the outliers in the master data set and improve the data quality.

2.2   Post-processing Approach

The essential contribution of this paper is that we have introduced a new *subjective* post-processing approach for discovering the interesting rules. When comparing with the interesting rules identified by the most objective measures, we believe that the interesting rules identified by our approach make more sense for the domain experts in practice. The core of the post-processing is to construct a classification model, which can learn the knowledge of domain experts and reflect the learned knowledge to classify the rules into classes "interesting" and "uninteresting". To reach this goal, three main parts of the post-processing are introduced: sampling, labeling and modelling.

Ideally, we want the domain experts to rate and label all of the discovered rules. These labelled rules are considered as training data to help us build the classification model. Unfortunately, asking the domain experts to label all the rules manually is often impractical. A more practical approach is to provide a smaller set of rules to the experts for labeling and use this information to construct the classifier. Therefore, we have proposed two different sampling methods to extract a "representative" sample set of rules. One is the so-called convex-hull-based sampling method, which can select the most diverse rules as training data for the classifier. The other is the clustering-based sampling method, which extends the method mentioned in [9]. With the help of the clustering-based sampling method, the most representative rules are selected from the non-redundant rules. However, in the paper, we use the convex-hull-based sampling method as an example to explain our post-processing approach. The clustering-based sampling method is not mentioned in the rest of the paper.

After the most diverse 100 rules are extracted from the non-redundant rules by the convex-hull sampling method, two domain experts (DM1 and DM2) rate and label these 100 rules with labels "interesting" and "uninteresting" respectively. Then, the labelled rules serve as training data to construct the classification model. In this paper, the classification algorithm Random-Forests [10] is applied to construct the classification model.

3   Results

In this section, we show the results of our proposed process in respect of each step in system design.

3.1   Results of association rule mining

Table 1 shows the results of the pre-processing, the application of association rule mining and removing redundant rules. It is observed that 199 trivial dimensions are deleted after the pre-processing and 6739 non-redundant rules are produced for the post-processing in the end.

Input Data:	451452 materials, 227 dimensions
Result of pre-processing:	451452 transactions, 28 dimensions
Result of association rule mining:	9508 association rules
Result of non-redundant rules:	6739 non-redundant rules

Table 1. Results from pre-processing, association rule mining and removing redundant rules

3.2 Results of post-processing

Using the convex-hull sampling method, we select the most diverse 100 rules for the domain experts. In addition, 50 rules are extracted by the random sampling method, which are used as test data to evaluate the classifier in section 4. Then, we give all of the 150 rules to two domain experts separately, and they label these rules with the labels "interesting" and "uninteresting". After that, we use these 100 labelled rules by one of the domain experts to construct the classifier, while the labelled rules of the other domain expert are used for the quality test. Initially, we assume that both of the domain experts have the same knowledge of the domain of material master data.

Labeling of domain experts Figure 2 present the results of labeling by DM1 and DM2 respectively. By the results of DM1, it is shown that the interesting class is rarer than the uninteresting class (interesting: uninteresting = 1:3). In our case, we take the results of DM1 as an example to construct our classifier.

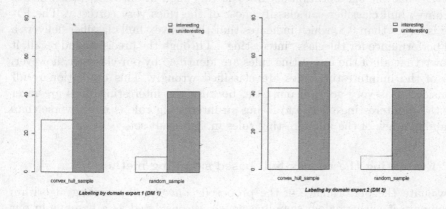

Fig. 2. Results of the class distribution of the labeled rules by domain expert 1 (DM1) and 2 (DM2)

Interestingness Binary Classification Model Based on the labelled 100 rules of DM1, a classifier is constructed using the Random-Forest-algorithm [10], that is, the so-called convex-hull classifier. In the following, the constructed convex-hull classifier of the Random-Forest-algorithm is displayed in Table 2. OOB-rate, F1-Score and AUC-value [11] are used to measure the performance of the convex-hull classifier. We evaluate the convex-hull classifier through the test data in the next section.

Classifier:	Convex-Hull Based Classifier
OOB Error Rate:	0.03
F1-Score:	0.94
AUC:	0.9899

Table 2. OOB-rate, F1-Score und AUC-value for convex-hull classifier built by random-forests

After training the convex-hull classifier, we have also applied the classifier to classify the remaining 6549 non-redundant rules. Different labels are rated by the domain experts, which lead to different results by the classifier.

4 Evaluation

To evaluate the performance of the constructed convex-hull classifier, we use a test data set containing 50 rules produced by the random sampling method. The classification accuracy, F1-Score and AUC-value are used to measure the performance of the classifier. Table 3 shows the results of this evaluation. It is observed that the classification accuracy is very high in our case, which means the convex-hull classifier can classify most of the rules very correctly. The F1-Score is larger than 0.8, which indicates that the convex-hull classifier delivers a good performance for the class "interesting". Through the precision and recall, it is shown that all of the interesting rules are identified by our classifier, however, some of the uninteresting rules are classified wrongly. This evaluation result is considered as very good in our case, because the interesting rules are rarer than the uninteresting ones, and losing an interesting rule is much worse than containing a few of the uninteresting rules in the result set.

4.1 Evaluating the convex-hull based sampling method

To evaluate the performance of the proposed convex-hull sampling algorithm we compare it with random sampling which is considered as a baseline in our case. A good sampling method can produce some of the "most informative and diverse" training data so that the classifier can learn more from the labels of the

Classifier:	Convex-Hull Based Classifier
Accuracy	0.94
F1-Score	0.8421
Precision	0.7273
Recall	1.0
AUC	0.9762

Table 3. Accuracy, F1-Score and AUC-value of the convex-hull classifier regarding a test data set (50 rules)

selected data and greatly improve the classification performance [12]. In order to evaluate our proposed sampling method, the performance of the correspondent classifiers, which are constructed based on the different training data selected by the different sampling methods, is evaluated instead. Two classifier are built, that is, the convex-hull-sampling-based classifier and the random-sampling-based classifier. Three test data sets are applied to measure the performance of these classifiers.

Figure 3 shows the evaluation results. As has been observed, during the comparison of the accuracy between the two classifiers, the convex-hull-sampling-based classifier has a better performance than the random- sampling-based classifier with regard to all of the three test data sets. This means that the convex-hull sampling method performs better than the random sampling method to construct the classifier with better performance.

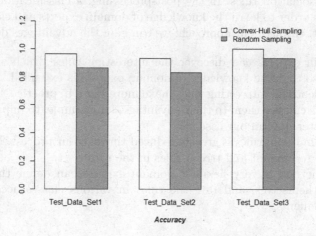

Fig. 3. Comparing the accuracy of the convex-hull classifier and the random sampling classifier regarding 3 test data sets.

Table 4 displays the results of the quality test. It shows that the classifier of DM1 has classified 62% of the rules in test data set 1 correctly and 84% of the rules in test data set 2. This means that the two domain experts (DM1 and DM2) have the same opinion on a minimum of 62% of the rules. We treat this result as acceptable in our case, in other words, the labels of DM1 are considered as reliable. However, it also indicates that the constructed convex-hull classifier cannot represent the knowledge of two domain experts, because they have a different understanding of the interestingness of the rules. To solve this problem, we suggest that a classifier can be trained based on the 62% rules with the same labels of the two experts.

	Test Dataset 1	Test Dataset 2
Accuracy	0.62	0.84
F1-Score	0.2916	0.5555

Table 4. Results of the quality test regarding the 2 test data sets

5   Conclusion

In this paper, we introduce a complete process which applies association rule mining in the area of master data. A system including three main parts: data pre-processing, application of association rule mining and post-processing for discovering the interesting rules is implemented. The main contribution of this paper is that we propose a new subjective method of post-processing the discovered association rules. In the post-processing, a classification model is constructed in order to learn the knowledge of domain experts and classify the rules. As a result, through our approach, we can gain the advantages described below:

- Domain experts can discover the interesting rules which are unknown to them so that the knowledge of domain experts is extended.
- The identified interesting rules make more sense in practice, because domain experts can use them to their advantage, for example, to improve the quality of master data in our case.
- The number of rules is greatly reduced through our process. Moreover, there are no redundant and trivial rules in the result set.
- Our method is very flexible. Domain experts can define the labels of the rules themselves, and rate the rules in various classes according to their requirements.

References

1. J. Hipp, M. Müller, J. Hohendorff, and F. Naumann, "Rule-based measurement of data quality in nominal data." in *ICIQ*, 2007, pp. 364–378.
2. B. Liu, W. Hsu, S. Chen, and Y. Ma, "Analyzing the subjective interestingness of association rules," *IEEE Intelligent Systems and their Applications*, vol. 15, no. 5, pp. 47–55, 2000.
3. "The main important sap material master tables (data & customizing)," http://sap4tech.net/sap-material-master-tables/, (Accessed on 02/20/2017).
4. S. Jaroszewicz and D. A. Simovici, "Pruning redundant association rules using maximum entropy principle," in *Pacific-Asia Conference on Knowledge Discovery and Data Mining*.   Springer, 2002, pp. 135–147.
5. R. Srikant and R. Agrawal, "Mining quantitative association rules in large relational tables," in *Acm Sigmod Record*, vol. 25, no. 2.   ACM, 1996, pp. 1–12.
6. R. Agrawal, H. Mannila, R. Srikant, H. Toivonen, A. I. Verkamo *et al.*, "Fast discovery of association rules." *Advances in knowledge discovery and data mining*, vol. 12, no. 1, pp. 307–328, 1996.
7. "What is pal? — sap hana platform," https://help.sap.com/viewer/2cfbc5cf2bc14f028cfbe2a2bba60a50/2.0.00/en-US, (Accessed on 02/20/2017).
8. M. J. Zaki, "Generating non-redundant association rules," in *Proceedings of the sixth ACM SIGKDD international conference on Knowledge discovery and data mining*.   ACM, 2000, pp. 34–43.
9. A. Strehl, G. K. Gupta, and J. Ghosh, "Distance based clustering of association rules," in *Proceedings ANNIE*, vol. 9, no. 1999.   Citeseer, 1999, pp. 759–764.
10. L. Breiman, "Random forests," *Machine learning*, vol. 45, no. 1, pp. 5–32, 2001.
11. T. Fawcett, "Roc graphs: Notes and practical considerations for researchers," *Machine learning*, vol. 31, no. 1, pp. 1–38, 2004.
12. Z. Xu, K. Yu, V. Tresp, X. Xu, and J. Wang, "Representative sampling for text classification using support vector machines," in *European Conference on Information Retrieval*.   Springer, 2003, pp. 393–407.

MapReduce and Spark-based Analytic Framework Using Social Media Data for Earlier Flu Outbreak Detection

Al Essa, Ali[1]
Faezipour, Miad [1]

[1] Computer Science and Engineering and Biomedical Engineering Departments
School of Engineering
University of Bridgeport, CT 06604, USA
aalessa@my.bridgeport.edu
mfaezipo@bridgeport.edu

Abstract. Influenza and flu can be serious problems, and can lead to death, as hundred thousands of people die every year due to seasonal flu. An early warning may help to prevent the spread of flu in the population. This kind of warning can be achieved by using social media data and big data tools and techniques. In this paper, a MapReduce and Spark-based analytic framework (MRSAF) using Twitter data is presented for faster flu outbreak detection. Different analysis cases are implemented using Apache Spark, Hadoop Systems and Hadoop Eco Systems to predict flu trends in different locations using Twitter data. The data was collected using a developed crawler which works together with the Twitter API to stream and filter the tweets based on flu-related keywords. The crawler is also designed to pre-process and clean the unintended attributes of the retrieved tweets. The results of the proposed solution show a strong relationship with the weekly Center for Disease Control and Prevention (CDC) reports.

1 Introduction

Seasonal influenza and flu can be a serious problem that may lead to death. About 250,000 to 500,000 deaths occur worldwide each year because of flu [1]. Public health care providers aim to know about the seasonal flu as soon as possible in order to take the required actions for their communities. Getting an early warning will help to prevent the spread of flu in the population.

Presently, a very large number of people use social media networks on a daily basis to share their news, events, and even their health status [2]. This leads to the idea of using commonly used social media networks for flu detection; that can provide an early flu warning to public health care providers to take the right action at the perfect time. Users of social media networks can be used as sensors to predict the flu trend in

© Springer International Publishing AG 2017
P. Perner (Ed.): ICDM 2017, LNAI 10357, pp. 246–257, 2017.
DOI: 10.1007/978-3-319-62701-4_19

a specific area and time. The social media network used for this project is Twitter. It is one of the most widely used social networks. It has over 271 million active users monthly [3]. The retrieved data from Twitter is enormous and contains large number of attributes, most of which will not be used for this study. Since the Twitter data is huge, Apache Spark, Hadoop systems and Hadoop Eco systems can be used for the analysis and for building a good prediction model for flu trend. This can help health care industry to provide high quality services at the right time by providing timely warnings.

2 Problem Definition

Reducing the impact of seasonal flu is important for public health. Studies show that it is possible to prevent of the spread of flu in the population if an early detection is made. Most of health care providers take the required action to the public after getting reports of flu from the Center for Disease Control and Prevention (CDC). This center collects data from health care providers to monitor Influenza –Like Illness (ILI) and publishes the reports. This takes one to two weeks delay, causing the required warning to come late to the provider's attention [1]. The providers need to be warned at the earliest time in order to take the appropriate actions to prevent the spread of flu.

Many solutions have been proposed to provide the warning as early as possible. These include monitoring web search queries like Google Flu Trend, monitoring call volume to advice lines, and monitoring the sales of drugs and flu shots taken by patients. In addition, textual and structural data mining techniques [4] have been used.

In this study, it is shown that early warnings can be achieved by monitoring Twitter posts. Twitter has become a very popular social media network. People use it to share news, events and even for their health status on a daily basis.

The flu prediction could be faster by taking the advantages of Apache Spark, MapReduce programming and Hadoop Eco Systems. These tools and techniques can use the available Twitter data to predict the flu trend at the earliest time. The proposed solution can help to produce analysis results from various scenarios.

To measure the quality of the proposed solution, the CDC reports were used as a ground truth. The prediction results showed a strong relationship with the CDC reports.

3 Data Collection

The data set for this study is a collection of Twitter posts obtained from Twitter social media network. The data is collected from Twitter website using a developed crawler which works with the Twitter API to stream tweets. The crawler is designed to filter the tweets based on flu-related keywords (flu and influenza) and also does the required pre-processing. Twitter API returns tweets in a JASON format which includes too many attributes. Most of these attributes are not needed for the flu trend prediction. The crawler cleans the retrieved tweets and only keeps the required attributes for the flu trend prediction analysis.

The data set consists of 896,905 tweets. They were collected over the period of November 2015, December 2015, October 2016, November 2016 and December 2016.

4   Background

An early warning about flu can be achieved by employing features of big data tools such as Apache Spark, Hadoop MapReduce programming and Hadoop Eco Systems. These tools together, with the available Twitter data can be used to detect flu trends at the earliest time.

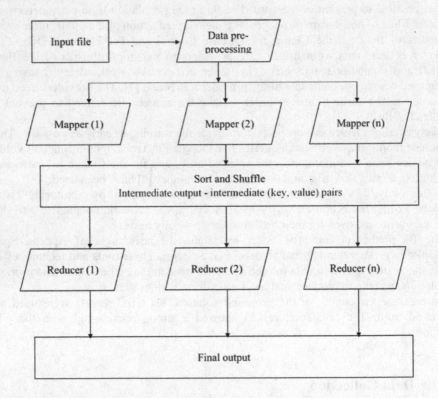

Figure 1. General flow of MapReduce Programming approach

4.1   Hadoop MapReduce Programming

MapReduce is an approach used to process a large dataset. It consists of two main functions: Map and Reduce. The Map function takes an input as a pair (key and value), groups all the values associated with the same key, and generates intermediate pairs to be passed to the Reduce function. The Reduce function merges all the pairs with the same intermediate key after processing the associated values such as summing them up. This programming model is automatically parallelized which allows programmers to utilize the resources of large distributed systems [5,6]. Figure 1 shows a general overview of the flow of MapReduce programming approach.

4.2   Hadoop Eco Systems

Hadoop Eco Systems such as Hive are data analytic tools to manage and query large datasets. They are built on top of Hadoop to provide an easy way to query and manage data. Hive allows users to query large datasets using a SQL-Like script (HiveQL) instead of MapReduce programming. The performance of queries written in HiveQL are similar to the ones written in MapReduce framework [7].

4.3   Spark

Spark was developed in 2009. It supports real time streaming data and fast queries. Spark runs on top of Hadoop to replace the data batch process of the traditional MapReduce model in order to support real time streaming data processes. Spark performs tasks based on two concepts. The first concept is the Resilient Distributed Dataset (RDD), which is an abstract collection of an element that can be processed in parallel. It is a read-only collection of objects partitioned across a set of nodes. RDD supports two kinds of operations: Transformations and Actions. Transformation operations take RDDs and only return new RDDs and nothing evaluated. Transformation functions include map, filter and reduceByKey. Action operations are also applied on RDDs which include evaluation and returning new values. Action functions include reduce, collect and take. The second concept is regarding the Directed Acyclic Graph (DAG) which is an engine that supports cyclic data flow. Spark creates a DAG for each job that consists of task stages (map and reduce) to be performed on a cluster [8].

5   Related Work

Many studies have been conducted to discover knowledge and predict flu trends from Twitter data.

5.1   Auto Regression Moving Average (ARMA) / SNEFT Framework

ARMA is a stochastic model which is composed of two forms: Auto Regression (AR) model and Moving Average (MA) model. The AR model is a prediction model. Its output depends linearly on the past values, a random value as an error, and a constant value. The MA model is used to represent the correlation between the past values and the white noise using linear regression.

Based on the ARMA model, Harshvardhan Achreckar et al. [1] proposed a framework called Social Network Enabled Flu Trends (SNEFT) that utilizes the ARMA model and the data obtained from CDC. Both are used in collaboration for better flu prediction trends. The architecture of the SNEFT consists of two main parts. The first part is used to predict influenza and Influenza Like Illness (ILI) using CDC data. The second part is used to provide flu warnings using Twitter data. The Auto regression Moving Average (ARMA) model is used to predict ILI incidence as a linear function of current and old Social Network data and historical ILI data (CDC data). Results show that Twitter data improves the output of the statistical models that were used for prediction. The SNEFT framework was tested with and without Twitter data together with CDC reports. It has been found that the Twitter data improves the accuracy of the prediction model. It has been shown that Twitter can provide real time measurement of influenza activity in the population.

5.2   Topic Models (Ailment Topic Aspect Model (ATAM and ATAM+)

Ailment Topic Aspect Model (ATAM) is a topic model that associates words with their hidden topics.   Michael J. Paul et al.   [9] showed that the ATAM model can be used to discover health topics posted by users in Twitter. The model is designed to discover more than a single disease. ATAM can be used to associate symptoms, treatments, and general words with an ailment (disease). Also, it has been shown that the ATAM model can track disease rate which matches the statistics published by the government (CDC).

Paul et al. [10] proposed a variant version of ATAM model called ATAM+. It is an enhanced model that can be used based on what can be learned from Twitter for public health to predict specific diseases such as influenza among other things. The results of the improved model show high quantitative correlation with government data (CDC) in detecting the flu trend using social media.

5.3   Autocorrelation Function (ACF)

ACF finds the correlation of the values of the same variables at different times (x_i, x_{i+1}).   Therefore, this method can be used for disease outbreak predictions. Disease outbreak trends in social media networks can be monitored by tracking a sudden high frequency of disease-content posts using ACF. It compares the averaged disease-related posts per day with the actual number of the same disease posts of that day. Courtney D Corley et al. [11] proposed a method to track ILI in social media using ACF and to identify possible web and social media communities [11]. The method tracks a sudden high frequency of flu-content posts using ACF. The method defines a seven day period as a period cycle for better accuracy and anomaly detection. It starts on Sundays and ends on Saturdays.

The results of this methodology show strong correlation with CDC reports. The Pearson correlation coefficient is used for evaluation. The value of r is 0.767 with a confidence level of 95%.

6   Proposed Framework

Figure 2 shows the components of the proposed framework which includes data collection of social media networks and big data tools for better detections.

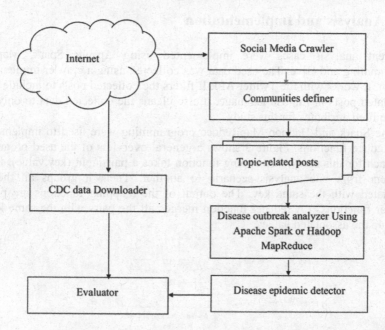

Figure 2. Components of the proposed framework

The crawler block collects data from social media networks. For this study, the data set is a collection of Twitter posts obtained from Twitter social media network. The crawler also filters and cleans the collected data.

Social media networks are very diverse. Therefore, community identification is an important issue in social media network analysis. The communities definer part helps to study only the targeted communities to get more accurate results. Many studies have been conducted for better user location detections such as [12][13]. It has been shown that a user location can be detected even in the lack of geo-location information by applying machine learning techniques on the user's post (tweet) [13].

The filtered disease-related posts for the only targeted community are passed to the analyzer component to be analyzed. This part of the framework utilizes the use of big data tools and techniques.

The results of the analyzer is passed to the detector. Then, the outbreak decision is made by considering the results of the analyzer.

Depending on the problem, a ground truth could be downloaded from different places. Government health-related data could be used as a ground truth. For this study, we used the available reports from the CDC ILINet surveillance system. CDC ILINet system monitors the spread of flu and flu-like illnesses and provides related reports on a weekly basis. Finally, results of the detector and the ground truth are both passed to the evaluator for the evaluation process.

7   Analysis and Implementation

Different analysis cases were implemented using Apache Spark, MapReduce Programming and Hive. The used data was collected using a crawler implemented in Python. It works with the Twitter API. It filters the collected posts to include only the flu related posts in English language. It also cleans the collected data to only consist the required attributes for this study.

Apache Spark and Hadoop MapReduce programming were used to implement Map and Reduce functions. Figure 3 shows a general overview of the used programming approach for this study. A mapping function takes a pair input (key,value). Keys are different from one analysis scenario to another. Then, it groups all the values associated with the same key. The output of the mapping functions are passed to reducer functions.   A reducer function merges all the pairs with the same keys and sums up the associated values.

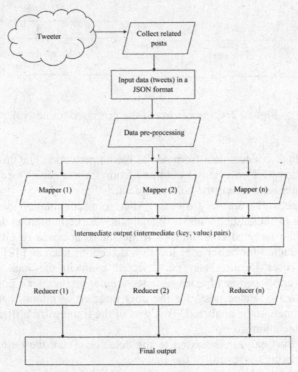

Figure 3. MapReduce approach using social media data

Hive was also used to query the collected data. It provides a SQL-Like script (HiveQL) to query and manage data.

8   Results

This section shows results of various analysis scenarios implemented using Apache Spark, Hadoop MapReduce Programming and Hive. It shows the ability of detecting flu disease outbreaks using social media data.

8.1   Flu Trend Based on Location

This analysis shows the flu trend in different locations. Figure 4 shows the results of this analysis. Results reflect the flu trend in different locations around the world and how it varies from one location to another.

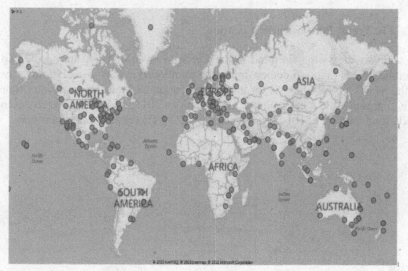

Figure 4. Flu trend based on location

8.2   Flu Trend Based on Time

This analysis shows all the English flu-tweets in November 2016 (week 44 to week 48). Figure 5 demonstrates the results of this analysis. This analysis shows how flu trend changes within a particular time.

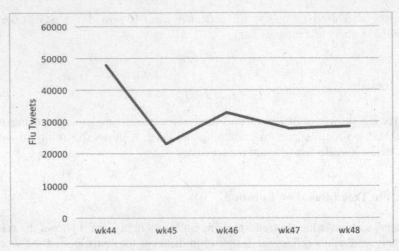

Figure 5. Flu trend based on time (2016)

8.3   Flu Trend for a Specific Location

This analysis shows the flu trend in the Central region of USA during the period of November 2015. Figure 6 demonstrates the number of doctor visits provided by CDC and the results of this analysis for the same location and the same period of time. The x axis represents the week number of November 2015, and the y axis represents the weekly total number of ILI cases reported by CDC/flu related tweets per week.

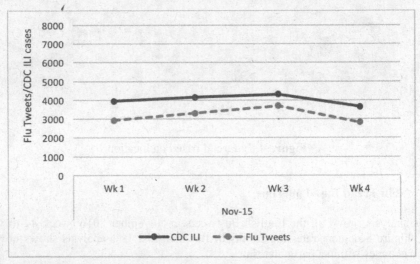

Figure 6. USA central region Flu trend during the period of Nov. 2015

9   Evaluation

An available ground truth can be used to evaluate the quality of the results of the proposed prediction model using big data techniques and tools.

Governments use surveillance systems such as ILINet to monitor the spread of flu and flu-like illnesses and provide related reports to the public. In Canada, the used surveillance system is called fluWatch. It also monitors the spread of flu and flu-like illnesses and provides the related reports.   Google Flu Trend is another system which monitors and counts the number of search queries about flu. It can be used to provide such warnings.   For this study, we used the CDC reports as the ground truth to compare with the proposed solution. Figure 7 shows the total number of ILI obtained from CDC and the total number of flu related posts during the period of November and December of 2015 and 2016. For the evaluation purpose, we used the Pacific Region which includes California, Oregon, Washington, Alaska and Hawaii states. The x axis represents the month, and the y axis represents the flu related tweets per month for the same region. Also, the y axis represents the actual total number of ILI cases reported by CDC excluding the cases for children less than four years old. The results show a strong relationship between the output of the proposed solution and the CDC reports. The Pearson correlation coefficient r is used to evaluate how the two datasets match. Its value ranges between 1 and -1. Let y_i be the observed value of the ground truth (CDC ILINet data), x_i be the predicted value by a proposed model, and y and x be the average values of $\{y_i\}$ and $\{x_i\}$, respectively. Using these notations, Pearson Correlation value r is defined as shown in Equation 1 [14].

$$r = \frac{\sum_{i=1}^{n}(y_i - \overline{y})(x_i - \overline{x})}{\sqrt{\sum_{i=1}^{n}(y_i - \overline{y})^2}\sqrt{\sum_{i=1}^{n}(x_i - \overline{x})^2}} \tag{1}$$

The two datasets match when $r = 1$. The correlation value between the results of the proposed solution and the CDC reports is high ($r=0.98$).

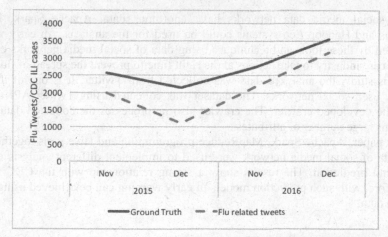

Figure 7. Relationship between the proposed solution and CDC ILI

Table 1 shows performance of some models that were conducted to discover knowledge and predict flu trends from Twitter data. Details of the proposed models is discussed in [1][9][10][11]. As shown in Table 1, the performance of the proposed solution is comparable with the performance of the existing methods.

Table 1. Performance of existing methods

Method name	r	Reference
SNEFT	0.9846	[1]
ATAM	0.934	[9]
ATAM+	0.958	[10]
ACF	0.767	[11]
Proposed Framework	0.98	

10   Conclusion

Most of health care providers need to get an early warning of the flu season. Early detection will help them to take the appropriate actions in order to prevent the spread of flu in a given population. Currently, the required warning is obtained from Center of Disease Control and Prevention (CDC). It takes about 1 to 2 weeks to be published. For that reason, it is important to come up with faster prediction models. With the benefit of big data tools and techniques together, with the use of social media networks, an early warning can be achieved. Presently, people use social media networks on a daily basis to share their news, events, and their daily health status. People can be used as sensors and a big data analysis system can be used to predict flu trend.

Since social media data networks have enormous data, Apache Spark, Hadoop systems and Hadoop Eco systems could be used for the analysis. An early warning obtained by these tools and techniques using data of social media networks can help health care industry to take actions at the right time to prevent the spread of flu.

For this study, the analyzed data was collected from Twitter. It is one of the most widely used social networks. The dataset was collected using Twitter API together with the developed crawler. The crawler also preprocesses the collected data, which contains large number of attributes.

In this paper, Apache Spark, MapReduce programming and Hive tools together, with the data of social media network were used to implement different analysis cases of flu trend prediction. The results show a strong relationship with the CDC reports. Therefore, with such prediction model, an early warning can be achieved using Social Media data.

References

1. Achrekar, H., Gandhe, A., Lazarus, R., Yu, S. H., & Liu, B. (2011, April). Predicting flu trends using Twitter data. *In Computer Communications Workshops (INFOCOM WKSHPS), 2011 IEEE Conference on pp. 702-707, IEEE.*
2. Murthy, D., Gross, A., & Longwell, S. (2011, June). Twitter and e-health: a case study of visualizing cancer networks on twitter. *In Information Society (i-Society), 2011 International Conference on pp.* 110-113, *IEEE.*
3. Nambisan, P., Luo, Z., Kapoor, A., Patrick, T. B., & Cisler, R. (2015, January). Social Media, Big Data, and Public Health Informatics: Ruminating Behavior of Depression Revealed through Twitter. *In System Sciences (HICSS), 2015 48th Hawaii International Conference on pp.* 2906-2913. *IEEE.*
4. Corley, C. D., Cook, D. J., Mikler, A. R., & Singh, K. P. (2010). Text and structural data mining of influenza mentions in web and social media, *International journal of environmental research and public health, 7(2), pp.* 596-615.
5. Dean, J., & Ghemawat, S. (2008). MapReduce: simplified data processing on large clusters, *Communications of the ACM,* 51(1), *pp.* 107-113.
6. Mohammed, E. A., Far, B. H., & Naugler, C. (2014). Applications of the MapReduce programming framework to clinical big data analysis: current landscape and future trends. BioData mining, *7:22, pp. 1-23.*
7. Haryono, G. P., & Zhou, Y. (2016, January). Profiling apache HIVE query from run time logs. *In 2016 International Conference on Big Data and Smart Computing (BigComp) pp.* 61-68, *IEEE.*
8. Verma, A., Mansuri, A. H., & Jain, N. (2016, March). Big data management processing with Hadoop MapReduce and spark technology: A comparison. *In Colossal Data Analysis and Networking (CDAN), Symposium on pp. 1-4, IEEE.*
9. Paul, Michael J., and Mark Dredze. "A model for mining public health topics from Twitter." Health 11 (2012): 16-6.
10. Paul MJ, Dredze M: You are what you Tweet: Analyzing Twitter for public health. *ICWSM 2011, 20:265-272.*
11. Corley C, Mikler AR, Singh KP, Cook DJ: Monitoring Influenza Trends through Mining Social Media. *In: BIOCOMP: 2009; 2009: 340-346.*
12. Jurgens, D. (2013). That's What Friends Are For: Inferring Location in Online Social Media Platforms Based on Social Relationships. *ICWSM, 13, 273-282.*
13. Cheng, Z., Caverlee, J., & Lee, K. (2010, October). You are where you tweet: a content-based approach to geo-locating twitter users. *In Proceedings of the 19th ACM international conference on Information and knowledge management(pp. 759-768). ACM.*
14. Santillana, M., Nguyen, A. T., Dredze, M., Paul, M. J., Nsoesie, E. O., & Brownstein, J. S. (2015). Combining search, social media, and traditional data sources to improve influenza surveillance. *PLoS Comput Biol, 11(10), e1004513.*

An Integrated Approach using Data Mining and System Dynamics to Policy Design: Effects of Electric Vehicle Adoption on CO$_2$ Emissions in Singapore

Bohao Zhang[1], Francis E. H. Tay[1]

[1] Department of Mechanical Engineering, Faculty of Engineering, Block EA, #02-17, National University of Singapore, 1 Engineering Drive 2, Singapore 117576
a0111069@u.nus.edu, mpetayeh@nus.edu.sg

Abstract. This study aims to demonstrate the utility of System dynamics (SD) thinking and data mining techniques as a policy analysis method to help Singapore achieve its greenhouse gases (GHG) emission target as part of the Paris climate agreement. We have developed a system dynamics model called *Singapore electric vehicle and transportation (SET)* and analyzed the long-term impacts of various emission reduction strategies. Data mining techniques were integrated into SD modelling, to create a more evidence-based decision-making framework as opposed to the prevalent intuitive modelling approach and ad hoc estimation of variables. In this study, data mining was utilized to aid in parameter fitting as well as the formulation of the model. We discovered that the current policies put in place to encourage electric vehicle (EV) adoption are insufficient for Singapore to electrify 50% of its vehicle population by the year 2050. Despite not achieving the electric vehicle target, the projected CO$_2$ emission still manages to be significantly lower than the year 2005 business as usual scenario, mainly because of switching to a cleaner fossil fuel for power generation as well as curbing the growth of vehicle population through the Certificate of Entitlement (COE). The results highlighted the usefulness of SD modelling not just in policy analysis, but also helping stakeholders to better understand the dynamics complexity of a system.

Keywords: System Dynamics modelling · Data mining · Bass diffusion model ·Diffusion of innovation · Non-linear least squares regression · Electric Vehicle·

1   Introduction

Policies made by authorities or senior management often have significant implications on those who are affected by the policies. Thus, it is essential for any governing body to be equipped with a toolkit that can systematically analyze the proposed policies and simulate it to test out the desired effects and understand the side effects.

As part of the Paris climate agreement, Singapore pledged to reduce its emissions intensity by 36% from 2005 levels by 2030 ("Singapore's Submission to the United Nations Framework Convention on Climate Change (UNFCCC)," n.d.) The case for

© Springer International Publishing AG 2017
P. Perner (Ed.): ICDM 2017, LNAI 10357, pp. 258–268, 2017.
DOI: 10.1007/978-3-319-62701-4_20

electric cars is now more compelling than ever, given that the transport sector will contribute 14.5% of greenhouse gas (GHG) emissions in Singapore by 2020 (Kuttan, 2016). Electric vehicles, especially cars and buses, are key enablers to help close the gaps in Singapore's commitments to the Paris Agreement, by reducing the use of Conventional vehicles on Singapore's roads. Singapore is aiming to increase usage of electric vehicles to help meet emission reduction targets agreed to at the Paris climate talks. A Land Transport Authority (LTA) led study has found that increasing the use of electric vehicles to half of all cars on the road by 2050 would reduce greenhouse gas emissions by 20 to 30% compared to a business as usual scenario (Marks, 2016).

If such a scenario were to happen, it will have significant long-term impacts on the energy sector in Singapore. The goal of this study is to employ System Dynamics thinking to explore the complex long-term effects of electric vehicle(EV) adoption in the transport sector in Singapore. This study is helpful to gain a better understanding of Singapore's land transport sector and to develop a policy analysis tool to investigate long-term impacts of various emission reduction strategies through a system dynamics simulation modelling. A land transport sector sub-model and EVs diffusion sub-model are developed in this study.

In addition, most of the existing models on the diffusion of EVs present limited interactions with another sector (Massiani, 2013). As many of the consequences of EV policies will transit through other sectors, such as the energy sector, this makes the inclusion of these other sectors necessary for a consistent Cost Benefit Analysis of EV policies (Massiani, 2013). This project aims to fill in this gap. Data mining techniques such as non-linear least square regression are also incorporated to facilitates the model creation process and the integration of their respective elements.

2 System Dynamics(SD) modelling approach

In this section, we briefly illustrate the system dynamics modelling approach. System dynamics is a discipline emerged in the late 1950s, which attempts to explore complex long-term policies in both public and private domains (Sterman, 2000). Jacobsen (1984) states several features of SD methodology that make it suitable for testing the social theory. First, it is possible to handle many variables simultaneously and study their fluctuations over time. Secondly, we can take account of multiple feedback loops in the system under investigation and study their mutual influences over time. Furthermore, we do not have to stick to a linear hypothesis, and can readily model any nonlinear relationship posited by the theory.

SD modelling allows the researcher to analyze complex systems from a cause-and-effect perspective, as compared to from a statistical standpoint. Furthermore, SD modelling allows us to track the various movements (such as money, virus population and people) as well as any buildups that may occur throughout the system (Gil, Matsuura, Monzon, and Samothrakis, 2005). It is also a method for developing management flight simulators, to help us learn about dynamic complexity, understand the sources of policy resistance, and design more effective policies.

Singapore's energy sector is highly integrated and SD was used to capture the interactions and feedback between various sub-systems. To capture the essence of EV adoption in Singapore, a system dynamics model called *Singapore Electric Vehicle and Transportation (SET)* was developed.

System dynamics thinking has been widely applied for policy analysis in several areas in the past. The purpose of a system dynamics (SD) intervention is to identify how structure and decision policies generate system behaviour identified as problematic, so that structural and policy-oriented solutions can be identified and implemented (Forrester, 1961; Sterman, 2000). The approach relies on formal simulation models to capture the detailed complexity of the problem situation and to make reliable behavioural interaction. SD have been used in a wide array of situations such as to examine sustainable water resources management in Singapore (Xi & Poh, 2014), to assess Singapore's healthcare affordability (Ng, Sy, & Jie Li, 2011, to study the complex issues involving Nuclear Energy in Singapore (Chia, Lim, Ng, & Nguyen, 2014), to evaluate different recycling scenarios in Singapore (Giannis, Chen, Yin, Tong, & Veksha, 2016), as well as to design hospital waste management (Chaerul, Tanaka, & Shekdar, 2008).

Nevertheless, it is important to point out that the expected outcomes are not necessarily quantitative point predictions for a variable, but rather a measure of the dynamic behaviour pattern of the system, given the inputs and conditions in the model (Gil, Matsuura, Monzon, and Samothrakis, 2005).

3   *SET* System Dynamics model

A four-step system dynamics modelling process introduced by Sterman (2000) and Ford (1999b) is used in this study, which includes problem articulation, model formulation, model testing, and scenario design and simulation.

On July 3 2015, Singapore submitted its Intended Nationally Determined Contribution (INDC), with a target of reducing the emissions intensity of GDP by 36% below 2005 levels by 2030 and stabilizing emissions, aiming for them to peak around 65 MtCO$_2$ in 2030 (NCCS, 2012)). The multiple energy efficiency measures are expected to improve energy and emission intensity but not to compensate for the increasing energy demand from the industry and buildings sectors, which will result in rising emissions (NCCS, 2012) By contrast, the transport sector energy demand and associated emissions are expected to stagnate as a result of multiple measures to promote public transport and improve the emissions intensity of road transport.

This study aims to employ the *SET* model to examine the current emission reduction measures pertaining to the land transport sector. The effects of three measures were looked at in this study, namely: promoting EV adoption, Encouraging use of public transport, and stricter vehicle ownership. The effectiveness will be evaluated based on two metrics in this study, derived from the policies intended goals:

1.   The transport sector associated CO$_2$ emissions at the year 2050. Two scenarios were compared, the business as usual scenario and the alternate scenario.

This is to assess if Singapore manages to reduce its CO_2 emission intensity from the land transport sector by 2030.

2. The number of EV by the year 2050. Singapore aimed to increase electric vehicle adoption to half of all cars on the road by 2050.

The *SET* model is an integration of 2 sub-models, namely the energy sector and the EVs adoption model. Since the number of electric vehicles is to be determined in 2050, the model will run from the year 2006 to 2050

3.1   Land transport sector model description

The land transport sector model aims to simulate the total CO2 emission from all land vehicles in Singapore. It comprised of five branches namely cars, motorcycles, goods and other vehicles, public buses, and the mass rapid transit (MRT). Figure 1 illustrates the casual loop diagram of the land transport sector. The positive and negative signs represent the reinforcing and balancing casual relationships respectively.

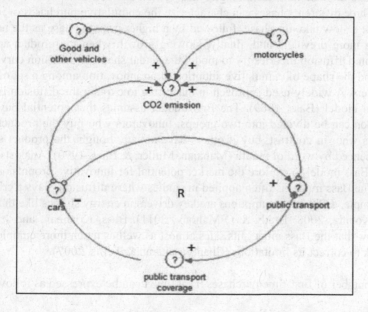

Figure 1: Casual loop diagram of land transport sector

3.2   EV diffusion model

To create the sub-model for the adoption of EVs, the Bass Diffusion Model was adopted, which is shown in Figure 2 below.

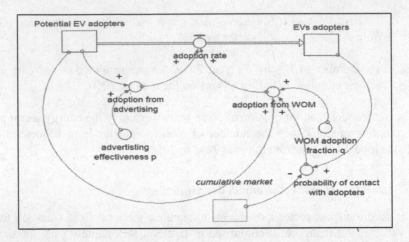

Figure 2: EV diffusion model

The adoption of new technologies has often been found to follow an S-shaped curve. Three different phases can characterize the cumulative purchases of innovative products: a slow take-up phase, followed by a higher growth phase as the technology becomes more prevalent and, finally, slowing growth when the product approaches saturation. Diffusion theories try to model the actual shape of diffusion curves i.e. the speed and the shape of cumulative adoption of an innovation among a set of prospective buyers. A widely-used approach in marketing to explain the diffusion is the Bass diffusion model (Bass, 1969). The Bass model assumes that potential buyers of an innovation can be divided into two groups: Innovators who buy the product first and Imitators who, in contrast, buy if others have already bought the product since they are influenced by word of mouth (Mahajan, Muller, & Bass, 1990). Analysts routinely use the Bass model to explore the market potential for innovative propulsion technologies. The Bass model is often applied in studies where diffusion plays a critical role. For example, diffusion assumptions are key drivers in energy studies like that of (Garrett & Koontz, 2008; Brady & O'Mahony, 2011). Bass, Krishnan, and Jain (1994) also show that the Bass model fits sales almost as well as much more complex models that seek to correct its limitations (Chandrasekaran & Tellis, 2007).

The number of first-time purchases n_t at time t can be expressed as follows:

$$n_t = \frac{dN_t}{dt} = p\,(M - N_t) + q\frac{1}{M}N_t(M - N_t)$$

Where:
n_t: product purchases in period t
N_t: cumulative product purchases until beginning of period t
M: cumulative market potential on the whole product's life cycle
p: coefficient of innovation
q: coefficient of imitation

Decisive for the implementation of the Bass model are the values assigned to the two parameters, p and q, which mathematically describe innovation and imitation mechanisms. In this study, values of p and q were estimated based on a time series of annual electric vehicle sales data from 2006 to 2016 by the method of non-linear least squares regression. The resulting value of p is 0.0000219 and q is 0.02. This result is consistent with a detailed study made by Cao (2004) and Lamberson (2008), who also supports a low p value as compared to other products. The low value of p indicates a slower diffusion in the beginning by innovators and subsequently peak sales at a much later time. This can be illustrated in Figure 3 below, where a smaller p value corresponds to a later peak (Ilonen, 2003).

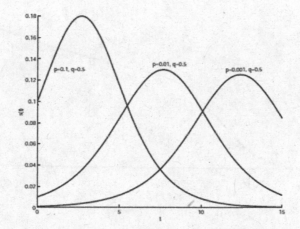

Figure 3: Graph of the Bass model with different values of innovators p

4   Results and Discussions

Two scenarios were simulated using *SET*.

Scenario 1 (SC1): The business as usual scenario, where no actions were taken to curb emissions and

Scenario 2 (SC2): The alternate scenario, where policies are put in place to reduce the emission from the transport sector.

4.1   Number of Electric Vehicles

Figure 4 shows the number of Electric vehicles adopted in Singapore from both scenarios. Both showed an increasing trend with time but differ greatly in magnitude. In SC1, where no efforts were put in place to spur the adoption of electric vehicles, it is shown that the population of EV grew from 1 in the year 2006 to 220 in the year 2050. This is significantly lesser compared to SC2, where the Singapore government introduced one thousandelectric vehicles between the year 2017 to 2020 as part of a vehicle sharing scheme. In SC2, the number of EV grew from 1 to 8267 across the same period. This suggests that the government's plan in introducing an additional one thousand EVs as well as two thousand charging stations from 2017 to 2020 does have some positive effects in encouraging the uptake of electric vehicles.

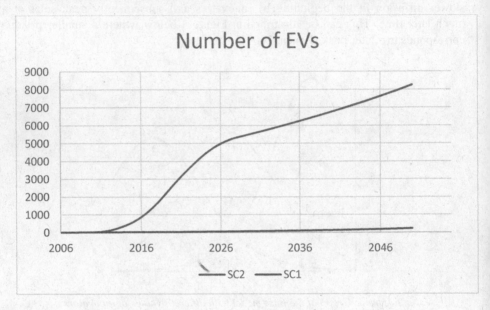

Figure 4: Cumulative number of EV adopters

4.2   Carbon dioxide (CO_2) emission

Figure 5 shows the carbon dioxide emission from both scenarios. SC1 shows a constant growth in emission while SC2 shows a decreasing growth, tending towards a limit. The emission at the year 2050 for SC2 is 9.5 Mt, 30% lower than SC1 at 14Mt. Contrary to the study led by LTA and NTU, the 30% reduction in CO2 emission was achieved even without electrifying 50 % of the vehicle population. Analyzing the simulation of *SET*, it was found that mainly 2 areas, the reduction in vehicle population and decreased emission intensity of power generation, were responsible for this 30% reduction. Since the year 2005, the government has been curbing the vehicle growth rate from 2% in the year 2005 to 1% in the year 2012 and subsequently 0.25% in the year 2015. This has drastically reduced the projected vehicle population by 30% from 1.56 million vehicles to 1.06 million vehicles in the year 2050.

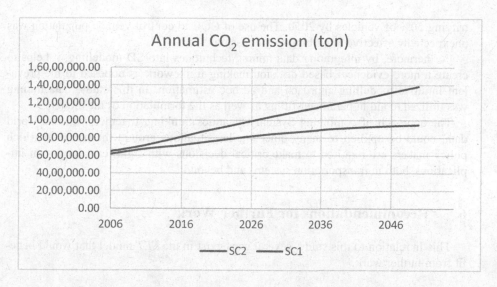

Figure 5: Annual carbon dioxide emission (ton)

Similarly, Singapore has been opting for cleaner energy sources to fuel electricity demand, moving away from petroleum products such as coal and fuel oil to the more environmentally-friendly fossil fuel alternative: natural gas. Natural gas now makes up 95.5% of Singapore's fuel mix, up from 74.4% in 2005 (EMA 2016). The electricity grid emission factor decreased from 0.5255 in the year 2005 to 0.4313 in the year 2015 (EMA, 2016). This greatly reduced the projected CO2 emission from power generation by 32% from 2.29Mt to 1.56Mt in the year 2050.

The results also suggest that the effectiveness of electric vehicle on CO_2 emission is not very evident in the near term. Figure 4 suggest that at the year 2050, EV adoption is still in its first stage. Generally, there is a very low penetration of pure EV in the Singapore market even in the long term till the year 2050. Referencing SC2, EVs only represent 0.8% of the vehicle population in the year 2050, far from the 50% mark that the government aims to achieve. Nonetheless, a 30% reduction in CO_2 emission was still achieved.

5   Conclusion

The aim of this study was to suggest a toolkit for policy analysis and this was largely demonstrated using System Dynamics thinking as a suitable decision support tool for Singapore's electro-mobility roadmap study. Driven by real-world data, the *SET* SD model adequately captures the essence of the land transport sector in Singapore. It is capable of developing management flight simulators to aid stakeholders in understanding dynamic complexity, identifying sources of policy resistance, and design more effective policies. The simulation results have given us insights into managing Singapore's land transport sector. We discovered that the current policies to encourage electric vehicles have limited effectiveness to realize the intention of electrifying 50% of vehicles by 2050. The use of COE to control vehicle population was unexpectedly effective in curbing CO_2 emission.

Furthermore, by integrating data mining techniques into SD modelling, it helps to create a more evidence- based decision making framework as opposed to the prevalent intuitive modelling approach and ad hoc estimation. In this study, data mining was utilized to aid in parameter fitting as well as the formulation of the model.

The approach of combining System Dynamics thinking, together with real-world data, could be applied to many other similar situations around the world in which policy makers are required to make critical decisions with significant long-term implications, both in transportation systems and beyond.

6   Recommendations for Further Work

This In relation to this study, several gaps exist in the *SET* model that would benefit from further work.

1. Emerging technology the likes of Uber and Grab are changing the transportation landscape and reshaping the future of mobility. It gave rise to a highly technology-driven sharing economy platform, where transportation is supplied "on demand', often very promptly. This gave users of such platforms more modes of transportation and could affect the effectiveness of policies that encourages public transportation. A Limits to Success system archetype as show in figure 6 between the emerging technology and public transport could be included in *SET* to deliver a more precise simulation.

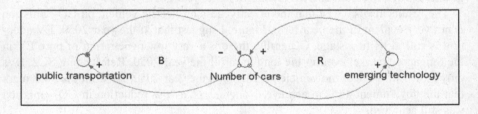

Figure 6: Limits to Success system CLD of public transportation and technology

2. The *SET* model could be broadened to include energy usage and the associated GHG emission from the whole country. The usage of EV has a direct impact on the energy sector. Several studies have attempted to consider the implications of different projections of future EV ownership on the demand and peak load of electricity under a range of charging assumptions (Electric Power Research Institute, 2007; Center for Entrepreneurship & Technology, 2009; and Dawar, Lesieutre, & Argonne, 2011). This helps to extend the capabilities of the model from a policy analysis tool to a more holistic decision support tool for transport policy planning.

The effects of price on EV adoption could be explored and subsequently included in the EV adoption sub-model to model its effects. Krishnan and Jain (1994) suggested an extended version of the original Bass diffusion model which includes the effects of advertising and price changes. The inclusion of such features can better assist the Singapore government in reviewing its Carbon Emissions-based Vehicle Scheme (CEVS),

References

1. Alcantara Gil, Benigno R. Matsuura, Masahiro. Monzon, Carlos Molina. Samothrakis, Ioannis. (2005). The use of system dynamics analysis and modeling techniques to explore policy levers in the fight against Middle Eastern terrorist groups. Monterey, California: Naval Postgraduate School.
2. Bass, F. M. (1969). A New Product Growth for Model Consumer Durables. *Management Science, 15*(5), 215-227. doi:10.1287/mnsc.15.5.215
3. Brady, J., & O'Mahony, M. (2011). Travel to work in Dublin: The potential impacts of electric vehicles on climate change and urban air quality. *Transportation Research Part D: Transport and Environment, 16*(2), 188-193. doi:10.1016/j.trd.2010.09.006
4. Cao, J., & Mokhtarian, P. L. (2004). The future demand for alternative fuel passenger vehicles: A diffusion of innovation approach.
5. Chaerul, M., Tanaka, M., & Shekdar, A. V. (2008). A system dynamics approach for hospital waste management. *Waste Management, 28*(2), 442-449. doi:10.1016/j.wasman.2007.01.007
6. Chandrasekaran, D., & Tellis, G. J. (2007). A Critical Review of Marketing Research on Diffusion of New Products. *Review of Marketing Research*, 39-80. doi:10.1108/s1548-6435(2007)0000003006
7. Chandrasekaran, D., & Tellis, G. J. (2010). Diffusion of Innovation. *Wiley International Encyclopedia of Marketing*. doi:10.1002/9781444316568.wiem05015
8. Chia, E. S., Lim, C. K., Ng, A., & Nguyen, N. H. (2014). The System Dynamics of Nuclear Energy in Singapore. *International Journal of Green Energy, 12*(1), 73-86. doi:10.1080/15435075.2014.889001
9. Dawar, V., Lesieutre, B. C., & Argonne, T. P. (2011). Impact of Electric Vehicles on energy market. *2011 IEEE Power and Energy Conference at Illinois*. doi:10.1109/peci.2011.5740487
10. Garrett, V., & Koontz, T. M. (2008). Breaking the cycle: Producer and consumer perspectives on the non-adoption of passive solar housing in the US. *Energy Policy, 36*(4), 1551-1566. doi:10.1016/j.enpol.2008.01.002
11. Giannis, A., Chen, M., Yin, K., Tong, H., & Veksha, A. (2016). Application of system dynamics modeling for evaluation of different recycling scenarios in Singapore. *Journal of Material Cycles and Waste Management*. doi:10.1007/s10163-016-0503-2
12. Ilonen, J. (2013). Predicting diffusion of innovation with self-organisation and machine learning.
13. Josh Marks. (2016, August 7). Singapore aims for 50% electric vehicles by 2050 | Inhabitat - Green Design, Innovation, Architecture, Green Building. Retrieved from http://inhabitat.com/singapore-aims-for-50-electric-vehicles-by-2050
14. Krishnan, T. V., Bass, F. M., & Jain, D. C. (1999). Optimal Pricing Strategy for New Products. *Management Science, 45*(12), 1650-1663. doi:10.1287/mnsc.45.12.1650

15. Lamberson,, P. J. (2008). The diffusion of hybrid electric vehicles. Future research directions in sustainable mobility and accessibility. In Sustainable mobility accessibility research and transformation (SMART) at the University of Michigan center for advancing research and solutions for society (CARSS).

16. Mahajan, V., Muller, E., & Bass, F. M. (1991). New Product Diffusion Models in Marketing: A Review and Directions for Research. *Diffusion of Technologies and Social Behavior*, 125-177. doi:10.1007/978-3-662-02700-4_6

17. Massiani, J. (2013). The Use of Stated Preferences to Forecast Alternative Fuel Vehicles Market Diffusion: Comparisons with Other Methods and Proposal for a Synthetic Utility Function. *SSRN Electronic Journal*. doi:10.2139/ssrn.2275756

18. Massiani, J., & Gohs, A. (2015). The choice of Bass model coefficients to forecast diffusion for innovative products: An empirical investigation for new automotive technologies. *Research in Transportation Economics*, *50*, 17-28. doi:10.1016/j.retrec.2015.06.003

19. NCCS. (2012). National Climate Change Secretariat (NCCS) - Climate Change and Singapore: Challenges. Opportunities. Partnerships. Retrieved from https://www.nccs.gov.sg/nccs-2012/sectoral-measures-to-reduce-emissions-up-to-2020.html

20. Ng, A. T., Sy, C., & Jie Li. (2011). A system dynamics model of Singapore healthcare affordability. *Proceedings of the 2011 Winter Simulation Conference (WSC)*. doi:10.1109/wsc.2011.6147853

21. Sanjay C Kuttan. (2016, August 2). Level the playing field for adopters of electric vehicles in Singapore, AsiaOne Singapore News. Retrieved from http://news.asiaone.com/news/singapore/level-playing-field-adopters-electric-vehicles-singapore

22. Singapore. Energy Market Authority. (2011). *Singapore energy statistics: Energising our nation*. Singapore: Research and Statistics Unit, Energy Market Authority.

23. Singapore's Submission to the United Nations Framework Convention on Climate Change (UNFCCC). (n.d.). Retrieved from https://www.mfa.gov.sg/content/mfa/overseasmission/geneva/press_statements_speeches/2015/201507/press_20150703.html

24. Sterman, J. D. (2000). Business dynamics: Systems thinking and modeling for a complex world. Boston: Irwin/McGraw-Hill.

25. Xi, X., & Poh, K. L. (2014). A Novel Integrated Decision Support Tool for Sustainable Water Resources Management in Singapore: Synergies Between System Dynamics and Analytic Hierarchy Process. *Water Resources Management*, *29*(4), 1329-1350. doi:10.1007/s11269-014-0876-8

26. Zhuang, Y., & Zhang, Q. (2014). Evaluating Municipal Water Management Options with the Incorporation of Water Quality and Energy Consumption. *Water Resources Management*, *29*(1), 35-61. doi:10.1007/s11269-014-0825-6

Using the Results of Capstone Analysis to Predict a Weather Outcome

Anthony G Nolan[1] and Warwick J. Graco[2]

[1] G3N1U5 Sydney Australia
tony@g3n1u5.com
[2] ATO Canberra Australia
warwick.graco@ato.gov.au

Abstract. In this paper the results of capstone analysis is applied to predict a weather outcome using a decision-tree model. It examines weather data of the capital cities of Australia in a 12 month period to see if the decision-tree models can predict rain in Sydney the next day. It produces a decision-tree model with the raw weather data for each capital city to make the predictions about this outcome. It also aggregates the raw data to provide a combined-city dataset. Finally, it combines and compresses the raw data for each city using capstone modelling. The capstone data for the cities is used to train another decision-tree model to see if this provides better predictions of rain in Sydney than those obtained from using single-city models and the combined-city model. The results of these comparisons and details of how capstoning works are provided in the paper

Keywords: Weather forecasting. Capstone Modelling, Classification and Regression Trees. Rattle

1   Introduction

Weather is a key factor that affects citizens in a variety of ways including as examples leisure activities, work attendance and moving from one location to another. It also impacts on shopping and use of health services. Adverse weather can be a risk to the community by threatening life and property.

In this paper weather is viewed as the current and predicted near future state of the atmosphere for a particular geographic area or location. In practicable terms, weather describes a series of environmental systems characterized by variables and'measurements. These can be categorized according to their impact. For instance, the level of temperature, the speed and direction of the wind, the amount of light, moisture and the amount of solar radiation received at any time influence the weather in a particular place. This in turn can affect the behaviours of citizens.

This paper reports the results of an investigation into weather as a large-scale, complex system and how it influences meteorological conditions in a specific geographic location. Specifically it models the weather conditions in the capital cities in Australia to predict whether it will rain the next day in Sydney, New South Wales,

© Springer International Publishing AG 2017
P. Perner (Ed.): ICDM 2017, LNAI 10357, pp. 269–277, 2017.
DOI: 10.1007/978-3-319-62701-4_21

Australia. The weather data for the capital cities used in this study include Perth West Australia, Adelaide South Australia, Darwin Northern Territory, Brisbane Queensland, Canberra Australian Capital Territory, Melbourne Victoria and Hobart Tasmania.

By using weather measurements such as wind, air pressure, sun shine hours, cloud cover, rain and evaporation in these various cities it enabled these datasets to be employed in a decision-tree model to predict the percentage probability of rain the next day in Sydney. This was done in three ways. The first was that the raw weather data from each capital city was used to predict this output. The second was to combine the raw weather data from all the capital cities bar Sydney to predict the output. The third was to combine and compress the data across these cities to predict the same output using a method called 'capstone analysis'. How capstone works is explained below.

The remainder of this paper reports the results of using single-city data, combined-city and the capstone data to predict whether it would rain the next day in Sydney. The data used in the analysis is described first.

2   Methodology

2.1   Data

The data set was taken from the Australian Bureau of Meteorology (BOM) website. This was for the period 1st November 2008 to 31st October 2009. The dataset contained 12 months of data, and has the following variables. Date, Day, Min Temp, Max Temp, Rain, Evaporation, Sunshine hours, Max wind gust (direction, Speed, Time), Then Temp, relative humidity, cloud cover, wind direction, wind speed and air pressure for 9am, and for 3pm (see Figure 1 for an extract of the data that can be obtained from the BOM website).

		Temps		Rain	Evap	Sun	Max wind gust			9 am						3 pm					
Date	Day	Min	Max				Dir	Spd	Time	Temp	RH	Cld	Dir	Spd	MSLP	Temp	RH	Cld	Dir	Spd	MSLP
		°C	°C	mm	mm	hours		km/h	local	°C	%	8th		km/h	hPa	°C	%	8th		km/h	hPa
1	We	20.6	27.2	26.4	7.2	2.2	E	43	06:34	21.4	87	6	SE	9	1021.0	25.0	67	7	ESE	22	1019.4
2	Th	20.5	26.3	3.2	3.6	7.6	E	26	13:03	23.0	81	3	WNW	13	1018.1	25.0	69	5	E	15	1016.0
3	Fr	20.7	24.8	37.8	5.2	1.4	SE	63	14:39	21.7	89	7	S	6	1016.9	22.2	85	8	SE	17	1016.0
4	Sa	19.9	23.3	27.6	4.0	0.2	SSE	54	12:51	21.5	87	7	SE	19	1015.2	20.9	86	7	S	20	1013.1
5	Su	20.1	23.0	7.6	2.4	0.0	SW	56	19:30	20.4	78	7	W	20	1010.2	22.1	76	7	SW	24	1009.6
6	Mo	19.6	25.7	2.4	4.0	7.7	SSW	65	11:41	20.5	63	5	WSW	19	1012.8	25.5	49	3	SSW	33	1013.2

Fig. 1. An Extract from the Australian Bureau of Meteorology Weather Data

2.2   Capital-City Models

The variables used for each capital city decision-tree model to produce the predictions for Sydney are listed in Table 1. These variables were converted to their raw values. For example, temperatures are converted from Celsius to Kelvin. The idea behind this was that where possible that numeric variables should have a ratio scale with a true zero so all variations between observations are a true representation of their movements.

Table 1. Weather Variables

Heading	Meaning	Units
Date	Day of the month	Day
Location	Location of observations	Name
Min Temps	Minimum temperature in the 24 hours to 9am.	degrees Celsius
Max Temp	Maximum temperaure in the 24 hours from 9am.	degrees Celsius
Rainfall	Precipitation (rainfall) in the 24 hours to 9am.	millimeters
Evaporation	Class A pan evaporation in the 24 hours to 9am	millimeters
Sunshine	Bright sunshine in the 24 hours to midnight	hours
WindGustDir	Direction of strongest gust in the 24 hours to midnight	16 compass points
WindGustSpeed	Speed of strongest wind gust in the 24 hours to midnight	kilometers per hour
WindSpeed	Wind speed averaged over 10 minutes intervals	kilometers per hour
Humidity	Relative humidity	percent
Pressure	Atmospheric pressure reduced to mean sea level (MSL)	hectopascals
Cloud	Fraction of sky obscured by cloud	eighths
RainToday	Did it rain the day of the observation	Yes/No

A binary target variable was added to the extracted BOM datasets to indicate if there was rain on each day of the one year period that was covered for Sydney. These converted datasets were used as inputs in a decision-tree model to predict if it will rain the next day.

The level of moisture had to be > 1mm for it to be judged that it had rained that day. This cut-off was selected to allow for days where there were frosts or heavy dews but no rain.

2.3 Capstone Analysis

Capstone Model (see Figure 2 for an example with weather) is a hierarchy of networked models where a small change in a lower model can have a cascading effect on the models higher up the hierarchy. There are parent-child relationships between models in the hierarchy. The basic challenge is how to represent data in a cause and effect framework which has multiple sources, that exists at different levels, and has dynamic properties. The capstone model employed in this study enables these issues to be accommodated in the modelling process and be self-adjusting and dynamic across the entire model structure. This is the approach used to predict the weather in Sydney.

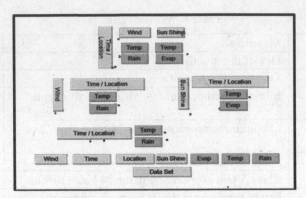

Fig. 2. Capstone Model of a Complex System of Events

In terms of how capstone analysis was applied to the weather data for the capital cities, the data for temperature, moisture and wind as examples for each day in each city was transformed using the peer relativity transformation formula described at Appendix 1 to this paper. This rescales the observations between 0 and 99. The rescale is determined by using the maximum value and the minimum value for that measurement within the Australian recorded weather observations. This provides a uniform rescale for all measurements.

The transformed scores were then converted to digital hash scores using a formula described in Appendix 2 to this paper. The resulting digital hash scores were then converted into peer relativity scores and were all combined and compressed into a single digital hash score in the middle layer. The resulting digital hash for each day was converted to a peer relativity score again and was provided as output in the top layer also called 'capstone layer'. This single score for each day contained all of the rolled-up weather data for the selected capital city. This process was repeated for the other 364 days of the chosen year. This provided a dataset consisting of capstone scores for eight cities for 365 days. This dataset provided the inputs for producing the decision-tree predictions of rain in Sydney next day.

2.4 Decision-Tree Model

The decision-tree model employed to make predictions with weather data is a version of the CART decision tree [1] sourced from the 'rpart' package of R. https://cran.r-project.org/web/packages/rpart/vignettes/longintro.pdf

This decision tree was chosen, as it is the most popular decision-tree in R and it is integrated into the program tools.

The model was accessed via the 'Rattle' package user interface. It has a special log window for producing the code and to cut and paste this into other software applications. More information on Rattle is found at https://cran.r-project.org/web/packages/rattle/rattle.pdf . The standard Rattle method was employed to run the decision-tree models where 70 percent of the data was used to train the models and the remaining 15 percent was employed to validate and a further 15 percent to test their performances [2].

The validation dataset is used to test different parameter settings or to test different choices of variables of the model. The testing dataset is only to be used to predict the unbiased error of the final results. It is important to note that this testing dataset is not used in any way in either building or fine tuning the models. This will provide an understanding of the level of accuracy of the model with previously unseen data. The model once developed and evaluated was applied to the whole dataset to predict the output of rain next day in Sydney.

An adjustment was also made for the distance of the furthest of the cities in Australia from Sydney to make allowances for time zone differences. This was done by altering the observations. Adelaide and Darwin in the middle of Australia were adjusted by one day and the most western city of Perth by two days.

Each city dataset containing 365 days of data was scored using the relevant decision-tree model to give a prediction of Sydney weather the following day. This gave 365 days of predicted rain for Sydney for each dataset for both individual city and combined cities. The predicted results were compared to the actual results and the agreement, precision and recall rates for each dataset were derived. The formula for deriving each of these performance measurements is listed in Table 2. The above steps were repeated for the capstone dataset to produce decision-tree predictions for Sydney weather and their performance results.

Table 2. Measurements used to Assess the Performance of the Decision-Tree Models

Performance Measurement	Formula	
		TP = True Positives or those that meet the criterion such as rainy day
		TN = True Negatives or those who do not meet the criterion such as dry day
Agreement Rate	=TP+TN/(TP+TN+FP+FN)	
Precision	=TP/(TP+FP)	FP = False Positives or those who appear to be positives but are negatives
Recall	=TP/(TP+FN)	FN = False Negatives or those who appear to be negatives but are positives

3   Results

The agreement, precision and recall rates for each dataset are shown Table 3 below. The results show that decision-tree model produced using capstone analysis produced the best results for predicting rain in Sydney next day compared to the results produced using the individual-city datasets and the combined-city dataset. It had a precision rate of 72.3 percent in predicting rain next day in Sydney. The precision rate for the combined city model was 46.3 percent and for the individual-city other models ranged from the low 30s through to the highest of 46.6 percent for the Melbourne, Canberra and Brisbane models.

The capstone decision-tree model also had the highest agreement rate of 87.9 percent indicating it was best predicting days with rain and days without rain the next day in Sydney. The next highest was the combined-city model of 75.6 percent. The individual-city models ranged from the high 60s to the mid 70s in their percentage agreements. The agreement rates were higher across the board for all datasets than the precision rates. This suggests the decision-tree models were better at predicting days that were dry rather those that were wet in Sydney.

The combined-city model had the very high recall rate of 91.7 percent followed by the capstone model of 82 percent. Recall rates for the individual-city models tended to be low with the best performing decision-tree model being for the Melbourne dataset. It had a recall rate of 50.3 percent.

4   Discussion

This analysis has focused on the performance of the decision-tree models using weather data obtained for capital cities in Australia to predict whether it would rain in Sydney next day. Single-city, combined-city and the capstone data were used for this purpose.

Attention was not focused on weather measurements such as wind, air pressure, sun shine hours, cloud cover, rain and evaporation that best predict the outcome. It is acknowledged that some of these will be more important than others in making the predictions. This issue is not discussed further in this paper.

Table 3. Performance of the Decision-tree Models for Predicting Rain in Sydney the Next Day

	Agreement Rate	Precision in Predicting Rain	Recall Rate
Perth Model	70.0%	36.3%	33.6%
Adelaide Model	70.6%	38.5%	33.7%
Darwin Model	67.7%	33.7%	34.0%
Hobart Model	73.7%	46.0%	44.9%
Melbourne Model	74.5%	46.6%	50.3%
Canberra Model	72.3%	40.6%	29.2%
Brisbane Model	72.3%	40.6%	29.2%
Sydney Model	73.4%	44.0%	37.1%
Combined-City Model	75.6%	46.3%	91.7%
Capstone Model	87.9%	72.3%	82.0%

This begs the question why did the model developed using the capstone data perform far better than the combined-city and individual-city models using the weather data in its raw form in predicting rain in Sydney. The following are suggested reasons as there has been no scientific analysis to ascertain what it is about capstoning that enabled it to perform better than the other models in making these predictions. This would require extensive analysis including testing various combinations of weather measurements to establish their impacts on the predicted outcomes.

Possible reasons why the capstone model performed better than the combined-city and the single-city models include:

- There is a 'boosting' process occurring inside the capstoning which helps to provide superior modelling results
- The digital hash formula used to compress data is a distance preserving transformation so that it brings like measurements together in a similar way that clustering algorithms bring like cases together in the same cluster

Capstone modelling provides a framework for the mapping of data points, the context for the results, and the means for comparing observations. It enables the building of an overarching model which is greater than the sum of its parts. Original data is transformed using the peer relativity formula and the digital hash formula into different combinations based on the analysis being undertaken. The digital hash gives a variable result while the peer relativity binning gives a grouped result. By grouping the hashed scores, an index which represents any number of observations can be created. The digital hashing of groups provides an index of grouped results. This allows comparisons between entities using the variables employed in the analysis. This also enables dynamic modelling where variables can be changed, inserted or deleted at any level. The adjustment has a cascading effect upon the entire model, and depending on the significance and sensitivity threshold of its contribution to the model.

In terms of an analogy, capstone analysis appears to operate in a similar manner to an artificial neural network where there is an input layer, a middle layer and an output layer. Instead of learning the patterns in the connections between the nodes of the neural net, capstone analysis transforms and combines scores for variables in the layers of the hierarchy to produce an output in the top layer. The capstone output in the top layer includes implicitly the complexities of the weather events in the cities that are used as inputs to this process.

There are many different uses of capstone analysis. It has been used with social networking analysis, decision making mapping, surveys analysis, sensor measurements, autonomous control, role play games, master character mapping and text mining. Though the evidence here is anecdotal it has been used to analyze populations to find very hard to identify cases of fraud and abuse – what are sometimes referred to as finding 'needles in the haystack'.

There are many potential uses of capstone analysis. It needs to be explored more extensively via the use of other datasets and its application to a broad range of issues to see how it fares. This would help confirm the usefulness of the process and whether it adds another capability to the 'armory' of those who analyze data to discover hard to detect cases and issues and those who seek to obtain superior modelling results.

5   Conclusion

Capstone analysis is a recently developed tool for analyzing data. Its major advantages are that it can combine qualitative and quantitative data together in compressed form that can capture the relationships between components that make a capstone and that it can be used for both supervised and unsupervised learning. It requires extensive testing and evaluation to confirm its efficacy and to assess its ubiquity. Until this is done any preliminary results that arise from the use of this technique are tentative. This applies to the results reported in this paper.

For those who would like to know more about the capstone methodology contact Tony Nolan, tony@g3n1u5.com

References

1. Breiman, L; Friedman, J. H.; Olshen, R. A.; Stone, C. J. <u>Classification and regression trees</u>. Monterey, CA: Wadsworth & Brooks/Cole Advanced Books & Software (1984)
2. Williams, G. <u>Data Mining with Rattle and R. The Art of Excavating Data for Knowledge Discovery.</u> New York: Springer. Section2.3 (2011)

Appendix 1

Nolan's Peer Relativity Transform

This is about taking an observation, which is defined by a series of characteristics. These characteristics define a subpopulation. Each subpopulation is extracted from the entire population, and the observation is then assigned a value which is relative to all the other members of that sub-population. For example, if an observation has five different characteristics to describe it, then that observation exists in five different subpopulations, and hence that value will be represented five different times, each representation ranging with values between 0 and 99.

The formula for doing this transformation of scores is:

$$X = \frac{(x - xmin)}{(xmax - xmin)} * n$$

Where $x \subseteq P$ and n = number of bins.

Appendix 2

Digital Hash Scores

This is a distance preserving discrete cosine transformation used to provide a digital fingerprint score for one or more variable scores. A digital fingerprint is an index value which can only be calculated by a unique combination of observations, either within the same variable, or across a number of variables. While it is not a perfect digital hash, it is reliable within the limitations defined by the matrix. To date the hash has worked across over 200 observations.

The formula for calculating a hash score is:

$$NDR\ Score = \sum_{i=1}^{n} X_i * Cos(7*i)$$

Where n = number of variables and X_i = ith variable score

Classification of Network Traffic Using Fuzzy Clustering for Network Security

Terrence P. Fries

Department of Computer Science, Indiana University of Pennsylvania, Indiana, PA USA
t.fries@iup.edu

Abstract. The use of computer networks has increased significantly in recent years. This proliferation, in combination with the interconnection of networks via the Internet, has drastically increased their vulnerability to attack by malicious agents. The wide variety of attack modes has exacerbated the problem in detecting attacks. Many current intrusion detection systems (IDS) are unable to identify unknown or mutated attack modes or are unable to operate in a dynamic environment as is necessary with mobile networks. As a result, it has become increasingly important to find new ways to implement and manage intrusion detection systems. Classification-based IDS are commonly used, however, they are often unable to adapt to dynamic environments or to identify previously unknown attack modes. Fuzzy-based systems accommodate the imprecision associated with mutated and previously unidentified attack modes. This paper presents a novel approach to intrusion detection using fuzzy clustering of TCP packets based upon a reduced set of features. The method is shown to provide superior performance in comparison to traditional classification approaches. In addition, the method demonstrates improved robustness in comparison to other evolutionary-based techniques.

Keywords: fuzzy clustering, fuzzy classification, intrusion detection, network security.

1   Introduction

As the use of computer networks becomes commonplace in society and users require more mobility, networks have been subjected to a dramatic increase in intrusions and malicious attacks. This necessitates finding more reliable and robust techniques to protect networks. Intrusion detection systems (IDS) provide network administrators with a tool to recognize normal and abnormal behavior of network traffic packets. An IDS is capable of identifying network intrusions, including malicious attacks, unauthorized access, and other anomalous behaviors. Due to the enormous volume of traffic now experienced by networks, application of big data analytics provides a promising approach to identifying intruders.

The majority of IDS utilize supervised or unsupervised pattern-recognition techniques to construct meta-classifiers which are then used for intrusion detection. These methodologies include statistical models, immune system approaches, protocol verifi-

© Springer International Publishing AG 2017
P. Perner (Ed.): ICDM 2017, LNAI 10357, pp. 278–285, 2017.
DOI: 10.1007/978-3-319-62701-4_22

cation, file and taint checking, neural networks, whitelisting, expression matching,
state transition analysis, dedicated languages, genetic algorithms, and burglar alarms
[1, 2]. These techniques function satisfactorily only in well-defined environments.
However, they are unable to identify complex or unknown attacks, or to adapt to dynamic environments such as mobile networks. Clustering provides a classification
technique, however, many of these approaches only identify elementary attacks and
fail when confronted with complex or unknown attacks [3].

Fuzzy clustering has been shown to be superior to traditional clustering and it
overcomes the limitation imposed by unknown attack modes and dynamic environments [4]. Despite its superior performance, most current methods experience limitations. Jiang [5] attempted to use a modified fuzzy c-means clustering algorithm, however, the system produces a poor 75-85% success rate. Other attempts suffer from a
high rate of false positives [6, 7, 8].

This research uses fuzzy c-means clustering to classify TCP packets as normal or
an intruder. However, this approach does not implement the modifications that hin-
dered Jiang's [5] success rate. The clusters are formed by analyzing very large datasets of TCP packets. Initial testing has shown it to provide excellent identification of
malicious packet without the high false positive rate.

Section 2 provides background on existing intrusion detection methods with a discussion of the advantages and limitation of each, an industry-standard test dataset, and
the concept behind fuzzy c-means clustering. The proposed method is described in
Section 3. Section 4 describes testing of the proposed fuzzy clustering IDS and the
test results. Conclusions and directions for future research regarding the fuzzy clustering IDS are in Section 5.

2   Background

2.1   Network Traffic Dataset

When comparing the performance of various intrusion detection techniques, it is
vital to use a common dataset for testing. The KDD Cup 1999 Dataset [21] has become the de facto standard for testing intrusion detection security systems. The KDD
dataset was developed by MIT Lincoln Labs from the 1998 DARPA Intrusion Detection Evaluation Program. The program simulated a typical U. S. Air Force network as
it was subjected to a variety of attacks. The dataset was created using a simulated
military network environment mimicking a typical U.S. Air Force local area network
subjected to normal traffic and simulated attacks. The dataset is comprised of 4GB of
compressed TCP dump data including approximately 5 million connection records
collected over a seven week period of network traffic. Each record contains forty-one
qualitative and quantitative features which were extracted from the TCP connection
data. The features include three categories: (1) basic features of the individual TCP
connections, (2) content features within a connection, and (3) features regarding error
rates using a two-second time window. Each record also includes a flag indicating

whether it is normal or an intrusion. Abnormal connections specify the type of attack represented by the connection.

The KDD dataset contains 24 known types of simulated attacks. In addition, 14 unknown types of attacks are included to test a systems ability to detect previously unknown novel attack modes or mutated attackers. The simulated attacks include all four of the most common categories of attack:

1. *Denial of service* (DoS). DoS attacks make the host computer or network unavailable for intended users. This is usually accomplished by flooding the host with TCP packets in order to overload the server. It may also be facilitated by targeting bugs in the system or exploiting flaws in the system configuration
2. *User to root* (U2R). In U2R attacks, the attacker gains access using a normal user account. Once access has been established, the intruder is able to exploit vulnerabilities in the system to gain root access. The two most common U2R attacks are to overflow the static buffer and to initiate program race conditions.
3. *Remote to user* (R2L). In R2L attacks, the attacker sends packets through a legitimate user's machine to the host. The intruder then gains access as a legitimate user by exploiting vulnerabilities in server software and/or security policies.
4. *Probing* (PROBE). Probe attacks are characterized by the attacker scanning a network to collect information in order to recognize known vulnerabilities. Probe attacks frequently use port scanning.

2.2 Fuzzy C-Means Clustering

Unlike traditional clustering, fuzzy c-means (FCM) clustering method allows one data item to belong to many clusters. The amount to which an item belongs to a cluster is expressed as a fuzzy set. FCM was developed by Dunn [10] and revised by Bezdek [11, 12].

The partitioning of data into c clusters is based on minimization of the generalized least-squared errors objective function

$$J(Z; U, V) = \sum_{k=1}^{N} \sum_{i=1}^{c} (u_{ik})^m \|z_k - v_i\|_A^2 \tag{1}$$

where Z is a vector containing the data such that

$$Z = \{z_1, z_2, \dots, z_n\} \in \mathbb{R}^n \tag{2}$$

Thus, z_k is the kth of the d-dimensional measured data and v_i is the d-dimensional centroid of the ith cluster. U is the fuzzy c-partition of Z containing the fuzzy membership functions corresponding to each data item defined as

$$U = [u_{ik}] \in M_{fc} \tag{3}$$

V is a vector of the cluster centers, or centroids,

$$V = \{v, v_2, \dots, v_c\} \in \mathbb{R}^n \tag{4}$$

The squared distance between any data item, z_k, and a cluster centroid, v_i, is determined as the inner-product distance using the A-norm as

$$d_{ikA}^2 = \|z_k - v_i\|_A^2 = (z_k - v_i)^T A(z_k - v_i) \tag{5}$$

The A-norm is a matrix that normalizes the data set. In practice, the A-norm may be a Euclidean, Diagonal, or Mahalonobois norm.

The weight associated with each squared error is $(u_{ik})^m$ where the weighting exponent, or fuzziness parameter, m controls the weight of the squared errors and, thus, determines the fuzziness of the clustering. Larger values of m result in more fuzziness in the memberships. While m may be any value greater than or equal to 1, testing has indicated that for most data, $1.5 \leq m \leq 3.0$ yields appropriate results [11].

Fuzzy clustering is accomplished by iterative optimization of the objective function, J. In each iteration, the membership functions u_{ik} and the cluster centroids v_i are updated using

$$u_{ik} = \frac{1}{\sum_{j=1}^{c}(d_{ikA}/d_{jkA})^{2/(m-1)}}, \quad 1 \leq i \leq c, \ 1 \leq k \leq N \tag{6}$$

and

$$v_i = \frac{\sum_{k=1}^{N}(u_{ik})^m z_k}{\sum_{k=1}^{N}(u_{ik})^m}, \quad 1 \leq i \leq c \tag{7}$$

Fuzzy c-means clustering (FCM) for this research is accomplished using the method shown in Figure 1.

3   Network Security Using Fuzzy Clustering

The network security system has three basic phases as shown in Fig. 2. The first two phases construct the classification system, and the third is the actual intrusion detection system. The phases are as follows:

1. *Feature subset selection.* Use dimension reduction techniques to reduce the number of TCP packet features necessary to classify a packet in a fuzzy cluster.
2. *Fuzzy clustering.* Apply fuzzy clustering to identify the centroid of n clusters. Label those clusters which are populated predominantly by malicious packets.
3. *Classify incoming TCP packets.* As packets arrive, determine the degree of membership in each cluster and flag those which demonstrate a strong association with clusters designated as malicious.

Initialize the following parameters:
- data set Z
- number of clusters $1 < c < N$
- weighting exponent m
- termination tolerance $\varepsilon > 0$
- norm-inducing matrix $\mathbf{A}$

Initialize the partition matrix randomly, such that $U^{(0)} \in M_{fc}$

Repeat for each iteration $l = 1, 2, \ldots$

1. Compute the cluster centroids:

$$v_i^{(l)} = \frac{\sum_{k=1}^{N} \left(u_{ik}^{(l-1)}\right)^m z_k}{\sum_{k=1}^{N} \left(u_{ik}^{(l-1)}\right)^m}, \quad 1 \le i \le c \tag{8}$$

2. Compute the distance of each data item to the new cluster centroids:

$$d_{ik\mathbf{A}}^2 = \left(z_k - v_i^{(l)}\right)^T A \left(z_k - v_i^{(l)}\right) \tag{9}$$

3. Update the partition matrix:

$$u_{ik}^{(l)} = \frac{1}{\sum_{j=1}^{c} \left(d_{ik\mathbf{A}}/d_{jk\mathbf{A}}\right)^{2/(m-1)}}, \quad 1 \le k \le N \tag{10}$$

4. If $\left\| U^{(l)} - U^{(l-1)} \right\| < \varepsilon$, then STOP,
   else go to step 1.

Fig. 1. Fuzzy c-means clustering (FCM) algorithm

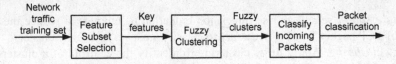

Fig. 2. Phases of the clustering security system

3.1 Feature Subset Selection

The fuzzy c-means clustering algorithm in Figure 1 computes the distance of each data item z_k to the each cluster centroid v_i. Both z_k and v_i are in d-dimensional space where d is the number of features in the TCP packet used for clustering. Therefore, it is desirable to reduce the number of features in order to reduce the complexity of the cluster, as well as for the classification in the final intrusion detection system in the third phase which performs the classification of incoming packets. The limitations of traditional dimension reduction techniques based on statistics and genetic clustering has been demonstrated [1]. Therefore, a genetic algorithm (GA) is used for feature subset selection with each chromosome corresponding to a candidate feature subset. Each chromosome is encoded as a string of 0's and 1's with the number of bits equal to the total number of features. Each bit represents a particular feature. If the bit is a '1', it indicates the attribute is to be used for training, while a '0' indicates the attribute is not to be used. A GA determines the optimal set of features to be used for training the rule set.

3.2   Creation of Fuzzy Clusters

The FCM algorithm is applied to a set of test data to compute the cluster centroids for N clusters. After the clusters have been established, label any clusters which are populated predominantly by malicious packets as attack clusters.

3.3   Using Fuzzy Clusters for Intrusion Detection

Once the clusters have been established and labeled as normal or malicious, the intrusion detection system is ready to identify new TCP packets. For each packet in the test data, determine its membership in each cluster, u_{ik}. If the membership exceeds a predetermined threshold, δ, for any malicious cluster, the system will block that packet and provide an alert identifying the packet and the mode of attack. Although the initial creation of the fuzzy clusters is computationally expensive, classification of individual packets does not hinder the speed of the intrusion detection system.

4   Testing and Results

The fuzzy clustering intrusion detection approach was tested using the KDD Cup 1999 Dataset [9]. The genetic algorithm was used for feature subset reduction. The GA used a 10% training subset of the KDD dataset. The training data contained approximately 500,000 connection records.

Feature subset selection in the first phase of the algorithm reduced the original 41 features available in a packet to a more manageable subset of 8 features:

duration	number of seconds of the connection
src_bytes	number of data bytes from source to destination
num_failed_logins	number failed login attempts
root_shell	1 if root shell is obtained; 0 otherwise
num_access_files	number of operations on access control files
serror_rate	percent of connections with "SYN" errors
same_srv_rate	percent of connections to same service
srv_count	number of connections to same service in past 2 seconds

Once the 8 features were identified, the FCM algorithm was applied to create 5 fuzzy clusters, one for each category of attack and one for normal packets.

The system was tested using the complete set of five million connection records in the KDD dataset. The full dataset contains 14 types of intrusion attacks not present in the training data. The new IDS successfully identified most of the intrusion attacks with a 98% success rate and a false positive rate of only 2%.

A second test was run to observe the effect of classifying each type of intrusion, rather than a single category for including all types of intrusions. In this test, 26 clusters were created: one for each of the 24 known types of simulated attacks, one for unknown types of attacks, and one for normal packets. This resulted in a 99% success rate and only 1.5% false positives. While this send test demonstrated a better accuracy, it suffered from excessive computation in determining the fuzzy membership of each packet in each of 26 clusters.

Table I provides a comparison of the fuzzy-genetic IDS with the other methods also using the KDD dataset. The proposed FCM system had the best intrusion detection rate and false positive rate of those In addition, the new system was able to correctly identify each of the 14 types of intrusions not in the training data. This demonstrates the robustness of the FCM approach

Table 1. Performance Comparison of IDS Methods

IDS Method	Intrusion Detection Rate	False Positive Rate
Fuzzy Clustering with Radial Basis Function [5]	90%	2%
Fuzzy Clustering with Simulated Annealing [6]	95%	11%
Fuzzy Clustering with Particle Swarm Optimization [7]	95%	2%
Proposed FCM IDS with 5 clusters	98%	2%
Proposed FCM IDS with 26 clusters	99%	1.5%

5   Conclusions and Future Research

Initial testing of the fuzzy clustering approach for network security provided promising results. However, using only 5 clusters allowed too many malicious packets to go unidentified. While increasing the clusters to 26 allowed improved identification of attacks, the computational time became prohibitive for real world operation.

Further research is required to determine the optimal number of clusters and how they should be divided to improve performance without impacting computational time. Additional Research to test various feature subset reduction techniques is planned. Reducing the number of features will reduce the computational complexity of the FCM approach. Further research is planned to test and evaluate the membership threshold in clusters for identifying attacks to determine the optimal membership for identification of malicious packets. Once these issues have been resolved, testing using datasets than the 4GB KDD dataset and in a real world environment will be conducted. There are also plans to incorporate the ability to learn new clusters or adjust cluster centroids based upon newly discovered attack modes

References

1. Y. Liu, K. Chen, X. Liao, and W. Zhang, "A genetic clustering method for intrusion detection," Pattern Recognition, vol. 37, pp. 927-942, May 2004.
2. T. Verwoerd and R. Hunt, "Intrusion detection techniques and approaches," Computer Communications vol. 25, pp. 1356-1365, September 2002.
3. B.A. Fessi, S. BenAbdallah, Y. Djemaiel, and N. Boudriga, "A clustering data fusion method for intrusion detection system," Proc. 11th IEEE Intl. Conf. on Computer and Information Technology, IEEE Press, 2011, pp. 539-545.
4. F. S. Gharehchopogh, N. Jabbai, and Z. G. Azar, "Evaluation of fuzzy k-means and k-means clustering algorithms in intrusion detection systems," Intl. J. of Scientific & Technology Research, vol. 1, 2012, pp. 66-72.
5. W. Jiang, M. Yao, and J. Yan, "Intrusion detection based on improved fuzzy c-means algorithm," Proc. 2008 Intl. Symp. Information Science and Engineering, 2008, pp. 326-329.
6. A. Ghadiri and N. Ghadiri, "An Adaptive Hybrid Architecture for Intrusion Detection based on Fuzzy Clustering and RBF Neural Networks," Proc. Ninth Annual Communication Networks and Services Research Conf., 2011, pp. 123-129.
7. J. Wu and G. Feng, "Intrusion detection based on simulated annealing and fuzzy c-means clustering," Proc. 2009 Intl. Conf. on Multimedia Information Networking and Security, IEEE Press, 2009, pp. 382-385.
8. R. Ensafi, S. Dehghanzadeh, and M. Akbarzadeh, "Optimizing fuzzy k-means for network anomaly detection using PSO," Proc. IEEE/ACS Intl. Conf. on Computer Systems and Applications (AICCSA 2008), IEEE Press, 2008, pp. 686-693
9. University of California, Irvine, "KDD99 Cup 1999 Dataset," 1999. [Online]. Available: http://kdd.ics.uci.edu/databases/kddcup99/kddcup99.html.
10. J. C. Dunn, "A fuzzy relative of the Isodata process and its use in detecting compact well-separated clusters," J. Cybernetics, vol. 3, 1973, pp. 32-57.
11. J. C. Bezdek, Pattern Recognition with Fuzzy Objective Function Algorithms, New York: Plenum Press, 1981.
12. J. C. Bezdek, R. Ehrlich, and W. Full, "FCM: The fuzzy c-means clustering algorithm," Computers & Geosciences, vol. 10, 1984, pp. 191-203.

Machine Learning and Pattern Recognition Techniques for Information Extraction to Improve Production Control and Design Decisions

Carlos A. Escobar[1,2], Ruben Morales-Menendez[2]

[1] General Motors, Research and Development, Warren MI 48092, USA,
`carlos.1.escobar@gm.com`
[2] Tecnológico de Monterrey, Monterrey, NL. 64849, México,
`rmm@itesm.mx`

Abstract In today's highly competitive global market, winning requires near-perfect quality. Although most mature organizations operate their processes at very low defects per million opportunities, customers expect completely defect-free products. Therefore, the prompt detection of rare quality events has become an issue of paramount importance and an opportunity for manufacturing companies to move quality standards forward. This paper presents the learning process and pattern recognition strategy for a knowledge-based intelligent supervisory system; in which the main goal is the detection of rare quality events through binary classification. The proposed strategy is validated using data derived from an automotive manufacturing systems. The l_1-regularized logistic regression is used as the learning algorithm for the classification task and to select the features that contain the most relevant information about the quality of the process. According to experimental results, 100% of defects can be detected effectively.

Keywords l_1-regularized logistic regression · Defect detection · Intelligent supervisory system · Quality Control · Model selection criterion

1 Introduction

In today's highly competitive global market, winning requires near perfect quality, since intense competition has led organizations to low profit margins. Consequently, a warranty event could make the difference between profit and loss. Moreover, customers use internet and social media tools (e.g., Google product review) to share their experiences, leaving organizations little flexibility to recover from their mistakes. A single bad customer experience can immediately affect companies' reputations and customers' loyalty.

In the quality domain, most mature organizations have merged business excellence, lean production, standards conformity, six sigma, design for six sigma, and other quality-oriented philosophies to create a more coherent approach [1]. Therefore, the manufacturing processes of these organizations only generate a

© Springer International Publishing AG 2017
P. Perner (Ed.): ICDM 2017, LNAI 10357, pp. 286–300, 2017.
DOI: 10.1007/978-3-319-62701-4_23

Acronyms	Definition
AI	Artificial Intelligence
CM	Confusion Matrix
CV	Cross-Validation
FN	False Negatives
FP	False Positives
FS	Feature Selection
ISCS	Intelligent Supervisory Control Systems
KB	Knowledge-Based
LASSO	Least Absolute Shrinkage and Selection Operator
LR	Logistic Regression
LP	Learning Process
ML	Machine Learning
MPCD	Maximum Probability of Correct Decision
PR	Pattern Recognition
TN	True Negatives
TP	True Positives
UMW	Ultrasonically Metal Welding

Table 1: Acronyms Table

few defects per million of opportunities. The detection of these rare quality events represents not only a research challenge, but also an opportunity to move manufacturing quality standards forward.

Impressive progress has been made in recent years, driven by exponential increases in computer power, database technologies, *Machine Learning (ML)* algorithms, optimization methods and big data [2]. From the point of view of manufacturing, the ability to efficiently capture and analyze big data has the potential to enhance traditional quality and productivity systems. The primary goal behind the generation and analysis of big data in industrial applications is to achieve fault-free (defect-free) processes [3, 4], through *Intelligent Supervisory Control Systems (ISCS)* [5].

A *Learning Process (LP)* and *Pattern Recognition (PR)* strategy for a knowledge-based *ISCS* is presented, aimed at detecting rare quality events from manufacturing systems. The defect detection is formulated as a binary classification problem, in which the l_1-regularized logistic regression is used as the learning algorithm. The outcome is a parsimonious predictive model that contains the most relevant features.

The proposed strategy is validated using data derived from an automotive manufacturing systems; (1) *Ultrasonically Metal Welding (UMW)* battery tabs from a battery assembly process. The main objective is to detect low-quality welds (*bad*).

The rest of this paper is organized as follows. It starts with a review of the theoretical background of this research in section 2. Then, section 3 describes the pattern recognition strategy, followed by an empirical study in section 4. Finally, section 5 concludes the paper.

2   Theoretical Background

The theoretical background of this research is briefly reviewed.

2.1   Machine Learning and Pattern Recognition

As discussed by [6], "As an intrinsic part of *Artificial Intelligence (AI)*, *ML*
refers to the software research area that enables algorithms to improve through
self-learning from data without any human intervention". *ML* algorithms learn
information directly from data without assuming a predetermined equation or
model. The two most basic assumptions underlying most *ML* analyses are that
the examples are independent and identically distributed, according to an unknown probability distribution. On the other hand, *PR* is a scientific discipline
that "deals with the automatic classification of a given object into one from a
number of different categories (e.g., classes)" [7].

In *ML* and *PR* domains, generalization refers to the prediction ability of a
learning algorithm model [8]. The generalization error is a function that measures
how well a trained algorithm generalizes to unseen data.

In general, the *PR* problem can be widely broken down into four stages: (1)
Feature space reduction, (2) Classifier design, (3) Classifier selection, and (4)
Classifier assessment.

2.2   Feature Space Reduction

The world of big data is changing dramatically, and feature access has grown
from tens to thousands, a trend that presents enormous challenges in the *Feature
Selection (FS)* context. Empirical evidence from FS literature exhibits that discarding irrelevant or redundant features improves generalization, helps in understanding the system, eases data collection, reduces running time requirements,
and reduces the effect of dimensionality [9–14]. This problem representation
highlights the importance of finding an optimal feature subset. This task can be
accomplished by *FS* or regularization.

Feature Selection The *FS* methods broadly fall into two classes: filters and
wrappers [15].

Filter methods select variables independently of the classification algorithm
or its error criteria, they assign weights to features individually and rank them
based on their relevance to the class labels. A feature is considered good and
thus selected if its associated weight is greater than the user-specified threshold [9]. The advantages of feature ranking algorithms are that they do not over-fit
the data and are computationally faster than wrappers, and hence they can be
efficiently applied to big data sets containing many features [10].

ReliefF is a well-know rank-based algorithm, the basic idea for numerical
features is to estimate the quality of each according to how well their values distinguish between instances of the same and different class labels. *ReliefF* searches

for a k of its nearest neighbors from the same class, called nearest *hits*, and also a k nearest neighbors from each of the different classes, called nearest *misses*, this procedure is repeated m times, which is the number of randomly selected instances. Thus, features are weighted and ranked by the average of the distances (*Manhattan* distance) of all *hits* and all *misses* [16] to select the most important features [17], developing a significance threshold τ. Features with an estimated weight below τ are considered irrelevant and therefore eliminated. The proposed limits for τ are $0 < \tau \leq 1/\sqrt{\alpha m}$ [16]; where α is the probability of accepting an irrelevant feature as relevant.

Regularization Another approach for feature space reduction is l_1-regularization. This method trims the hypothesis space by constraining the magnitudes of the parameters [18]. Regularization adds a penalty term to the least square function to prevent over-fitting [19]. l_1-norm formulations have the advantage of generating very sparse solutions while maintaining accuracy. The classifier-fitted parameters θ_i are multiplied by a coefficient λ to shrink them toward zero. This procedure effectively reduces the feature space and protects against over-fitting with irrelevant and redundant features. The value of λ can be tuned through validation or CV. Regularization methods may perform better than FS methods [20].

2.3 Classifier Design, Selection and Assessment

A classifier is a supervised learning algorithm that analyzes the training data (e.g., data with a class label) and fits a model. In a typical PR analysis, the training data set is used to train a set of candidate models using different tuning parameters.

It is important to choose an appropriate validation or CV method to evaluate the generalization ability of each candidate model, and select the *best*, according to a relevant model selection criterion.

For information-theoretic model selection approaches in the analysis of empirical data refer to [21]. Common performance metrics for model selection based on recognition rates —correct decisions made— can be found in [22].

For a data-rich analysis, it is recommended the hold-out validation method, an approach in which a data set is randomly divided into three parts: training set, validation set, and test set. As a typical rule of thumb, 50 percent of the initial data set is allocated to training, 25 percent to validation, and 25 percent to testing [23].

Once the best candidate model has been selected, it is recommended that the model's generalization performance be tested on a new data set before the model is deployed. This can also determine whether the model satisfies the learning requirement [23]. The generalization performance can be efficiently evaluated using a *Confusion Matrix (CM)*.

Confusion Matrix In predictive analytics, a *CM* [22] is a table with two rows and two columns that reports the number of *False Positives (FP)*, *False Negatives (FN)*, *True Positives (TP)*, and *True Negatives (TN)*. This allows more detailed analysis than just the proportion of correct guesses since it is sensitive to the recognition rate by class.

A type-I error may be compared with a *FP* prediction, and it is denoted by the greek letter α. On the other hand, a type-II error may be compared with a false *FN*, and it is denoted by the greek letter β [24]. Alpha, and beta are estimated by:

$$\alpha = \frac{\text{FP}}{\text{FP} + \text{TN}}, \tag{1}$$

$$\beta = \frac{\text{FN}}{\text{FN} + \text{TP}}. \tag{2}$$

2.4 Logistic Regression

Logistic Regression (LR), which uses a transformation of the values of a linear combination of the features, is widely used in classification problems. It is an unconstrained convex problem with a continuous differentiable objective function that can be solved either by the Newton's method or the conjugate gradient. *LR* models the probability distribution of the class label y, given a feature vector x [25].

$$P(y = 1|x; \theta) = \sigma(\theta^T x) = \frac{1}{1 + \exp(-\theta^T x)}. \tag{3}$$

where $\theta \in \mathbb{R}^N$ are the parameters of the *LR* model and $\sigma(\cdot)$ is the sigmoid function. The sigmoid curve (logistic function), maps values in $(-\infty, \infty)$ to $[0, 1]$. The discrimination function itself is not linear anymore, but the decision boundary is still linear.

Under the Laplacian prior $p(\theta) = (\lambda/2)^N \exp(-\lambda||\theta||_1)(\lambda > 0)$, the *Maximum A Posteriori (MAP)* estimate of the parameters θ is:

$$\min_\theta \sum_{i=1}^M -\log p(y^{(i)}|\mathbf{x}^{(i)}; \theta) + \lambda||\theta||_1. \tag{4}$$

This optimization problem is referred to as l_1-regularized *LR*. This algorithm is widely applied in problems with small training sets or with high dimensional input space. However, adding the l_1-regularization makes the optimization problem computationally more expensive to solve. For solving the l_1-regularized *LR* [26], the *Least Absolute Shrinkage and Selection Operator (LASSO)* is an efficient method.

2.5 Intelligent Supervisory Control Systems

ISCSs are computer-based decision support systems that incorporate a variety of *AI* and non-*AI* techniques to monitor, control, and diagnose process variables to assist operators with the tasks of monitoring, detecting, and diagnosing process anomalies, or in taking appropriate actions to control processes [27]. Developing and deploying an *ISCS* requires a lot of collaborative intellectual work from different engineering disciplines.

There are three general solution approaches for supporting the tasks of monitoring, control, and diagnosis: (1) data-driven, for which the most popular techniques are Principal Component Analysis, Fisher discriminant analysis, and Partial Least-Squares analysis; (2) analytical, an approach founded in first principles or other mathematical models; and (3) *Knowledge-Based (KB)*, founded in *AI*, specifically Expert Systems, Fuzzy Logic, *ML*, and *PR* [27, 28].

Due to the explosion of industrial big data, *KB-ISCS*s have received great attention. Since the scale of the data generated from manufacturing systems cannot be efficiently managed by traditional process monitoring and quality control methods, a *KB* scheme might be an advantageous approach.

3 Learning Process and Pattern Recognition Strategy

The proposed *LP* and *PR* strategy for a *KB-ISCS* considers the l_1-regularized *LR* as the learning algorithm. Fig. 1 displays the proposed strategy, which uses the hold-out data partition method (framed into a 4-stage approach). The input is a set of candidate features, the outcome is a parsimonious predictive model that contains the most relevant features to the quality of the product. This model is used to detect rare quality events in manufacturing systems. The candidate features can be derived from sensor signals following typical feature construction techniques [29] or from process physical knowledge.

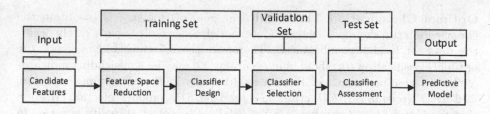

Fig. 1: Learning process and pattern recognition framework.

Two main assumptions that must be satisfied are: (1) the faulty events must be generated during the manufacturing process and captured by the signals; (2) since the *LR* learning algorithm is a linear classifier, the decision boundaries between the two classes must be linear. Due to the dynamic nature of manufacturing systems, the predictive model should be updated constantly to maintain its generalization ability.

3.1   Feature Space Reduction

The goal of this stage is to eliminate irrelevant features from the analysis. For manufacturing processes, massive amounts of data and the lack of a comprehensive physical understanding may result in the development of many irrelevant features. This problem representation highlights the importance of preprocessing the data. The *ReliefF* algorithm is used to obtain the feature ranking, and the associated weight of each feature is compared with τ to eliminate the irrelevant ones.

3.2   Classifier Design

The main goal of this stage is to design the classifier and to identify which features contain the most relevant information to the quality of the product. While the classifier is aimed to detect rare quality events, the features included in the predictive model may provide valuable engineering information. Although feature interpretation is out of the scope of this approach, analyzing the data-derived predictive model from a physics perspective may support engineers in systematically discovering hidden patterns and unknown correlations that may guide them to identify root causes and solve quality problems.

The training set is used to fit n-candidate l_1-regularized *LR* models by varying the penalty value λ. It is recommended to start with the largest value of λ that gives a non-null model (i.e., a model with the intercept only), and from that point decrease the value of λ to develop more candidate models with more features. The rationale behind this approach is that the form of the model is not known in advance; therefore, it can be approximated by generating a set of candidate models. This analysis can be computationally performed using the *LASSO* method in MATLAB or R.

Optimal Classification Threshold   The goal of this step is to obtain the classification threshold of each candidate model. Since faulty events rarely occur, the data set is highly unbalanced. Therefore, the 0.5 threshold may not the optimal classification threshold, and accuracy [22] may be a misleading indicator of classification performance. To address this scenario, the *Maximum Probability of Correct Decision (MPCD)* criterion is used [30, 31]. A model selection criterion that tends to be very sensitive to *FN*s — failure to detect a quality event — in highly unbalanced data. *MPCD* is estimated by:

$$\text{MPCD} = (1 - \alpha)(1 - \beta). \tag{5}$$

Since *MPCD* is used as a model selection criterion, *gamma* (γ), the optimal classication threshold with respect to *MPCD* of each candidate model is obtained. γ is found by enumerating all candidate classification thresholds (midpoints between two consecutive examples), and estimating the *MPCD* at each threshold. γ is the maximum value of all candidate classification thresholds, a graphical representation of this procedure is shown in Fig. 4.

3.3 Classifier Selection

In the context of *PR*, the primary purpose of this stage is to select the *best* candidate model with respect to generalization (*MPCD*). The validation data set is used to estimate the *MPCD* of each candidate model, and the model with the highest value should be selected.

3.4 Classifier Assessment

The generalization performance of the selected model is evaluated on the testing set. The classifier must be assessed without the bias induced in the validation stage. This stage ensures that the model satisfies the learning target; due to the nature of the problem, *FN*s are the main concern. The target can be simplified to develop a model that produces zero or nearly zero *FN*s with the least possible number of *FP*s.

4 Empirical study

To show the effectiveness of the proposed strategy, an automotive case study is presented.

4.1 Ultrasonic Metal Welding

UMW is a solid state bonding process that uses high frequency ultrasonic vibration energy to generate oscillating shears between metal sheets clamped under pressure. It is an ideal process for bonding conductive materials such as copper, aluminum, brass, gold, and silver, and for joining dissimilar materials. Recently, ultrasonic metal welding has been adopted for battery tab joining in the manufacturing of vehicle battery packs. Creating reliable joints between battery tabs is critical because one low-quality connection may cause performance degradation or the failure of the entire battery pack. It is important to evaluate the quality of all joints prior to connecting the modules and assembling the battery pack [13].

The data used for this analysis is derived from an *UMW* of battery tabs. A very stable process, that only generates a few defective welds per million of opportunities. However, all the welds in the battery must be good for the unit to function. This problem representation not only highlights the engineering intellectual challenge but also the importance of a zero-defects policy.

The collected data set contains a binary outcome (*good/bad*) with 54 features derived from signals (e.g., acoustics, power, and linear variable differential transformers) following typical feature construction techniques [29]. The data set is highly unbalanced since it contains only 36 bad batteries out of 40,000 examples (0.09%). The data set is partitioned following the hold-out validation scheme (including *bads* in each data set): training set (20,000), validation set (10,000), and testing set (10,000).

Feature Space Reduction In order to eliminate irrelevant features, the data set is initially preprocessed using the *ReliefF* algorithm. *ReliefF* is run with $k = 5$ nearest neighbors and a significance threshold of $\tau = 0.031622$, (calculated based on $1/\sqrt{am}$ — $\alpha = 0.05$, and $m = 20,000$). According to the algorithm, feature 26 is the most important feature, while feature 14 is the lowest quality feature. Fig. 2 summarizes the feature ranking and which features are selected based τ. According to *ReliefF*, 43 features —out of 54— should be selected.

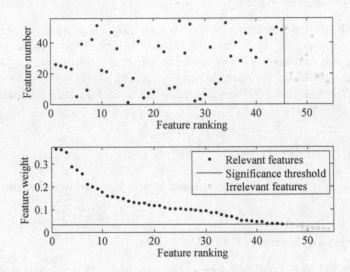

Fig. 2: Feature ranking and selection using ReliefF.

Classifier Design The training set was used to fit 100 regularized *LR* models. The *LASSO* method was applied to estimate the fitted least-squares regression coefficients for a set of 100 regularization coefficients λ, starting with the largest value of λ that gives a non-null model (i.e., a model with the intercept only). However, the non-null model is not included in the analysis since its estimated *MPCD* equals zero. Fig. 3a displays each candidate model's associated value of λ, Fig. 3b the number of features, and Fig. 3c the associated values of γ. As shown by Fig. 3a and Fig. 3b, the number of features decreases as the value of λ increases, therefore, selecting the right model is one of the main challenges.

Optimal Classification Threshold Figure 4 shows the optimal classification threshold search of candidate model 69.

Classifier Selection The goal is to select the candidate model with the highest *MPCD*. In the context of the problem that is being solved, the goal is to detect low-quality welds. Due to the relevance of failing to detect a potential defect, the

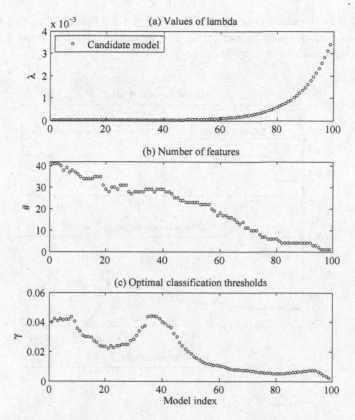

Fig. 3: Candidate model information.

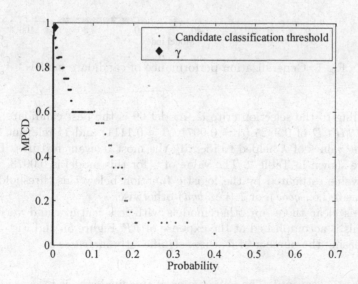

Fig. 4: Optimal classification threshold search of candidate model 69.

type-II error is the main concern of this analysis; for this reason, the *MPCD* is also used as a model selection criteria. The estimated *MPCD*, α, and β of each model are summarized in Fig. 5.

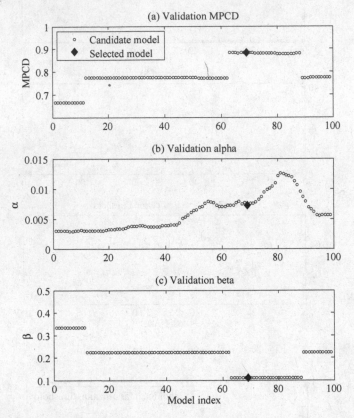

Fig. 5: Generalization performance of candidate models.

According to the selection criteria, model 69 is the best candidate, with an estimated *MPCD* of 0.8825 ($\alpha = 0.0071$, $\beta = 0.1111$) and 11-relevant features, varying the values of λ helped to identify the most relevant features. The coefficients are shown in Table 2. The value of γ for this model is 0.0073, meaning that any value estimated by the logistic function below this threshold will be classified as 0 (i.e., *good*), or 1 (i.e., *bad*) otherwise.

While is clear there are other models with less features and very similar *MPCD*, this is accomplished at the expense of *FP*. Figure 5b and Fig. 3b, show that decreasing the number of features significantly increases α.

Classifier Assessment The importance of this final step is to assess the classifier without the bias induced in the validation stage, and to ensure the model

Coef	Value	Coef	Value	Coef	Value	Coef	Value	Coef	Value	Coef	Value
θ_0	372.24	θ_4	-0.0077	θ_5	29.68	θ_9	3.20	θ_{21}	0.0019	θ_{22}	-0.0015
θ_{26}	-0.0122	θ_{30}	-0.0018	θ_{31}	-0.0035	θ_{32}	0.0536	θ_{36}	0.0108	θ_{48}	1.7122

Table 2: Coefficients of model 69.

satisfies the learning target; due to the nature of the problem, FNs are the main concern. Therefore, the goal can be simplified to develop a model that produces zero or nearly zero FNs with the least possible number of FPs.

The estimated $MPCD$ of the final model on the testing data is 0.9956, with an estimated $\beta = 0$ and $\alpha = 0.0044$. The testing set includes approximately 10,000 records, with seven bad batteries. The classifier correctly classified the seven bad units and only misclassified 44 good units. Recognition rates are summarized on Table 3. According to model assessment results, LR not only shows high prediction ability, but also did not commit any type-II error.

	Declare good	Declare bad
good	9949	44
bad	0	7

Table 3: Confusion Matrix

5 Conclusions and Future Work

Today's business environment sustains mainly those companies committed to a zero-defects policy. This quality challenge was the main driver of this research, where a LP and PR strategy was developed for a $KB\text{-}ISCS$. The proposed approach was aimed at detecting rare quality events in manufacturing systems and to identify the most relevant features to the quality of the product. The defect detection was formulated as a binary classification problem and validated in a data set derived from an automotive manufacturing system. The main objective was to detect low-quality welds (*bad*) from the UMW of battery tabs from a battery assembly process.

The l_1-regularized LR was used as the learning algorithm for the classification task and to identify the most important features. Since the form of the model was not known in advance, a set of candidate models were developed —by varying the value of λ— as an effort to approximate the true model. Chosen model exhibited high capacity to detect rare quality events, since 100% of the defective units on the testing set were detected.

The l_1 penalty term helped to identify the features most relevant to the quality of the product. Although the identification of relevant features may support engineers in systematically discovering hidden patterns and unknown correlations, it was beyond the scope of this research to use a physics-based perspective to analyze the influence of these features over the manufacturing system.

The proposed strategy used $MPCD$, a model selection criterion very sensitive to FNs and developed γ, an optimal classification threshold with respect to $MPCD$.

The proposed approach can be adapted and widely applied to manufacturing processes to boost the performance of traditional quality methods and potentially move quality standards forward, where soon virtually no defective product will reach the market.

Since $MPCD$ is founded exclusively on recognition rates, future research along this path could focus on adding a penalty term for model complexity.

Acknowledgments Authors would like to express our deepest appreciation to Dr. Debejyo Chakraborty, Diana Wegner and Dr. Xianfeng Hu, who helped us to complete this report. A special gratitude is given to Dr. Jeffrey Abell, whose ideas and contributions illuminated this research.

Bibliography

[1] A. S. of Quality, "Emergence - 2011 Future of Quality Study," *ASQ: The Global Voice of Quality*, 2011.

[2] K. Schwab, "The Fourth Industrial Revolution: What It Means, How to Respond," *World Economic Forum*, 2016.

[3] S. Yin and O. Kaynak, "Big Data for Modern Industry: Challenges and Trends [Point of View]," *Proc of the IEEE*, vol. 103, no. 2, pp. 143–146, 2015.

[4] S. Yin, X. Li, H. Gao, and O. Kaynak, "Data-based Techniques Focused on Modern Industry: An Overview," *IEEE Trans on Industrial Electronics*, vol. 62, no. 1, pp. 657–667, 2015.

[5] V. Venkatasubramanian, R. Rengaswamy, S. Kavuri, and K. Yin, "A Review of Process Fault Detection and Diagnosis: Part III: Process History based Methods," *Computers & Chemical Eng*, vol. 27, no. 3, pp. 327–346, 2003.

[6] P. Ghosh, "A Comparative Roundup: Artificial Intelligence vs. Machine Learning vs. Deep Learning," June 2016. [Online]. Available: www.dataversity.net/ai-vs-machine-learning-vs-deep-learning

[7] S. Theodoridis and K. Koutroumbas, "Pattern Recognition and Neural Networks," in *Machine Learning and its Applications*. Springer, 2001, pp. 169–195.

[8] Z. Zhou, "Ensemble Learning," in *Encyclopedia of Biometrics*. Springer, 2009, pp. 270–273.

[9] L. Yu and H. Liu, "Feature Selection for High-Dimensional Data: A Fast Correlation-based Filter Solution," in *ICML*, vol. 3, 2003, pp. 856–863.

[10] M. Hall, "Correlation-based Feature Selection of Discrete and Numeric Class Machine Learning," in *Proc of the 17th Int Conf on Machine Learning*. University of Waikato, 2000, pp. 359–366.

[11] K. Nicodemus and J. Malley, "Predictor Correlation Impacts Machine Learning Algorithms: Implications for Genomic Studies," *Bioinformatics*, vol. 25, no. 15, pp. 1884–1890, 2009.

[12] F. Wang, Y. Yang, X. Lv, J. Xu, and L. Li, "Feature Selection using Feature Ranking, Correlation Analysis and Chaotic Binary Particle Swarm Optimization," in *5th IEEE Int Conf on Software Engineering and Service Science*. IEEE, 2014, pp. 305–309.

[13] C. Shao, K. Paynabar, T. Kim, J. Jin, S. Hu, J. Spicer, H. Wang, and J. Abell, "Feature Selection for Manufacturing Process Monitoring using Cross-Validation," *J. of Manufacturing Systems*, vol. 10, 2013.

[14] S. Wu, Y. Hu, W. Wang, X. Feng, and W. Shu, "Application of Global Optimization Methods for Feature Selection and Machine learning," *Mathematical Problems in Eng*, 2013.

[15] A. Ng, "On Feature Selection: Learning with Exponentially Many Irrevelant Features as Training Examples," in *Proc of the 15th Int Conf on Machine*

Learning.   MIT, Dept. of Electrical Engineering and Computer Science, 1998, pp. 404–412.

[16] M. Robnik-Šikonja and I. Kononenko, "Theoretical and Empirical Analysis of ReliefF and RReliefF," *Machine Learning*, vol. 53, no. 1-2, pp. 23–69, 2003.

[17] K. Kira and L. Rendell, "The Feature Selection Problem: Traditional Methods and a New Algorithm," in *AAAI*, vol. 2, 1992, pp. 129–134.

[18] C. Bishop, *Neural Networks for Pattern Recognition.*   Oxford University Press, 1995.

[19] A. Ng, "Feature Selection L1 vs L2 Regularization and Rotational Invariance," in *Proc. of the 21st Int Conf on Machine Learning.*   ACM, 2004, p. 78.

[20] E. Xing, M. Jordan, R. Karp *et al.*, "Feature Selection for High-Dimensional Genomic Microarray Data," in *ICML*, vol. 1, 2001, pp. 601–608.

[21] M. Peruggia, "Model Selection and Multimodel Inference: A Practical Information-Theoretic Approach," *J of the American Statistical Association*, vol. 98, no. 463, pp. 778–779, 2003.

[22] T. Fawcett, "An Introduction to ROC Analysis," *Pattern Recognition Letters*, vol. 27, no. 8, pp. 861–874, 2006.

[23] J. Friedman, T. Hastie, and R. Tibshirani, *The Elements of Statistical Learning.*   Statistics Springer, Berlin, 2001, vol. 1.

[24] J. Devore, *Probability and Statistics for Engineering and the Sciences.*   Cengage Learning, 2015.

[25] S. Lee, H. Lee, P. Abbeel, and A. Ng, "Efficient L_1 Regularized Logistic Regression," in *Proc of the National Conf on Artificial Intelligence*, vol. 21, no. 1.   Cambridge, MA, 2006, p. 401.

[26] R. Tibshirani, "Regression Shrinkage and Selection via the LASSO," *J. of the Royal Statistical Society. Series B (Methodological)*, pp. 267–288, 1996.

[27] V. Uraikul, W. Chan, and P. Tontiwachwuthikul, "Artificial Intelligence for Monitoring and Supervisory Control of Process Systems," *Eng Applications of Artificial Intelligence*, vol. 20, no. 2, pp. 115–131, 2007.

[28] L. Chiang, R. Braatz, and E. Russell, *Fault Detection and Diagnosis in Industrial Systems.*   Springer Science & Business Media, 2001.

[29] L. Huan and H. Motoda, "Feature Extraction, Construction and Selection: A Data Mining Perspective," 1998.

[30] J. A. Abell, J. P. Spicer, M. A. Wincek, H. Wang, and D. Chakraborty, "Binary Classification of Items of Interest in a Repeatable Process," *US Patent* , no. US8757469B2, June 2014. [Online]. Available: www.google. com/patents/US20130105556

[31] J. A. Abell, D. Chakraborty, C. A. Escobar, K. H. Im, D. M. Wegner, and M. A. Wincek, "Big data driven manufacturing — process-monitoring-for-quality philosophy," *ASME Journal of Manufacturing Science and Engineering (JMSE) on Data Science-Enhanced Manufacturing*, vol. 139, no. 10, October 2017.

Real-time Prediction of Styrene Production Volume based on Machine Learning Algorithms

Yikai Wu, Fang Hou, Xiaopei Cheng

Accenture, 21/F West Tower, World Financial Center, No. 1, East 3rd
Ring Middle Road, 100020 Beijing, China
{Yikai.a.wu@accenture.com, fang.hou@accenture.com, peter.xiaopei.cheng@accenture.com}

Abstract. Due to wide application of styrene and complex process of modern dehydrogenation of ethylbenzene, traditional methods usually spend much more time on chemical examinations and tests for identification of the production volume. Generally, there are several hours or days of time lag for the information to be made available. In this article, the whole ethylene cracking plants are investigated. The generalized regression neural network model is designed to timely predict the styrene output after the high-dimensional reduction. The usefulness of the model will be demonstrated by specific cases. The appropriate data mining techniques and implementation details will also be depicted. Finally, the simulation results show that this model can monitor the styrene output per hour with high accuracy.

Keywords: styrene monomer; high-dimensionality; real-time prediction

1   Introduction

Styrene, primarily prepared from dehydrogenation of ethylbenzene, is a derivative of benzene used in a wide range of applications such as automobiles, electrical and electronics etc. As the second most important monomer in the chemical industry, its annual consumption is more than 30 MT/year. The global market of styrene and related derivatives will grow at a CAGR of 4.82% over the period of 2014-2019 [1]. Therefore, styrene production plays an important role and has great economic value for petrochemical factories. In this paper, in the real petrochemical factories, the most widely used adiabatic dehydrogenation of the ethylbenzene will be discussed and analyzed.

In fact, there are several studies regarding the process optimization and stability conditions for dehydrogenation of ethylbenzene, such as adjusting the steam oil ratio or temperature for styrene production for higher productivity [2]. Shell and Crowe (1969) were the first to apply the optimal modeling for styrene reactors to improve the existing reactor operations with steam tem-

© Springer International Publishing AG 2017
P. Perner (Ed.): ICDM 2017, LNAI 10357, pp. 301–312, 2017.
DOI: 10.1007/978-3-319-62701-4_24

perature, steam rate and other parameters [3], etc. Most of their studies are based on the petrochemical mechanism processes and parameters.

In this work, a new predictive model is used to analyze and explain the timely styrene monomer output. There are two existent practical problems. The first one is to find out the related parameters with yield of styrene monomer using the high dimensional reduction method; the second one is to provide timely information for workers, and construct a model for timely monitoring of the styrene monomer production with algorithm filtered parameters despite the period of chemical examination for the styrene monomer production rate is relative long (max. three times a day). The structure is organized as follows. the second section describes the preparation work, domain process under-standing, parameter selection for modeling, and the noise reduction method, and presents the specific results of data processing. The third section con-structs a real time predictive model and analyzes the predictive value. Finally, all the findings and results are discussed to draw conclusions.

2   Data Generation and Preprocessing

2.1   Domain process

In the process, the core part of the device is the ethylbenzene dehydrogena-tion system, of which the operating conditions play an important role in sty-rene monomer production. The ethylbenzene catalytic dehydrogenation sys-tem consists of a heating furnace, two adiabatic radial reactors and many heat exchangers, as shown in Figure 1. The raw material of ethylbenzene was first mixed with steam. After passing the heat exchanger and heating furnace, the material will put into the two-stage reactor reaction. Then, the mixture of styrene monomer and related styrene derivatives, such as methylbenzene, will be produced.

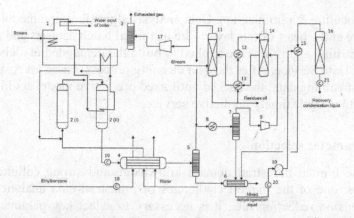

Fig. 1. Schematic diagram for ethylbenzene dehydrogenation, 1-furnance; 2-cascade reactor; 3, 5, 7, 14-separating tower; 4, 6, 9, 12, 13, 15-heat exchangers; 9, 17-compressor; 10-pump; 11, 16-gas stripper; 18, 19, 20, 21-sampling points

2.2 Conceptual Industrial Analytics & Modeling Process

In the real factory, to determine the accurate styrene monomer output, the engineer will periodically extract the mixture at the sampling points in Fig. 1. The styrene content can be calculated by the empirical formula following the chemical examination. There are several hours or a few days of delay before the engineer obtain the helpful information. We leveraged multi-data sources and big data techniques to build a predictive model for addressing the delay problem. The conceptual industrial analytics & modeling process is an integrated process for industrial customized real-time analytics. Modeling optimization is sketched in Fig. 2

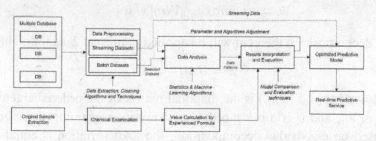

Fig. 2. Conceptual industrial analytics & modeling process

- Data preprocessing. In general, flow-oriented process data and subject-oriented operational data are integrated in a data preparation layer. Then, various data cleaning methods, such as noise reduction (for typical industrial data problem [4]), abnormal point processing etc.

- Data modeling & optimization. Data analysis algorithms and model evaluation are at the heart of this layer. The historical batch datasets and machine learning algorithms are applied to build the demanded models. Industrial Data Service. After the data cleaning process, based on Apache Kafka streaming data, the backed optimized predictive model could be deployed to provide timely predictive service.

2.3   Parameter selection

Due to the limited industrial domain knowledge and strong collinearity of parameters, one of the biggest challenges of industrial data analytics is the high-dimension reduction, i.e., it is necessary to select key parameters that could present valuable information for the model building. In statistics, there are many dimension reduction methods, such as Principal Component Analysis (PCA), Random Forest Algorithms (RFA) [5], Multi-dimensional Scaling (MDS) [6], etc. In this paper, only PCA and RFA are applied due to their popularity and strong theoretical backgrounds.

Principal Component Analysis (PCA)

PCA model could reduce the dimension of data by selecting a few orthogonal linear combinations of the original parameters with the largest variance [7].

Assume there are n_i linear components in the domain process, and $i = 1, 2, ..., n_i$ is the linear combination with the largest variance, $s_i = x^T w_i$, where the p dimensional coefficient vector $w_i = (w_{1,1}, ... w_{1,p}, ..., w_{n_i,p})^T$, therefore,

$$w_i = argmax_{\|w_i\|} Var\{x^T w\} \tag{1}$$

spectral decomposition theorem will be applied to present Σ as following,

$$\Sigma = U \Lambda U^T \tag{2}$$

where $\Lambda = diag(\lambda_1, ..., \lambda_p)$ is the diagonal matrix of the ordered eigenvalues $\lambda_1 \leq \cdots \leq \lambda_p$, and U is a $p \times p$ orthogonal matrix containing the eigenvectors. Based on eigenvalue decomposition, the total variation is equal to the sum of the eigenvalue of the variance matrix,

$$\sum_{i=1}^{p} Var(LPC_i) = \sum_{i=1}^{p} \lambda_i = \sum_{i=1}^{p} trace(\Sigma) \tag{3}$$

The selected LPCs are defined as the "new variables". Here, considering that most petrochemical parameters are strong collinearity, the $S_j, j \in (1, \dots, n_j)$ LPCs are first calculated. The scree plot is studied to determine the number of important parameters in each S_j; secondly, for determining the eigenvector which has the smallest eigenvalue (the least important LPC), the second smallest eigenvalue and so on are calculated. This process is repeated until only q variables remain. Then the following function could be obtained,

$$S_j = w_{j,1}X_1 + \dots + w_{j,q}X_q \tag{4}$$

where $j \in (1, \dots, n_j)$ and $q \in (1, \dots, p)$, in the jth LPC S_j, q parameters are determined to represent this component. In our use case, the results of each selected LPCs S_j, where $j = 8$ are sown in Fig.3. Higher correlation coefficient parameters are extracted and determined through PCA applying,

$$X^{PCA} = (X_{S_1,1}, X_{S_1,2}, \dots, X_{S_1,q_1}, X_{S_2,1}, X_{S_2,2}, \dots, X_{S_2,q_2}, \dots, X_{S_j,q_j})^T \tag{5}$$

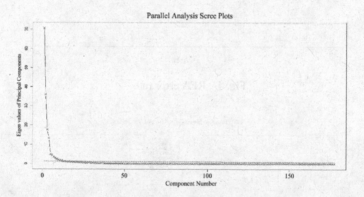

Fig. 3. the LPCs S_j with Parallel Analysis Scree Plots

Random Forest Method (RFA)

Compared with PCA, a random forest classification tree is constructed to filter high dimensional parameters using different bootstrap sample data. Here, we assume m_{tree} bootstrap samples from the raw data for each of the purposed bootstrap samples, and grow an unpruned classification or regression tree, with the modification that at each node, rather than choosing the best split among all predictors[8]. m_{try} is sampled from all predictors. The best split is selected from these sensor parameters. Then, new data are predicted by aggregating the prediction of the $m'_{tree}, m' \leq m$ trees. The RFA error rate and importance ranking of parameters are presented in Fig. 4 and Fig.5.

The random forest algorithm estimates the importance of variables by look-ing at how much the prediction error increases when the sample data for that variable is permuted while all others are left unchanged. The necessary calcu-lations are carried out tree by tree as the random forest is constructed. The corresponding calcualtion results are shown in the Fig. 5). The proximity matrix can be used to identify the structure in the data. The (i, j) element of the proximity matrix produced by the random forest is the fraction of trees in which elements i and j fall in the same terminal node.

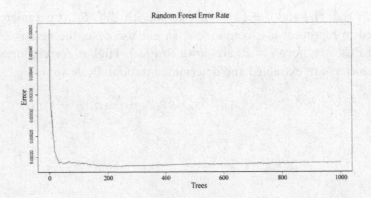

Fig. 4.   RFA error rate

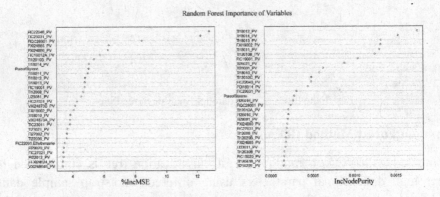

Fig. 5. Importance ranking of parameters using 500 decision trees

Based on random forest importance ranking key parameters could be ob-tained as following,

$$X^{RFA} = (X_1, X_2, \dots, X_{m'_{tree}})^T \tag{6}$$

Extracting common key parameters X^* from X^{PCA} in (6) and X^{RFA} in (7), and defined X^*,

$$X^* = X^{PCA} \cap X^{RFA} \tag{7}$$

There are a total of 33 overlapped parameters in X^* which are prepared to fit the predictive model in third chapter.

2.4 Noise reduction

Due to the instability of the styrene production from the chemical test data, the raw data contain a lot of noise which is also one of the special problems of the sensor data. In this stage, the Kalman filter method is used to deal with noise data. The Kalman filter is essentially a set of mathematical equations that implement a predictor-corrector type estimator through vector model building. The recursive method is used to search the optimal solution in the minimum variance. It has been widely used in communication, stochastic control, signal processing and other fields [9]. Here, we extracted the training datasets from July 2014 to April 2015 to design the model and set the data from April.2015 to October 2015 as test dataset for evaluating the models (Fig. 6). Meanwhile, the Kalman filter is used to reduce the noise of styrene monomer production. The simplified mathematical functions are as following,

$$\hat{x}_t = F_t\hat{x}_{t-1} + B_tu_t \tag{8}$$

Where F_t is the transition matrix, B for control matrix. The above formula $\hat{x}_t$ is only an estimate, not the best value.

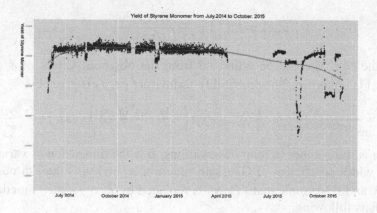

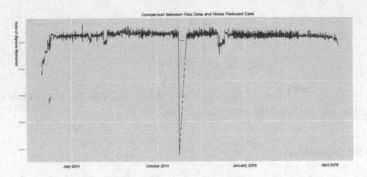

Fig. 6. Comparison of raw data and noise reduced data

After adjusting the algorithm parameters, we obtain a relatively flat styrene yield curve. The two-styrene monomer production will be set as input to fit the predictive model in the third chapter.

3   Real-time Predictive Model Fitting

3.1   Generalized Regression Neural Network

The generalized regression neural network (GRNN) is often applied to approximate the predictive function. The key idea is to estimate a linear or nonlinear regression surface on independent variables. Its basic theory is nonlinear regression analysis. Assume the joint probability density function of random variable X and Y is $f(x, y)$, then the expected value is given by,

$$y(x) = E\left(\frac{y}{x}\right) = \frac{\int_{-\infty}^{\infty} yf(x,y)dy}{\int_{-\infty}^{\infty} f(x,y)dy} \tag{9}$$

Where the $f(x, y)$ could be estimated from the observed values of X and Y, when that is unknown. The probability estimator $\hat{f}(x, y)$ is given by,

$$\hat{f}(x,y) = \frac{1}{(2\pi)^{(p+1)/2}\sigma^{(p+1)}} \cdot \frac{1}{n} \cdot \sum_{i=1}^{n} exp\left\{-\frac{(x-x_i)^T(x-x_i)}{2\sigma^2}\right\} \cdot exp\left\{-\frac{(y-y_i)^2}{2\sigma^2}\right\} \tag{10}$$

Where n is the number of total observations, p is the dimension of variable X, σ is the width coefficient of Gaussian function, x_i and y_i is the i-th observation of X and Y. Assume that $f(x, y)$ is replaced by $\hat{f}(x, y)$, then function (9) is shown as following,

$$\hat{y}(x) = E\left(\frac{y}{x}\right) = \frac{\sum_{i=1}^{n} y_i \cdot \exp\left\{-\frac{(x-x_i)^T(x-x_i)}{2\sigma^2}\right\}}{\sum_{i=1}^{n} \exp\left\{-\frac{(x-x_i)^T(x-x_i)}{2\sigma^2}\right\}} \tag{11}$$

Based on above function, when the width coefficient σ is relative large, $\hat{y}(x)$ approximates the mean of all sample dependent variables. However, when the width coefficient σ approaches 0, $\hat{y}(x)$ and the training sample are very close. Therefore, there are a lot of width coefficients and methods for σ selection to determine the optimal width coefficient for the GRNN model.

There are four layers in the GRNN, the input layer, the hidden layer, the summation layer and the output layer. The generalized regression neural network architecture is shown in Fig. 7.

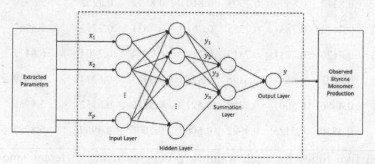

Fig. 7. Generalized regression neural network architecture

3.2 Results and Discussion

Here, we selected the GRNN to fit the styrene monomer predictive model due to its excellent prediction power. In the second chapter, 33 extracted operational parameters served as inputs in the GRNN. There is only the smoothing factor σ to be determined as it is not necessary to do the priori method of parameters selection. The smoothing factor σ usually ranges from 0.1 to 1 with optimal value.

To obtain the optimal predictive model, we compared three models, GRNN, GRNN with after Kalman-filter reduced data, and the random forest regression model. Meanwhile, several methods could be used to evaluate the performance of the models. For instance, the accuracy rate (difference between the actual and estimated value), SSE (sum square error), RMSE (root-mean-square error), MAPE (mean absolute percentage error) and others, most frequently the MSE and RMSE are used, and defined as following,

$$MSE = \frac{1}{N}\sum_{n=1}^{N}(y_n - \hat{y}_n)^2 \qquad\qquad (12)$$

$$RMSE = \sqrt{MSE} \qquad\qquad (13)$$

Where y is actual value, $\hat{y}$ is estimated value, N is total number of observation.

Table 1. GRNN results with different smoothing factor

σ	R^2 of Actual vs. Predicted					Avg. R^2	Elapsed Time
0.05	0.9402	0.9146	0.9336	0.9249	0.9471	0.9311	5.2921
0.1	0.9188	0.8965	0.9227	0.9194	0.9013	0.9125	5.3322
0.15	0.8342	0.8596	0.7903	0.8101	0.8192	0.8252	5.8930
0.2	0.7442	0.7831	0.7555	0.7619	0.7831	0.7713	5.5967
0.3	0.7120	0.7231	0.7415	0.7142	0.7014	0.7204	6.8475
0.5	0.6232	0.6543	0.6196	0.6282	0.6303	0.6334	5.6739
0.8	0.4986	0.5001	0.5328	0.5213	0.5510	0.5132	5.9936
1	0.4840	0.5125	0.4922	0.5003	0.5006	0.4969	5.8795

We tried five times for each training processes using different smoothing factor values, and the results of GRNN modeling are shown in the table 1. Meanwhile, the statistical performance is used to compare the different modeling results in table 2.

Table 2. Comparison of predictive model results

Model	R^2	MSE	RMSE
GRNN	0.9311	3203.88	56.60
GRNN-Kalman filter	0.9740	742.40	27.24
Random forest regression	0.8343	16705.60	129.25

In Table 2, all models had a large R^2 value (from 83.43% to 97.40%). Meanwhile, the GRNN-Kalman filter model has excellent predictive results with highest R^2 value (97.40%) and with lower MSE and RMSE value. It means that the GRNN-Kalman filter predictive model is much better than other selected models. Therefore, the GRNN-Kalman filter model could effectively and timely predict the styrene monomer production.

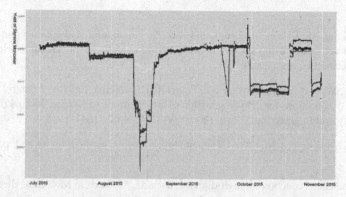

Fig. 8. Comparison of three models prediction and actual results. Here, black curve represents actual production, red curve represents prediction by GRNN-Kalman filter model, blue curve representsprediction by GRNN model, and yellow curve represents prediction by random forest regression model.

4   Conclusion

Statistical algorithms and machine learning methods are considered being in the infancy and limitedly used in the industrial enterprise. In this study, the generalized regression neural network and random forest regression methods are applied to explore the complex and nonlinear relationships between the styrene monomer production and the 33 extracted operational parameters. The results have shown that we perform successful data preparation and the GRNN Kalman filter model prediction have the highest accuracy, the model could be deployed to timely predict the styrene output. Meanwhile, compared with other neural network models, only one parameter σ needs to be determined.

ACKNOWLEDGMENT

I would like to acknowledge everyone who have participated in this client sides project and in the development of this paper. This work is carried out within projects supported by Accenture Beijing Lab and Accenture Resources Group in Great China.

Reference

1. Global Styrene Market 2015-2019, Research and Markets, Globe Newswire, Dublin, June 10. 2015

2. Lim, H., Kang, M., Chang, M., Lee, J. and Park, S., "Simulation and optimization of a styrene monomer reactor using a neural network hybrid model", 15th Triennial World Congress, Barcelona, Spain, (2002).

3. Bin Zhang, Weimin Yang, Zhiyong Wu, Feng Qian, Predictive functional control algorithm and stability conditions for dehydrogenation of ethylbenzene to styrene. Journal of Chemical Industry and Engineering (China) (2008), Vol. 59, No.7.: 1641-1645

4. Batini, C., Scannapieca, M. (2006). Data quality: concepts, methodologies and techniques. Springer.

5. L. Li, Dimension reduction for high-dimensional data. Methods in Molecular Biology. 620: 417-34, 2010.

6. J. B. Kruskal and M. Wish, Multidimensional Scaling. Beverly Hills, CA, U.S.A.: Sage Publication Inc., 1978.

7. S. van de Geer and P. Buehlmann, Statistics for High-Dimensional Data: Methods, Theory and Application, Springer Series in Statistics, DOI 10.1007/978-3-642-20192-9_6, 2002.

8. L. Breiman. Bagging predictors. Machine learning, 24 (2):123-140, 1996.

9. Greg Welch and Gary Bishop. 1995. "An Introduction to the Kalman Filter," University of North Carolina, Department of Computer Science, TR 95-041.

A graph-based ranking model for automatic keyphrases extraction from Arabic documents

Mohamed Salim EL BAZZI

IRF-SIC Laboratory, Faculty of sciences
Ibn Zohr University, Agadir, Morocco
elbazzi.mohamedsalim@edu.uiz.ac.ma

Driss MAMMASS

IRF-SIC Laboratory, Faculty of sciences
Ibn Zohr University, Agadir, Morocco
mammass@uiz.ac.ma

Taher ZAKI

IRF-SIC Laboratory, Faculty of sciences
Ibn Zohr University, Agadir, Morocco
t.zaki@uiz.ac.ma

Abdelatif ENNAJI

LITIS Laboratory, university of Rouen
Rouen, France
abdel.ennaji@univ-rouen.fr

Abstract. Automatic keyphrases extraction is to extract a set of phrases that are related to the main topics discussed in a document. They have served in several areas of text mining such as information retrieval and classification of a large text collection. Consequently, they have proved their effectiveness. Due to its importance, automatic keyphrases extraction from Arabic documents has received a lot of attention. For instance, the KP-Miner system was proposed to extract Arabic keyphrases, and demonstrates through experimentation and comparison with other systems its effectiveness. In this paper, we introduce TextRank, a graph-based ranking model, used successfully in many tasks of text processing, to compute term weights from graphs of documents. Vertices represent the document's terms, and edges represent term co-occurrence within a fixed window. It is an innovative unsupervised method that we have adapted to extract Arabic keyphrases, and assess its effectiveness. The obtained results with TextRank are compared with those obtained with KPMiner, owing to the fact that both systems do not need a training step.

Keywords: Keyphrases Extraction; KPMiner; TextRank; Arabic Documents

© Springer International Publishing AG 2017
P. Perner (Ed.): ICDM 2017, LNAI 10357, pp. 313–322, 2017.
DOI: 10.1007/978-3-319-62701-4_25

1   Introduction

Automatic extraction of keyphrases is an important task that allows the synthetic representation and highlighting of important topics in a document. This task facilitates access to relevant documents to ensure better use of the document.

Automatic methods of keyphrases extraction have the particularity to be independent from the domain and the language of the documents to be analyzed. This abstraction is due to the fact that terms are analyzed according to the given document. Many approaches are proposed to this aim, some rely on statistics only, while others combine them with more complex representations of documents.

The task of extracting keyphrases from Arabic documents is becoming increasingly important. In fact, keyphrases can be used to tag digital libraries, to valorize web contents and news streaming, and many more textual data collections. Nevertheless, keyphrases extaction can be a step to improve several text mining applications, including but not limited to classification [1] and information retrieval.

For instance, KP-Miner system [2] is used to extract keyphrases from Arabic documents and demonstrates through experimentation and comparison with widely used systems its effectiveness and efficiency.

Otherwise, Graph ranking algorithms, such as the TextRank [3] have been used successfully in English keyphrases extraction to compute term weights from graphs of a given document, where vertices represent the document's terms, and edges represent term co-occurrence within a fixed window.

Unlike other existing keyphrase extraction systems, the KPMiner and TextRank do not need to be trained in order to achieve their tasks. For this reason, we propose to compare the KPMiner system, considering its outperformance of other keyphrases extraction systems, and TextRank algorithm that we have adapted to extract Arabic keyphrases.

The rest of this paper is organized as following. The second part introduces related works. The third part is dedicated to explain the methods that will be compared in this paper. In the fourth section, we explain the conducted experiments and their results. Finally, we conclude by highlighting some ideas that have been introduced in this work.

2   Related works

Several approaches attempt to define what a keyphrase is by relying on certain statistical methods and studying the importance of the relationship to the notion of a candidate term. The more important a candidate term is to the analyzed document, the more relevant it is as a keyphrase.

Shishtawy et al. in their work [4], add linguistic knowledge to the extraction process, rather than relying only on statistics. They claim that using this approach may provide better results. Their work is based on combining the linguistic knowledge and the machine learning techniques to extract keyphrases from Arabic documents. The proposed system is based on three main steps: Linguistic pre-processing, candidate phrase extraction by generation of N-grams based on their Pos Tags, and feature vector calculation.

Najadat [5] consider the whole phrase which will be nominated as a candidate phrase if its frequency exceeds 2 (exist at least twice within a single paragraph). Phrase frequency (PF), summation of phrase terms frequencies (Tf), PF×IDF (Phrase Frequency–Inverse Document Frequency), Phrase Position, Title Threshold and phrase distribution are the used features in Najadat's approach. The score of each phrase is computed.

Combining linguistic methods with statistical methods could lead to better performance. Thus, Nabil [6] introduces his work using a stemmer and Part of Speech tagger (POS). The stemmer and the POS tagger generate a list of candidate keyphrases. These keyphrases are then weighted by different methods such as TFIDF which still have high performance, according to Nabil, compared to most of the current used methods. The linguistic aspect helps to boost the performance of text mining applications as they add abstract semantic representation to a given document.

Liu et al. [7] consider that the best keyphrase must be: understandable, semantically relevant with the document and have high coverage of the whole document. Their method relies on four steps. First, Candidate terms are selected after filtering out the stop words. Second, term relatedness is calculated. They use some measures to calculate the semantic relatedness of candidate terms. The third step is term clustering. Based on term relatedness, they group candidate terms into clusters and find the exemplar terms of each cluster. Finally, they use these exemplar terms to extract keyphrases from the document.

According to EL Baltagy [2], The number and frequency of compound terms in any given document is usually less than that of single keywords. Yet, Effective keyphrase extraction depends on the determination of an appropriate boosting feature for candidate keyphrases. This boosting feature is related to the ratio of single to compound terms in each document. They consider the position of the first occurrence of any given phrase because it is related to the fact that the more important a term is, the more likely it is to appear 'sooner' in the document than later. Nonetheless, they consider also the hypothesis that after a given threshold is passed in any given document, phrases appearing for the first time are highly unlikely to be keyphrases. These observations have been translated into a set of heuristics that are implemented by KPMiner system in three main steps: candidate keyphrase selection, candidate keyphrase weight calculation and final keyphrase refinement.

In the work of Ali et al [8]., the authors proceed by document preprocessing including Tokenization, Stop word removal and Stemming. Then, they extract candidate terms by selecting all noun phrases from the Arabic text as candidate Keyphrases. The next step is to extract N- gram phrases from the Arabic phrase. Accordingly, they used POS patterns to identify the candidate noun phrases. Candidate features are extracted by ranking the candidate keyphrase using Term Frequency (TF-IDF). Finally, they introduce a classification process. The idea is to enrich statistical and linguistic information. The keyphrase are used as input to know whether it classifies the term as keyphrase or not. Linear Logistic Regression (LLR), Linear Discriminant Analysis (LDA), Support Vector Machines (SVM) are used for comparison purpose.

The approach of Sarkar [9] is based on four phases. After document preprocessing, candidate keyphrase are identified. A candidate keyphrase is considered as a sequence of words containing no punctuations and stop words, then this sequence is broken into smaller phrases. Afterward, scores are assigned to candidate keyphrases. The weight of a candidate keyphrase is computed using three important features: phrase frequency, inverse document frequency and domain specificity. next, K top-ranked candidate keyphrases are selected as the final list of keyphrases. The value of K is specified by the user.

Mihalcia and tarau present TextRank [3], a Graph-based ranking algorithm to extract keyphrases. The graph based algorithms are a way of deciding the importance of a vertex within a graph. The idea of implementing a graph-based ranking model comes from the concept of recommendation. When one vertex is linked to another one, this connection is considered as a recommendation between those vertices. To fulfill this target, four steps must be accomplished. First of all, to identify text units that best define the document using a Pos tagger and add them as vertices in the graph. Second of all, to identify relations that connect text units, and use these relations to draw edges between vertices in the graph. Edges can be directed or undirected, weighted or unweighted. Third of all, to iterate the graph-based ranking algorithm until convergence. Finally, to sort vertices based on their final score.

Reference	Used methods	Comment
Nabil (2015)	Stemming, Pos Tagging, TFIDF	Combining linguistic methods with statistical methods
Ali(2014)	Preprocessing, TF-IDF	enrich statistical and linguistic information
Mihalcea (2004)	Pos Tagging, TextRank	A Graph based ranking algorithm
El-Beltagy (2009)	KPMiner, frequency	Consider the position of the first occurrence of any given phrase
Elshishtawy (2009)	Linguistic preprocessing, N-grams, Pos Tags, and feature vector calculation.	Combining the linguistic knowledge and the machine learning techniques
Najadat (2016)	frequency	Phrase frequency (PF), summation of phrase terms frequencies (Tf), PFIDF (Phrase FrequencyInverse Document Frequecy), Phrase Position, Title Threshold and phrase distribution

Liu (2009)	Preprocessing, term relatedness, clustering.	The best keyphrase must be: understandable, semantically relevant with the document and have high coverage of the whole document
Sarkar (2013)	phrase frequency, inverse document frequency and domain specificity	A candidate keyphrase is considered as a sequence of words that does not contain neither punctuations nor stop words, and then this sequence is broken into smaller phrases

Table 1. Overview of related works

The extraction of candidate terms is a preliminary step in the extraction of keyphrases. If a term of the analyzed document is not present in the set of candidate terms, then the keyphrasee can not be extracted. Used with unsupervised methods, candidate terms are ordered according to an importance score obtained either from themselves or from the importance of the words that compose them. If a method is based only on words, then the score of a candidate term is generally calculated as the sum of the words composing it. However, this is not always fair, so it is an important disadvantage of methods working on words to extract key terms. Indeed, the summation may favor words that contain many important words non terms vis-à-vis containing very few words, but important.

3 Arabic keyphrases extraction

The unsupervised methods have the particularity to abstract the specific nature of the processed data. This abstraction is due to approaches which valorize semantic importance, degree of informativness, and structure.

3.1 KPMiner

The KPMiner system uses TF-IDF for weighing terms. The use of IDF information for compound phrase weight calculation would improve the extraction of phrases low-frequency phrases. This would build a general rather than a domain-specific exemplar keyphrases extractor, since combinations are much larger.The use of linguistic features will help to extract meaningful keyphrases.

$$TFIDF_{doc}(term) = tf \times idf \tag{1}$$

with $$idf = \log (N/n_t) \tag{2}$$

tf represents the term frequency, **N** is the total number of documents in the corpus and **n**$_t$ the number of documents where the term **t** appears KPMiner is used to extract Arabic keyphrases and have been compared with KEA extractor [10].

In the experiment, seven, fifteen and twenty keywords were extracted by the system from a document set consisting of 100 Arabic articles. The three systems have been applied to the same dataset to extract extract keyphrases. (table 2 and 3 show the obtained results).

	15 keyphrases		20 keyphrases	
	Kpminer	*KEA*	*KPMINER*	*KEA*
Precision	0.143	0.112	0.124	0.092
recall	0.358	0.303	0.376	0.326

Table 2. Comparison between developed KP-Miner system and KEA-3.0 [10]

	7 keyphrases		15 keyphrases		20 keyphrases	
	KPMINER	*KEA*	*KPMINER*	*KEA*	*KPMINER*	*KEA*
F-Measure	0.241	0.198	0.205	0.17	0.186	0.143

Table 3. Comparison between the average F-measure of the developed KP-Miner system and KEA-3.0 when 7, 15, and 20 keyphrases are extracted

3.2 TextRank

TextRank is an adaptation of PageRank for automatic keywords extraction. It is an algorithm proposed by Mihalcea and Tarau [3] that represent text as graph. Each vertex represent a term and point to its predecessors and successors following the reading flow. The score of the vertex v, denoted S (vi), is defined as follows:

$$S(v_i) = (1 - d) + d \times \sum_{v_j \in \text{Adj}(v_i)} \frac{w_{ji}}{\sum_{v_k \in Adj(v_j)} w_{jk}} S(v_j) \qquad (4)$$

Where **Adj (v**$_j$**)** represents the neighbors of **v**, and **d** is a damping factor set at 0.85. **w**$_{ji}$ is a weight associated to each edge, and represents the frequency of co-occurrence of two words within a window of 2 to10 words. This weight can be any other relationship between words such as, but not limited to, lexical or semantic relations, contextual overlap.

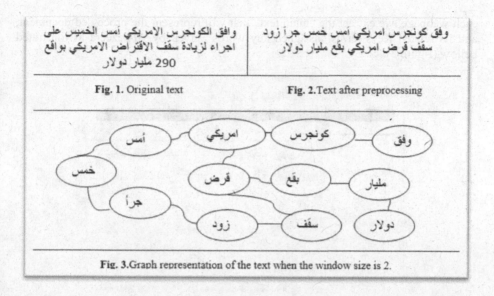

وافق الكونجرس الامريكي أمس الخميس على اجراء لزيادة سقف الاقتراض الامريكي بواقع 290 مليار دولار	وفق كونجرس امريكي أمس خمس جرأ زود سقف قرض امريكي بقع مليار دولار
Fig. 1. Original text	**Fig. 2.** Text after preprocessing

Fig. 3. Graph representation of the text when the window size is 2.

Fig. 1. Graph based representation of a document

For each node of the graph, a score is calculated iteratively to simulate the concept of recommendation of a term by its adjacent nodes. The score at each node grants a ranking value to the word which assesses its importance within the document. Then, the sorted list of words is used to extract keywords.
Experiments and

4   RESULTS

The use of a learning corpus implies that the learned models are specific to the subject and the language of it. This specificity can be advantageous when the domain and the language represented by the corpus are the same for the documents which will be analyzed, but if this is not the case, the results of the extraction may decrease.

In this paper, we propose an adaptation of textRank to extract Arabic Keyphrases. For comparison purpose, we introduce in this section two automatic methods. In one hand, we will give an overview of KPMiner which is so far an efficient system of Arabic keyphrases extraction. On the other hand, we will present TextRank algorithm and its adaptation to Arabic language.

4.1   DATASET

To validate the proposed algorithm, we have tested the two systems KPMiner and TextRank on a corpus of 100 documents extracted manually from Al-Jazeera[1]. From

---

[1]   http://www.aljazeera.net

each webpage, we extract the "main text" that will represent the processed document, and its related Keys terms "assigned key words" given by Al-Jazeera as illustrated bellow (figure 2).

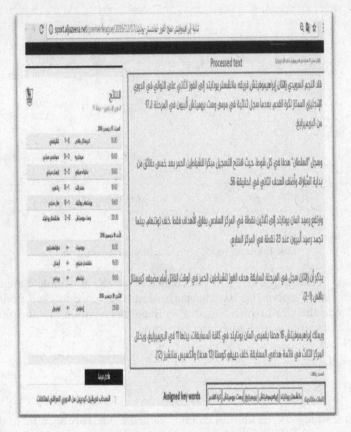

Fig. 2. Example of the constructed corpus

For an efficient validation we have tested the two systems on the same corpus. To evaluate the performance, three conventional metrics are used: precision, recall and F-measure.

4.2   RESULTS

First, we run each system on the main text (documents of corpus) to extract keyphrases. Then, we compare the extracted keyphrases with those assigned by Al-Jazeera to each document. According to KPMiner's results, the most the number of extracted is high, the less it is efficient. So our approach aim's to solve this problem. Thus, we select all extracted keyphrases without using any threshold. The following results are drown from the conducted experiments.

	Precision %	Recall %	Fmeasure
KPminer	7.293	1.162	2.005
TextRank	14.476	4.407	6.757

Table 4. Comparison between TextRank and KPMiner

Table 4 shows that our proposed approach outperforms the KPMiner approach. We can argue this outperformance by two remarks. Firstly, TextRank is a graph-based approach which introduces the concept of recommendation or voting. This concept grant the high score to the each term whitin a structural modeling. In contrast, KPMiner used the bag of word representation which consider that terms are independent in modeling space. Secondly, the used corpus contains a huge variation of topics which complicates the task and influence negatively the efficiency of keyphrases extraction systems when using statistical methods.

5   Conclusion

The keyphrases extraction is a task of analyzing a document and extracting its important aspects. These keyphrases use simple text units to construct a synthetic view of the document. A set of keyphrases can therefore be perceived as one of the text units which key elements are expressed using aggregation between them.

In this paper, we have conducted an experiment to assess the efficiency of adapted TextRank for Arabic document compared with KPminer system. The experimental results show that our proposed method, based on graph modeling of documents, outperformed the KPMiner approach.

In our future work, we will apply Automatic Keyphrases Exctraction from Arabic documents to improve many related fields to Arabic text mining, such as information retrieval. Nonetheless, we will continue to search more suitable approaches to extract Arabic keyphrases.

6   References

1. Mohamed Salim El Bazzi; Taher Zaki; Driss Mammass; Stemming versus multi-words indexing for Arabic documents classification. 11th International Conference on Intelligent Systems: Theories and Applications (SITA) 2016
2. Samhaa R El-Beltagy and Ahmed Rafea. Kp-miner: A keyphrase extraction system for english and arabic documents. Information Systems, 34(1):132–144, 2009.
3. Mihalcea R. et P. Tarau. Textrank: Bring- ing order into texts. In Proceedings of EMNLP, pages 404–41, 2004.

4. El-shishtawy T.A. & Al-sammak A.K   Arabic Keyphrase Extraction using Linguistic knowledge and Machine Learning TechniquesProceedings of the Second International Conference on Arabic Language Resources and Tools, 2009

5. Hassan M. Najadat, Mohammed N. Al-Kabi, Ismail I. Hmeidi, Maysa Mahmoud Bany Issa. Automatic   Keyphrase   Extractor   from   Arabic Documents. (IJACSA) International Journal of Advanced Computer Science and Applications,
Vol. 7, No. 2, 2016

6. Mahmoud Nabil, Mohamed Aly, Amir F. Atiya. New Approaches for Extracting Arabic Keyphrases. 2015 First International Conference on Arabic Computational Linguistics.

7. Liu, Z. , Peng, L. , Yabin, Z. et S. Maosong. 2009b. Clustering to find exemplar terms for keyphrase extraction. In Proceedings of the 2009 Conference on Empirical Methods in Natural Language Processing, pages 257–266.

8. Nidaa Ghalib Ali, Nazlia Omar. Arabic Keyphrases Extraction Using a Hybrid of Statistical and Machine Learning Methods. International Conference on Information Technology and Multimedia (ICIMU), November 18 – 20, 2014, Putrajaya, Malaysia

9. Kamal Sarkar. A Hybrid Approach to Extract Keyphrases from Medical Documents. International Journal of Computer Applications (0975 – 8887) Volume 63– No.18, February 2013

10. I.H. Witten, et al., KEA: Practical automatic keyphrase extraction, in: The Fourth ACM Conference on Digital Libraries, 1999

Mining Frequent Subgraph Pattern over a Collection of Attributed-Graphs and Construction of a Relation Hierarchy for Result Reporting

Petra Perner

Institute of Computer Vision and Applied Computer Sciences, IBaI
PSF 30 11 14
04251 Leipzig

e-mail: pperner@ibai-institut.de

Abstract. Graphs are compared to the standard attribute-value representations sophisticated data structures. Besides the description of an entity, a graph representation can also represent the relation of the entities to each other and by doing that it can be build up a complex network of knowledge pieces. These networks may be different kinds of networks such as telecommunication networks, computer networks, biological networks, and Web and social community networks. There are broad applications that require graph-based representations such as chemical informatics, bioinformatics, computer vision, video indexing, text retrieval, and Web analysis. Attributed graphs are a special form of graphs that describe the nodes and the edges of a graph by attributes. An important task of graph mining is mining frequent subgraph patterns. The summarization of graphs into groups of subgraphs are used for further characterization, discrimination, classification, and cluster analysis of a collection of graphs. In this paper, we introduce mining frequent subgraph pattern over a collection of attributed graphs. We describe the graph representation. We explain the fast graph-matching algorithm. How to deal with the similarity on the attributes on the nodes and the edges is described. Then, we explain the main part of our paper our algorithm for frequent subgraph mining. We describe how the found results can be reported in such a way that a human can easily overlook the results and how he can use the discovered knowledge for his application. The reporting is done by constructing a relation hierarchy over the discovered groups of subgraphs. The user gets the relation hierarchy of the subgraphs graphically displayed. We give results for our algorithm on an application of attributed image graphs. These image graphs have been obtained by automatic image analysis of ultra-sonic images of welding seams. The task was to classify the images into different defects such as pore, whole, and cracks.

Keywords: Mining Frequent Subgraph Pattern, Attributed-Graph Mining, Construction of a Relation Hierarchy, instance-based learning, graph-based learning, high dimensional data

© Springer International Publishing AG 2017
P. Perner (Ed.): ICDM 2017, LNAI 10357, pp. 323–344, 2017.
DOI: 10.1007/978-3-319-62701-4_26

1   Introduction

Graphs are compared to the standard attribute-value representations sophisticated data structures. Besides the description of an entity, a graph representation can also represent the relation of the entities to each other and by doing that it can be build up a complex network of knowledge pieces. These networks may be different kinds of networks such as telecommunication networks, computer networks, link mining [1], biological networks [5], and Web and social community networks [3]. There are broad applications that require graph-based representations such as chemical informatics, bioinformatics, computer vision, video indexing, text retrieval, and Web analysis. Attributed graphs are a special form of graphs that describe the nodes and the edges of a graph by attributes. An important task of graph mining is mining frequent subgraph patterns [2][4]. The summarization of graphs into groups of subgraphs are used for further characterization, discrimination, classification, and cluster analysis of a collection of graphs.

In this paper, we introduce mining frequent subgraph pattern over a collection of attributed graphs.

Determination of similarity between structural attributed graphs is time-consuming and requires a good matching procedure. Most applications have attribute assignments to the components of the structure and attributes that describe the relationships between the components. This requires a similarity measure, which can describe both structural similarity and the proximity of the attribute labels. The similarity measure should have the flexibility to view the similarity from these different perspectives.

We propose an algorithm that can incrementally learns the groups of subgraphs. The discovered subgraphs form a hierarchy and can be up-dated incrementally as soon as new graphs are available. The tentative underlying conceptual structure of the hierarchy is visually presented to the user. We describe two approaches for mining of frequent subgraphs. Both are based on approximate graph subsumption. The first approach is based on a divide-and-conquer strategy whereas the second is based on a split-and-merge strategy that better allows fitting the hierarchy to the actual structure of the application, but requires more complex operations. The first approach uses a fixed threshold for the similarity values. The second approach uses for the grouping of the graphs an evaluation function.

The paper is organized as follow: In Section 2 we will describe the concepts for indexing and learning. The definition of a graph, the similarity measure as well as matching are presented in Section 3. Our two approaches for learning the subgraphs are described in Section 4. We compare our methods to related work in Section 5. Finally, we give conclusions in Section 6.

2   Concepts for Indexing and Learning

2.1 Organization and Retrieval of Graphs

To speed up the findings of subgraphs, we use a sophisticated organization of the subgraph data base. This organization should allow separating the set of similar graphs from those graphs not similar to the recent problem at the earliest stage of the retrieval process. Therefore, we need to find a relation p that allows us to order our graphs:

Definition 1: A relation p on a set CB is called a partial order on CB if it is re-flexive, antisymmetric, and transitive. In this case, the pair $\langle CB, p \rangle$ is called a partial ordered set or partial-ordered set.

The relation can be chosen depending on the application. One common approach is to order the graphs based on the similarity value. The set of graphs can be reduced by the similarity measure to a set of similarity values. The relation $\leq$ over these similarity values gives us a partial order over these groups of subgraphs. The derived hierarchy consists of nodes and edges. Each node in this hierarchy contains a set of subgraphs that do not exceed a specified similarity value. The edges show the similarity relation between the nodes. The relation between two successor nodes can be expressed as follows: Let z be a node and x and y are two successor nodes of z, then x subsumes z and y subsumes z. By tracing down the hierarchy, the space gets smaller and smaller until finally a node will not have any successor. This node will contain a set of similar graphs. Among these set of graph has to be found the most similar graph to the query graph. Although we still have to carry out matching, the number of matches will have decreased through the hierarchical ordering. The nodes can be represented by the prototypes of the set of graphs assigned to the node. When the hierarchy is used to process a query, the query is only matched with the prototype. Depending on the outcome of the matching process, the query branches right or left of the node.

The problem is to determine the right relation p that allows organize the graph data base, a procedure for learning prototypes and graph classes, and a similarity measure.

2.2 Learning Over the Groups of Detected Subgraphs

Learning process aims to find groups of subgraphs in an incremental manner. That means not all graphs need to be available from scratch. They can arrive in temporal sequence and the new subgraph group can be detected or the already detected groups can be updated. Let X be a set of graphs collected in the data base CB. The relation between each group of subgraph in the data base is expressed by the similarity value sim. The data base can be partitioned into n graph classes $C: CB = \bigcup_{i=1}^{n} C_i$ such that the intra-class similarity is high and the inter-class similarity low. The set of graphs in each class C can be represented by a graph that generally describes the cluster. This representative can be the prototype, the median, or an a-priori selected graph. Whereas the prototype implies that the representative is the mean of the cluster, which can easily be calculated from numerical data, the median is the graph whose sum of all distances to all other graph in a cluster is minimal. The relation between the different graph classes C can be expressed by higher order constructs expressed e.g. as super classes that gives us a hierarchical structure over the data base.

There are different learning strategies that can take place over the hierarchy:

Learning takes place if a new graph x has to be stored into the data base such that:
$$CB_{n+1} = CB_n \cup \{x\}.$$

It may incrementally learn the graph classes and/or the prototypes representing the class.

The relationship between the different graph or graph classes may be updated according to the new graph class.

The system may learn the similarity measure.

2.2.1 Learning of Prototypes

The prototype or the representative of a graph class is the most general representation of a graph class. A class of graphs is a set of graphs sharing similar properties. The graph class is a set of graphs that do not exceed a boundary for the intra-class dissimilarity. Graphs that are on the boundary of this hyper-ball have a maximal dissimilarity value. A prototype can be selected a-priori by the domain user. This approach is preferable when the domain expert knows for sure the properties of the prototype. The prototype can be calculated by averaging over all graphs in a graph class, or the median of the graphs is chosen. When only a few graphs are available in a class and subsequently new cases are stored in the class, then it is preferable incrementally update the prototype according to the new detected subgraphs.

2.2.2 Learning of Higher Order Constructs

The ordering of the different subgraph classes gives an understanding of how these graph classes are related to each other. For two graph classes, which are connected by an edge similarity relation holds. Graph classes that are located at a higher position in the hierarchy apply to a wider range of graphs than those located near the leaves of the hierarchy. By learning how these graph classes are related to each other, higher order constructs are learned [6].

2.2.3 Learning of Similarity

For generality, we will also consider here that feature weights can be learned. By introducing feature weights, we can put special emphasis on some features for the similarity calculation. It is possible to introduce local and global feature weights. A feature weight for a specific attribute is called a local feature weight. A feature weight that averages over all local feature weights for a graph is called a global feature weight. This can improve the accuracy of the learnt hierarchy over the data base. By updating these feature weights we can learn similarity [7][8]. Learning feature weights in a structural representation is computationally complex and time consuming. Therefore, we will only propose a similarity measure for structural representations, which can handle feature weights, but we will not consider how these feature weights can be learned. Instead, we will assume that the user can specify this feature weight a priori.

3 Structural Representation and Structural Similarity Measure

Now, we introduce the basic definitions and notation that will be used in this paper.

3.1 Definition of a Graph

The structural representation of complex knowledge pieces of different applications can be described as a graph. If we assign attributes to the nodes and the edges, then we have an attributed graph defined as follow:

Definition 2:

W ... set of attribute values

A ... set of all attributes

$b:\ A \rightarrow W$ partial mapping, called attribute assignments

B ... set of all attribute assignments over A and W.

A graph $G = (N, p, q)$ consists of

N ... finite set of nodes

$p:\ N \rightarrow B$ mapping of nodes to attribute assignment

$q:\ E \rightarrow B$ mapping of nodes to attribute assignment, where $E = (N \cdot N) - I_N$ and

I_N is the Identity relation in N.

This representation allows us to consider only objects, the spatial relation between objects, or the whole graph itself (see Section 3.2.).

For generality, we will not consider specific attribute assignments to the nodes and the edges. To give the reader an idea what these attributes could be we will consider the graphs to be images such as the ultrasonic images [9]. The images may show defects such as cracks in a metal component taken by an ultra sonic image acquisition system called SAFT, see Figure 1a to Figure 1d. A defect comprises of several reflection points that are in a certain spatial relation to each other. Here the nodes could be the objects (reflection points) in the image and the edges are the spatial relation between these objects (e.g. right-of, behind, ...). Each object has attributes (e.g. size, mean gray level value,..) that are associated to the corresponding node within the graph. We will not describe the application in this paper, nor will we describe how these objects are extracted from the images, nor will we not consider how these objects are combined into a graph and labeled by symbolic terms. For these details we refer to [10].

Fig. 1a Original Image

Fig. 1b Binary Image

Fig. 1c Result

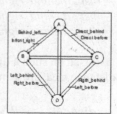
Fig. 1d Representation

3.2 Similarity Measure

We may define our problem of similarity as to find structural identity between two structures. However, structural identity is a very strong requirement. An alternative approach is to require only part isomorphism.

Definition 3

Two graphs $G_1 = (N_1, p_1, q_1)$ and $G_2 = (N_2, p_2, q_2)$ are in the relation $G_1 \leq G_2$ iff there exists a one-to-one mapping $f : N_1 \rightarrow N_2$ with

(1) $p_1(x) = p_2(f(x))$ for all $x \in N_1$ and

(2) $q_1(x, y) = q_2(f(x), f(y))$ for all $x, y \in N_1, x \neq y$.

Definition 3 and 4 require identity in the attribute assignments of the nodes and the edges. To allow proximity in the attributes labels, we introduce the following way to handle similarity.

In Definitions 3 and 4 we may relax the required correspondence of attribute assignment of nodes and edges to that we introduce ranges of tolerance:

If $a \in A$ is an attribute and $W_a \subseteq W$ is the set of all attribute values which can be assigned
to a, then we can determine for each attribute a a mapping:

$$dist_a : W_a \rightarrow [0,1].$$

The normalization to a real interval is not absolutely necessary but advantageous for the comparison of attribute assignments.

For example, let a be the attribute spatial_relation as it is in e.g. in the ultra sonic image the relation between the nodes and

$W_a = \{behind, right, behind\_left, in\_front\_right, ... \}$

then we could define:

$dist_a$ *(behind_right, behind_right) = 0*
$dist_a$ *(behind_right, infront_right) = 0.25*
$dist_a$ *(behind_right, behind_left) = 0.75*.

Based on such a distance measure for attributes, we can define different variants of distance measure as mapping:

$$dist : B^2 \rightarrow R^+$$

(R^+ is the set of positive real numbers) in the following way:

$$dist(x, y) = \frac{1}{D} \sum_{a \in D} dist_a(x(a), y(a))$$

with $D = domain(x) \cap domain(y)$.

Usually, in the comparison of graphs not all attributes have the same priority. Thus it is good to determine a weight factor w_a and then define the distance as follows:

$$dist(x, y) = \frac{1}{D} \sum_{a \in D} w_a \cdot dist_a(x(a), y(a))$$

For definition of part isomorphism, we get the following variant:

Definition 4

Two graphs $G_1 = (N_1, p_1, q_1)$ and $G_2 = (N_2, p_2, q_2)$ are in the relation $G_1 \leq G_2$ iff there exists a one-to-one mapping $f : N_1 \to N_2$ and thresholds C_1, C_2 with

(1) $dist(p_1(x), p_2(f(x)) \leq C_1$ for all $x \in N_1$

(2) $dist(q_1(x, y), q_2(f(x), f(y)) \leq C_2$ for all $x, y \in N_1 \ x \neq y$.

Obviously it is possible to introduce a separate constant for each attribute. Depending on the application, the similarity may be sharpened by a global threshold: If it is possible to establish a correspondence g according to the requirements mentioned above, then an additional condition should be fulfilled:

$$\sum_{(x,y) \in g} dist(x, y) \leq C_3$$

where C_3 is the global threshold. .

3.3 Mean of a Graph

The mean graph can be computed as follows:

Definition 5

A Graph $G_p = (N_p, p_p, q_p)$ is a prototype of a Class of Cases $C_i = \{G_1, G_2, ..., G_t(N_t, p_t, q_t)\}$ iff $G_p \in C_i$ and if there is a one-to-one mapping $f : N_p \to N_i$ with

(1) $p_p(x_i) = \frac{1}{n} \sum_{n=1}^{t} p_n(f(x_i))$ for all $x_i \in N$ and

(2) $q_p(x_i, y_i) = \frac{1}{n} \sum_{n=1}^{t} q_n(f(x_i), f(y_i))$ for all $x_i, y_i \in N$.

3.4 Graph Subsumption

On the basis of the part isomorphism, we can introduce a partial order over the set of graphs. If a graph G_1 is included in another graph G_2 than the two graphs are in the relation $G_1 \leq G_2$ and the number of nodes of G_1 is not higher than the number of nodes of G_2. We can also say G_1 subsumes G_2 and we can write $G_1 \succ G_2$. If we have three graphs $G_1, G_2,$ and G_3 with $G_3 \succ G_1$ and $G_3 \succ G_2$, then G_3 is the most specific common substructure mscs of the two graphs G_1 and G_2. We can also say that G_3 is the generalization or the general concept of the two graphs G_1 and G_2.

We can use these relations to learn the hierarchy over our subgraph classes and on the other hand it allows us to discover the underlying concept of the domain.

3.5 Graph Matching

Now consider an algorithm for determining the part isomorphism of two graphs. The main approach is to find a superset of all possible correspondences f and then exclude non-promising cases. In the following we assume that the number of nodes of G_1 is not greater than the number of nodes of G_2.

A technical aid is to assign to each node n a temporary attribute list $K(n)$ of all attribute assignments of all the connected edges:

$$K(n) = (a : q(n,m) = a, m = N - \{n\}) \text{ where } (n \in N).$$

The order of list elements has no meaning. Because all edges exist in a graph the length of $K(n)$ is equal to $2 \cdot (|N| - 1)$.

For demonstration purposes, consider the examples in Figures 3a-3b. The result would be:

$$K(X) = (bl, bl, br)$$
$$K(Y) = (br, br, bl).$$

In the worst case the complexity of the algorithm is $O(|N|^3)$.

In the next step we assign to each node of G_1 all nodes of G_2 that could be assigned by a mapping f. That means we calculate the following sets:

$$L(n) = \{m : m \in N_2, p_1(n) = p_2(m), K(n) \subseteq K(m)\}.$$

The inclusion $K(n) \subseteq K(m)$ shows that the list $K(m)$ contains the list $K(n)$ without considering the order of the elements. When the list $K(n)$ contains multiple times an attribute assignment, then the list $K(m)$ also contains multiple times this attribute assignment.

For the example in Fig. 3a and Fig. 3b we get the following L-sets:

$$L(X) = \{A\}$$
$$L(Y) = \{B_1\}$$
$$L(Z) = \{C\}$$
$$L(U) = \{D, B_2\}.$$

We did not consider in this example the attribute assignments of the nodes.
Now, the construction of the mapping f is prepared and if there exists any mapping then the following condition must hold:

$$f(n) \in L(n) \rightarrow (n \in N_1).$$

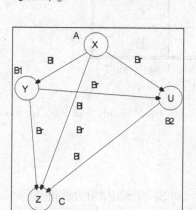

Fig. 3a Graph_1

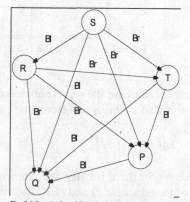

Fig. 3b Graph_2 and Result of Subgraph Isomorphism to Graph 1

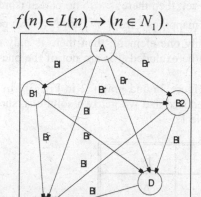

Fig. 3c Second Result of Subgraph Isomorphism of Graph_1 and Graph_2

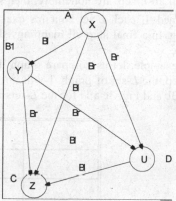

Fig. 3d Graph_3 and Results for Subgraph Isomorphism to Graph 2

The first condition for the mapping f regarding the attribute assignments of nodes holds because of the construction procedure of the L-sets. In case that one set $L(n)$ is empty, there is no partial isomorphism.

Also, if there are nonempty sets, in a third step it has to be checked if the attribute assignments of the edges match.

If there is no match, then the corresponding L-set should be reduced to:

> **for** all nodes n_1 of G_1
> > **for** all nodes n_2 of $L(n_1)$
> > > **for** all edges (n_1, m_1) of G_1
> > > **if** for all nodes m_2 of $L(m_1)$
> > > $p_1(n_1, m_1) \neq p_2(n_2, m_2)$
> > > **then** $L(n_1) := L(n_1) \setminus \{n_2\}$

If the L-set of a node has changed during this procedure, then the examinations already carried out should be repeated. That means that this procedure should be repeated until none of the L-sets has changed.

If the result of the third step is an empty L-set, then there is also no partial isomorphism. If all L-sets are nonempty, then some mappings f from N_1 to N_2 have been determined. If each L-set contains exactly only one element, then there is only one mapping. In a final step all mappings should be excluded that are not of the one-to-one type.

For example, let us compare graph_1 in Figure 3a and graph_2 in Fig. 3b. In the third step, the L-set of graph_1 will not be reduced and we get two solutions, shown in Fig. 3b and Fig. 3c and for the L-sets in table 1.

N_1	f_1	f_2
X	A	A
Y	B_1	B_1
Z	C	C
U	D	B_2

Table 1 Results for the L-Sets for Matching Graph_1
with Graph_2 and Graph_3

When we compare the representation of graph_1 and graph_3 in Fig. 3d, the L-set of graph_1 also contains two elements:

$$L(U) = \{T, P\}.$$

However, in the third step the element T will be excluded if the attribute assignments of the edges (U, Y) and (T, R) do not match when node U is examined.

If the L-set of a node has changed during the third step, then the examinations already carried out should be repeated. That means that step 3 is to be repeated until there is no change in any L-set.

This algorithm has a total complexity of the order $O\left(\left|N_2\right|^3, \left|N_1\right|^3 \cdot \left|M\right|^3\right)$. $\left|M\right|$ represents the maximal number of elements in any L-set $\left(\left|M\right| \leq \left|N_2\right|\right)$.

The similarity in the attribute labels can be handled by the way the L-sets are defined and particularly the inclusion of K-lists:

Given C a real constant, $n \in N_1$ and $n \in N_2$. $K(n) \sqsubseteq_c K(m)$ is true iff for each attribute assignment b_1 of the list $K(n)$ attribute assignment b_2 of $K(m)$ exists, such that $dist(b_1, b_2) \leq C$. Each element of $K(m)$ is to be assigned to a different element in list $K(n)$.

Obviously it is possible to introduce a separate constant for each attribute. Depending on the application, the inclusion of the K-lists may be sharpened by a global threshold:

If it is possible to establish a correspondence g according to the requirements mentioned above, then an additional condition should be fulfilled:

$$\sum_{(x,y) \in g} dist(x, y) \leq C_3 \ (C_3 - \text{threshold constant}) \ .$$

Then we get the following definition for the L-set:

Definition 6

$$L(n) = \left\{m : m \in N_2, dist(p_1(n), p_2(m)) \leq C_1, K(n) \sqsubseteq_c K(m)\right\}.$$

In step 3 of the algorithm for the determination of one-to-one mapping, we should also consider the defined distance function for the comparison of the attribute assignments of the edges. This new calculation increases the total amount of effort, but the complexity of the algorithm is not changed.

4 Organization and Learning over the Groups of Learnt Subgraphs

We propose a hierarchical organization schema of the subgraph classees that can be up-dated incrementally. Two approaches are described in the following section. Both are based on approximate graph subsumption. The first one is based on a divide-and conquer strategy whereas the second is based on an evaluation function and a split-and-merge strategy.

4.1 Approach I

4.1.1 Index Structure
The initial data base may be built up by existing graphs. Therefore, a nonincremental learning procedure is required in order to find the subgraph groups and to define the hierarchy over these groups. When new graphs arrive they may be

stored into the subgraph data base by searching for similar graphs or opening a new subgraph class. Therefore, we need an incremental learning procedure [8].

As an important relation between graphs we have considered similarity based on part isomorphism. Because of this characteristic, it is possible to organize the data base as a directed graph.

In terminological logics [11] the extension of the concept, with respect to a domain of interpretation, is the set of all things in the domain which satisfy the concept description. As we have shown before, the subsumption is a subset relation between the extensions of concepts. Subsumption allows us to observe the conceptual knowledge of the domain of interpretation. Therefore, we introduce a new attribute a which is called concept_description (in short: cd). To this attribute are assigned the names for the various subgraph classes if wanted by the user. They are not known before they should be learned during usage of the usage of the system. Thus the attribute gets the temporary value "unknown". We assume that the attribute values can be specified by the user of the system when a new subgraph class is found by the incremental learning procedure and this name is assigned to the attribute.

In the following, we will define the hierarchy of the subclasses as a graph that contains the graphs of the application in the nodes:

Definition 7

H is given, the set of all graphs.

A index graph is a tupel $IB = (N, E, p)$, with

(1) $N \subseteq H$ set of nodes and

(2) $E \subseteq N^2$ set of edges.

This set should show the partial isomorphism in the set of nodes, meaning it should be valid $x \leq y \Rightarrow (x, y) \in E$ for all $x, y \in N$.

(3) $p : N \rightarrow B$ mapping of class names to the index graph (also the attribute values for the sub-class).

Because of the transitivity of part isomorphism, certain edges can be directly derived from other edges and do not need to be separately stored. A relaxation of point 2 in definition 5 can be reduced storage capacity.

In the nodes of the index graph are stored the names of the classes and not the graph itself.

Note that graph identifiers are stored into the nodes of the directed graph, not the actual subclass. The root note is the dummy node.

It may happen that by matching two graphs we will find two solutions, as it is shown in Figures 3a-3d. This will result in an index structure as shown in Figure 4. The hypergraph will branch into two paths for one solution. That means at $n=4$ we have to match twice, once for the structure {A, B1, C, D, B2}and the other time for the structure {A, B1, C, B2, D}. Both will result in the same solution, but by doing so the matching time will double.

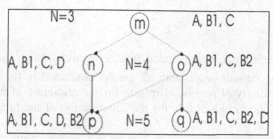

Fig. 4 Hypergraph and the Problem of two Solutions

The solution to this problem could be that in the hierarchy a link to point p will advise the matcher to follow this path. However, this will only be the right solution if the increase in the number of nodes from one step to another is one and not more. Otherwise, more than one solution will still be possible and should be considered during the construction and up-date of the hierarchy.

4.1.2 Incremental Learning of the Hierarchy

Now the task is to build up the graphs of IB into a supergraph by a learning environment.

Input:

Supergraph $IB = (N, E, p)$ and

graph $x \in H$.

Output:

modified Supergraph $IB' = (N', E', p')$

with $N' \subseteq N \cup \{x\}, E \subseteq E', p \subseteq p'$

At the beginning of the learning process or the process of construction of index graph N can be an empty set.

The attribute assignment function p' gives the values $(p'(x), (dd))$ as an output. This is an answer to the question: What is the name of the image name that is mirrored in the graph x?

The inclusion

$N' \subseteq N \cup \{x\}$

says that the graph x may be isomorphic to one graph y contained in the data base, so $x \le y$ and also $y \le x$ hold. Then no new node is created that means the data base is not increased.

The algorithm for the construction of the modified hierarchy IB' can also use the circumstance that no graph is part isomorphic to another graph if it has more nodes than the second one.

As a technical aid for the algorithm we introduced a set N_i. N_i contains all graphs of the data base IB with exactly i nodes. If the maximal number of nodes of the graph contained in the data base is k, then:

$$N = \bigcup_{i=k}^{k} N_i \ .$$

The graph which has to be included in the hierarchy has l nodes $(l > 0)$. By comparison of the current graph with all graphs contained in the hierarchy, we can make use of transitivity of part isomorphism for the reduction of the nodes that have to be compared. The full algorithm for the construction of the hierarchy is shown in Figure 5.

If we use the approach described in Section 3.2 for uncertainty handling, then we can use the algorithm presented in Section 3.5 without any changes. However we should notice that for each group of graphs that is approximate isomorphic, the graph that occurred first is stored in the hierarchy. Therefore, it is better to calculate for every instance and each new instance of a group a prototype and store it in the hierarchy of the data base.

Figure 5 illustrates this hierarchy. Suppose we have given a set of graphs, where the supergraph is the empty graph, then we open the first node in the supergraph for this graph at level n which refers to the number of nodes this graph has, and make a link to the root node which is the dummy node. Then a new graph is given to the supergraph. It is first classified by traversal of the tentative supergraph. If it does not match with a graph stored in the hierarchy then a new node is opened at the level which refers to the number of nodes this graph has and a link to the root node is made. If the node matches with the graph in the super-graph and if the graph is in the relation $G_1 < G_2$, then a new node is opened at the level k, the link between the root node and the recent node is removed and a link between the new node and the old node is inserted and another link from the new node to the root is installed. This procedure repeats until all graph of a data base are inserted into the supergraph.

Algorithm

E' := E;
Z := N;
for all y∈ N₁
if x ≤ y **then** [IB' := IB; return];
N' := N ∪ {x};
for all i with 0 < i < l;
for all y ∈ Nᵢ \ Z;
for all y ≤ x **then** [Z := Z \ {u | u ≤ y, u ∈ Z};
E' := E' ∪ { (y,x)}];
for all i with l < i ≤ k
for all y ∈ Nᵢ \ Z
if x ≤ y **then** [Z := Z \ {u _ y ≤ u, u ∈ Z };
E' := E' ∪ { (x,y)}];
p' := p ∪ { (x, (dd : unknown))};

Fig. 5 Algorithm of Approach I

4.1.3 Detection of Frequent Subgraph, Retrieval, and Result

Retrieval is done by classifying the current graph through the hierarchy until a node represented by a prototype having the same number of nodes as the query graph is reached. Then the most similar graph in this graph class is determined.

The output of the system are all graphs y in the database which are in relation to the query x as follows:

$$(x \leq y \vee x \geq y) \leq C$$

where C is a constant which can be chosen by the user of the system.

In the learnt hierarchy, Graphs that are grouped together in one subgraph class can be viewed by the user by clicking at the node which opens a new window showing all graphs that belong to the subgraph class. The node itself is labelled with the name of the graphs in this example but can be also labeled with a user given name.

A visualization component allows the user to view the organization of his subgraphs. It shows him the similarity relation between the graphs and the subgraph classes and by doing this gives him a good understanding of his domain.

4.2 Approach II

Whereas in Section 4.1, the hierarchy is built based on a divide-and-conquer technique, in this approach we use a strategy that is more flexible to fit the hierarchy dynamically to the different graphs. This approach is called conceptual clustering. For numerical data you may find the method and the algorithm in [12][13][14]. Graph conceptual clustering was introduced by Perner [15]. It does not only allow to incorporate new cases into the hierarchy and open new nodes by splitting the leaf nodes into two child nodes, it also allows to merge existing nodes and to split existing nodes at every position in the hierarchy.

4.2.1 The Hierarchy of Subgraphs

A concept hierarchy is a directed graph in which the root node represents the set of all input instances and the terminal nodes represent individual instances. Internal nodes stand for sets of instances attached to those nodes and represent a super concept. The super concept can be represented by a generalized representation of this set of instances such as the prototype, the median or a user-selected instance. Therefore a concept C, called a class, in the concept hierarchy is represented by an abstract concept description and a list of pointers to each child concept $M(C) = \{C_1, C_2, ..., C_i, ..., C_n\}$, where C_i is the child concept, called a subclass of concept C.

4.2.2 Learning

4.2.2.1 Utility Function

When several distinct partitions are incrementally generated over the hierarchy, a heuristic is used to evaluate the partitions. This function evaluates the global quality of a single partition and favors partitions that maximize potential for inferring information. In doing this, it attempts to minimize intra-class variances and to maximize inter-case variances. The employment of an evaluation function avoids the problem of defining a threshold for intra-class similarity and inter-class similarity. The threshold is to determine domain-dependent and it is not easy to define them a priori properly.

However, the resulting hierarchy depends on a properly chosen threshold. We will see later on by example what influence it has on the hierarchy.

$$SCORE = \frac{1}{m}\left|\sum_{j=1}^{m} p_j \left(G_{pj} - \overline{G}\right)^2 - \sum_{j=1}^{m} p_j G^2{}_{vj}\right| \qquad (2)$$

Given a partition $\{C_1, C_2, ..., C_m\}$, the partition which maximizes the difference between the subgroup-class variance s_B and the within-subgroup class variance s_W

$$SCORE = \frac{1}{m}\left|s_B^{*2} - s_W^{*2}\right| \Rightarrow MAX! \qquad (1)$$

is chosen as the right partition:

The normalization to m (m -the number of partitions) is necessary to compare different partitions.

If G_{pj} is the prototype of the j -th class in the hierarchy at level k, $\overline{G}$ the mean graph of all graphs in level k, and $G^2{}_{vj}$ is the variance of the graphs in the partition j, then:

where p_j is the relative frequency of cases in the partition j.

4.2.2.2 Learning of New Classes and Updating Existing Classes

It might happen that the reasoning process results in the answer: "There is no similar graph in the hierarchy". This indicates that such a graph has not been seen before. The graph needs to be incorporated into the hierarchy in order to close the gap in the hierarchy. This is done by classifying the graph according to the existing hierarchy and a new node is opened in the hierarchy at that position where the similarity relation holds. The new node represents a new subgraph class and the present graph is taken as class representative.

The evidence of new classes increases when new similar graphs arrive, where the evaluation measure holds. They are incorporated into the node and the concept description of the node is updated. This is considered as learning of the hierarchy over the subclasses of graphs.

4.2.2.3 Prototype Learning

When learning a class of graphs, graphs where the evaluation measure holds are grouped together. The graph that appeared first would be the representative of the class of these subgraphs and each new graph is compared to this subclass of graphs. Obviously, the first graph to appear might not always be a good graph. Therefore, it is better to compute a prototype for the class of graphs as described in Section 3.3.

In the same manner, we can calculate the variance of the graphs in one subgraph class. The resulting prototype is not a graph that exists in reality. Another strategy for calculating a prototype is to calculate the median of the graphs in a subgraph class.

4.2.2.4 Refinement and Abstraction of Subgraph Classes

The constructed subgraph classes and the hierarchy are sensitive to the order presentation of the graphs. It would result in creating different hierarchies from different orders of the same graphs. We have already included one operation to avoid this problem by learning of prototypes. Two additional operations [12] should help it recover from such non-optimal hierarchies. At each level of the classification process, the system considers merging the two nodes that best classify the new instance. If the resulting subgraph class is better according to the evaluation function described in Section 4.2.2.1 than the original, the operation combines the two nodes into a single, more abstract subgraph class. This transforms a hierarchy of N nodes into one having N+1 nodes, see Fig. 6. This process is equivalent to the application of a "climb-generalization tree" operator [13], with the property that the generalization tree is itself built and maintained by the system. Other merge strategies are described in [14].

The inverse operation is splitting of one node into two nodes as illustrated in Figure 7. This is also known as refinement. At each level, the learning algorithm decides to classify an instance as a member of an existing subgraph classes, it also considers removing this subgraph class and elevating its children.

If this leads to an improved hierarchy, the algorithm changes the structure of the hierarchy accordingly.

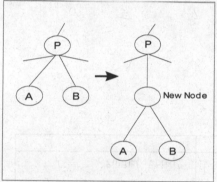

Fig. 7 Node Merging

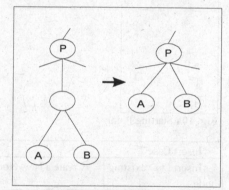

Fig. 8 Node Splitting

4.2.3 Algorithm

Now, that we have defined our evaluation function and the different learning levels, we can describe our learning algorithm. We adapt the notion of Fisher [14] for concept learning to our problem. If a new graph has to input into the hierarchy over the graphs, it is tentatively placed into the hierarchy by applying all the different learning operators described before. The operation which gives us the highest evaluation score is chosen. The new graph is entered into the hierarchy and the subgraph hierarchy is reorganized according to the selected learning operation. Figure 9 shows the learning algorithm in detail.

4.2.4 The Learning Algorithm in Operation: an Example

The learning process is illustrated in Fig. 10a-d for the four cases shown in Fig. 2a-d that were used to construct the hierarchy in Figure 6.

The learning algorithm first tests where the graph should placed into the hierarchy. Figure 10a shows how graph_1 is incorporated into the hierarchy. At this stage, the hierarchy is empty, and so the only possible operation is to *create a new node* for graph_1. Next, graph_2 is inserted into the hierarchy (see Fig. 10b). First, it is placed in the existing node and the score for this operation is calculated. Then the *"create a new node"* operation is performed and its score is calculated. The last operation which can be applied in this graph is *node merging*. The score for this operation is also calculated. Since all three scores have the same value, the first operation *"insert into existing node"* is selected on this occasion.

Figure 10c shows how graph_3 is incorporated into the hierarchy. Any of the operations *"insert to existing node"*, *"create a new node"*, and *"node merging"* can be applied to the existing hierarchy. The highest score is obtained for the operation *"create a new node"*. This operation is therefore used to update the hierarchy. Graph_4 can be inserted into either of the two existing nodes (see Fig. 10d). The operations *"create a new node"* and *"node merging"* can then be applied in sequence. The highest score is obtained for the operation *"insert to node_2"*. This is the final hierarchy resulting from the application of the four example graphs. Note that the *"node-splitting"* operation was not used in our example because the resulting hierarchy did not allow for this operation. The learning algorithm should results to more meaningful groupings of graphs since the algorithm does not rely to much on the order of the example.

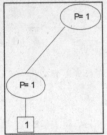

Fig. 10a Starting Point

Insert Case 2		
Insert to existing node	Create a New Node	Node Merging
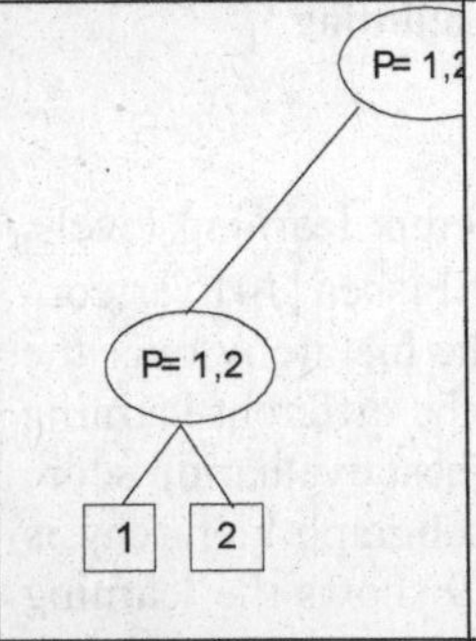	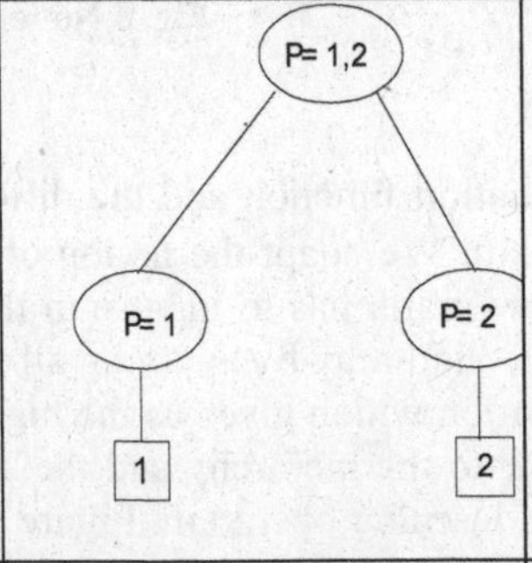	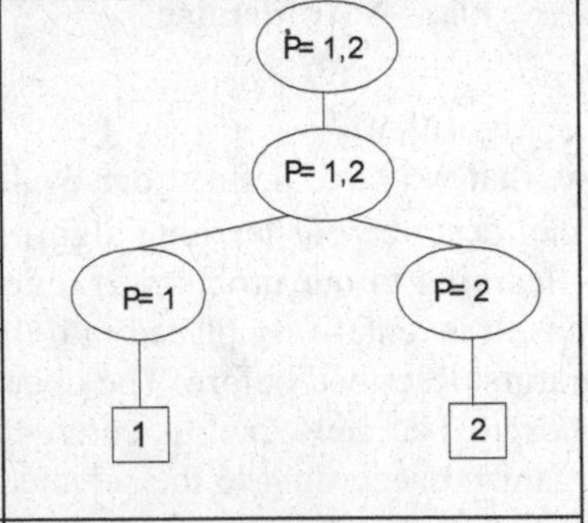
SB = 0	SB = 0,00018172	SB = 0,00018172
SW = 0,00018172	SW = 0	SW = 0 .
SCORE= 0,00018172	SCORE = 0,00018172	SCORE = 0,00018172
Resulting Case Base		

Fig. 10b Insert Case 2 into the case base

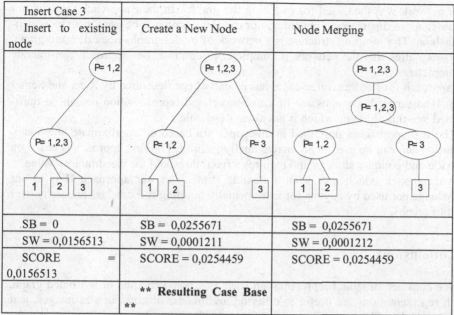

Insert Case 3		
Insert to existing node	Create a New Node	Node Merging
SB = 0	SB = 0,0255671	SB = 0,0255671
SW = 0,0156513	SW = 0,0001211	SW = 0,0001212
SCORE = 0,0156513	SCORE = 0,0254459	SCORE = 0,0254459
	** Resulting Case Base **	

Fig. 10c Insert Case 3 into Case Base

4.3 Discussion of the two Approaches

The results in Approach I shown in Figure 6 are two nodes containing a set of graphs each. The second node is highlighted in Figure 5 and shows that the two graphs (graph_3 and graph_4) are belonging to the same subgraph-class. The first node shows only the name of graph_1 but also contains graph_2. The partition {1,2} and {3,4} is reached by a threshold of 0.0025 for the similarity. Matching has to be performed three times for this structure. First, matching the prototypes is performed and afterwards is determined the closest graph in the subgraph-class of the prototype with the highest similarity. If we increase the threshold to 0.02 than we obtain the partition {1,2,4}{3}. This hierarchy still has two nodes but the time for matching is increased since now in one node there are three graph that should be matched.

The four graphs are placed into four separate nodes each when using a threshold less than 0.001. Thus, for this hierarchy matching has to be performed four times.

This example shows how the threshold effects the resulting hierarchy.

Algorithm II automatically selects the best partition {1,2} and {3,4}. The predetermination of a threshold is not necessary. This algorithm protects the user from a try-and-test procedure in order to figure out the best threshold for the similarity of his application and besides that it guarantees that the hierarchy will not grow too deep or too broad. The evaluation measure as well as the allowed operations over the hierarchy keep the hierarchy in balance according the observed cases.

5 Related Work

Bunke and Messmer [16] used a network of model graphs for matching graphs that is created based on the subgraph isomorphism relation. They calculated off-line each possible permutation from each model graph and used them to construct the network.

This network was then used for matching the graphs. In their approach they did not consider approximate graph matching, nor can their approach be used in an incremental fashion. The resulting structure is a network of model graphs not a directed graph. It also requires that the network is complete. In contrast, our approach can tolerate incompleteness.

An approach for tree structured cases has recently been described by Ricci and Senter [17]. It assumes that graphs are in a tree-structured representation or can be transformed into this structure which is not always possible.

The two approaches described in this paper are based on approximate graph subsumption and can be used in an incrementally fashion. The first approach is based on a divide-and-conquer strategy and requires a fixed threshold for the similarity value.

An approach which uses such a flexible strategy as our approach II is to our knowledge not used by anyone for incrementally building the subclass hierarchy over a set of graphs.

6 Conclusions

We consider structural representation such as a non-attributed or attributed graph. Such representations are useful to describe multimedia objects such as images, text documents, log-files or even in social networks, building design, software engineering, and timetabling. The similarity between these objects must be determined based on their structural relations as well as on the similarity of their attributes.

We described two approaches that can detect frequent subgraphs in a data base of graphs.

Our two approaches forming hierarchy over subclasses of graphs are based on approximate graph subsumption. The first approach requires an off-line defined threshold for the similarity value. The chosen threshold as well as the order of graphs presented to the system strongly influences the resulting hierarchy. Therefore, not the first stored graph in the group is taken as the class representative instead of that a prototype is incrementally calculated. However, this approach can only construct new subgraph classes. Once a subgraph class has been established the structure cannot be reversed. This is only possible in the second approach since it can merge and split nodes at any position in the hierarchy. Another advantage of the second approach is that it does not rely on a fixed threshold for similarity. Instead, an evaluation function is used to select the right partition. This gives the flexibility needed when incrementally building the hierarchy.

The performance of the two approaches was evaluated on a data set of ultrasonic images from non-destructive testing. The image description of the reflection points of the defect in the image is represented as an attributed graph. The learning approach for the subgraph detection was developed for real-time collection of ultra sonic images into the graph data base. It allows to up-date the hierarchical organization of the subgraphs at any time a new graph is collected.

Input:	Case Base Supergraph CB
	An unclassified case G
Output:	Modified Case Base CB´

Top-level call:	Case base (top-node, G)
Variables:	A, B, C, and D are nodes in the hierarchy
	K, L, M, and T are partition scores

Case base (N, G)
IF N is a terminal node,
        THEN Create-new-terminals (N, G)
          Incorporate (N, G)
     ELSE   For each child A of node N,
            Compute the score for placing G in A.
            Compute the scores for all other action with G
            Let B be the node with the highest score K.
            Let D be the node with the second highest score.
                 Let L be the score for placing I in a new node C.
            Let M be the score for merging B and D into one node.
            Let T be the score for splitting D into its children.

            IF K is the best score
                THEN Case base (P, G) (place G in case class B).
                   ELSE IF L is the best score,
               THEN Input a new node C
                          Case base (N, G)
             Else IF M is the best score,
               Then let O be merged (B, D, N)
                   Case base (N, G)
             Else IF T is the best score,
               Then Split (B, N)
                     Case base (N, G)

Operations over Case base

Variables: X, O, B, and D are nodes in the hierarchy.
           G is the new case

Incorporate (N, G)
    Update the prototype and the variance of case class N

Create new terminals (N, G)
    Create a new child W of node N.
    Initialize prototype and variance

Merge (B, D, N)
    Make O a new child of N
    Remove B and D as children of N
    Add the instances of P and R and all children of B and D to the node O
    Compute prototype and variance from the instances of B and D

Split (B, N)
    Divide Instances of Node B into two subsets according to evaluation criteria
    Add children D and E to node N
    Insert the two subsets of instances to the corresponding nodes D and E
    Compute new prototype and variance for node D and E
    Add children to node D if subgraph of children is similar to subgraph of node D
    Add children to node E if subgraph of children is similar to subgraph of node E

Figure 9 Algorithm II

References

1. AP Appel, LG Moyano, Link and Graph Mining in the Big Data Era - Handbook of Big Data Technologies, 2017 —Springer
2. Jiawei Han,Jian Pei,Micheline Kamber, Data Mining —Concepts and Techniques, Elsevier, 2011
3. PV Bindu, PS Thilagam, D Ahuja, Discovering suspicious behavior in multilayer social networks - Computers in Human Behavior, 2017 —Elsevier, 3rd Edition, 2011
4. Chapter 9: Graph Mining, Social Network Analysis, and Multirelational Data Mining
5. T Yu, D Yan, L Zhu, S Yang, The Semantic Web for Complex Network Analysis in Biomedical Domain - Information Technology in 2015 7th International Conference on Information Technology in Medicine and Education (ITME), 2015
6. P. Perner, Different Learning Strategies in a Case-Based Reasoning System for Image Interpretation, Advances in Case-Based Reasoning, B. Smith and P. Cunningham (Eds.), LNAI 1488, Springer Verlag 1998, S. 251-261.
7. D. Wettscherek, D.W. Aha and T. Mohri, A review and empirical evaluation of feature weighting methods for a class of lazy learning algorithms, in Artifical Intelligence Review (also available on the Web from http://www.aic.nrl.navy.mil/~aha.
8. A. Bonzano and P. Cunningham, Learning Feature Weights for CBR: Global versus Local
9. P. Perner, Using CBR Learning for the Low-Level and High-Level Unit of a Image Interpretation System, In: Sameer Singh (Eds.), Advances in Pattern Recognition, Springer Verlag 1998, pp. 45-54.
10. Perner, Ultra Sonic Image Interpretation, IAPR Workshop on Machine Vision and Applications MVA`96, Tokyo Japan, page 552-554
11. B. Nebel, Reasoning and Revision in Hybrid Representation Systems, Springer Verlag 1990
12. J.H. Gennari, P. Langley, and D. Fisher, "Models of Incremental Concept Formation," Artificial Intelligence 40 (1989) 11-61.
13. R.S. Michalski. A theory and methodology of inductive learning. In R. S. Michalski, J.G. Carbonell, and T.M. Mitchell, (Eds.), Machine Learning: Artificial Intelligence Approach. Morgan Kaufmann, 1983.
14. D.H. Fisher, "Knowledge Acquisition via Incremental Clustering," Machine Learning, 2: 139-172, 1987.
15. P. Perner, Case-base maintenance by conceptual clustering of graphs, Engineering Applications of Artificial Intelligence, vol. 19, No. 4, 2006, pp. 381-295
16. H. Bunke and B. Messmer. Similarity measures for structured representations. In S. Wess, K.-D. Althoff, and M.M. Richter (eds.), Topics in Case-Based Reasoning, Springer Verlag 1994, pp. 106-118
17. F. Ricci and L. Senter, Structured Cases, Trees and Efficient Retrieval, In Proc.: B. Smyth and P. Cunningham, Advances in Case-Based Reasoning, Springer Verlag 98, pp. 88-99.

Author Index

Printed in the United States
By Bookmasters